# Methods in Enzymology

## Volume XVI
## FAST REACTIONS

# METHODS IN ENZYMOLOGY

EDITORS-IN-CHIEF

Sidney P. Colowick    Nathan O. Kaplan

*Methods in Enzymology*

*Volume XVI*

# Fast Reactions

EDITED BY

*Kenneth Kustin*

DEPARTMENT OF CHEMISTRY
BRANDEIS UNIVERSITY
WALTHAM, MASSACHUSETTS

1969

ACADEMIC PRESS   New York and London

ACADEMIC PRESS, INC.
111 Fifth Avenue, New York, New York 10003

*United Kingdom Edition published by*
ACADEMIC PRESS, INC. (LONDON) LTD.
Berkeley Square House, London W1X 6BA

LIBRARY OF CONGRESS CATALOG CARD NUMBER: 54-9110

PRINTED IN THE UNITED STATES OF AMERICA

# Contributors to Volume XVI

Article numbers are shown in parentheses following the names of contributors.
Affiliations listed are current.

R. A. ALBERTY (2), *Massachusetts Institute of Technology, Cambridge, Massachusetts*

LEO C. M. DE MAEYER (4), *Max Planck Institut für Physikalische Chemie, Göttingen, West Germany*

FRIEDER EGGERS (3), *Max Planck Institut für Physikalische Chemie, Göttingen, West Germany*

THAYER C. FRENCH (1), *Department of Chemistry, Massachusetts Institute of Technology, Cambridge, Massachusetts*

QUENTIN H. GIBSON (6), *Biochemistry and Molecular Biology, Cornell University, Ithaca, New York*

H. GUTFREUND (7), *Department of Biochemistry, University of Bristol, The Medical School, Bristol, England*

GORDON G. HAMMES (1), *Department of Chemistry, Cornell University, Ithaca, New York*

H. E. JOHNS (8), *Department of Medical Biophysics, University of Toronto, Toronto, Canada*

KENNETH KUSTIN (3), *Department of Chemistry, Brandeis University, Waltham, Massachusetts*

H. RÜPPEL (9), *Technische Universität Berlin, Max Volmer Institut für Physikalische Chemie, Berlin, West Germany*

M. T. TAKAHASHI (2), *Department of Biochemistry, University of Massachusetts, Amherst, Massachusetts*

GREGORIO WEBER (10), *Department of Chemistry, University of Illinois, Urbana, Illinois*

H. T. WITT (9), *Technische Universität Berlin, Max Volmer Institut für Physikalische Chemie, Berlin, West Germany*

PETR ZUMAN (5), *Department of Chemistry, The University of Birmingham, Birmingham, England*

# Preface

The title of this book, "Fast Reactions" might imply to the reader a restricted view of chemical kinetics, invoking images of "instantaneous" processes and select methods for their measurement. Actually, quite the opposite picture is closer to the truth. The developments of modern solution kinetics were thrust at the broad problem of detecting and measuring individual steps in chemical reaction mechanisms. This goal was accomplished by expanding the hitherto restricted kinetic time range to new limits. New methods, capable of time resolution shorter than a millisecond, were therefore introduced. Ultimately, a resolution of the order of $10^{-10}$ to $10^{-11}$ seconds was achieved; at this point, chemical activation begins to yield to processes of a more physical nature, generally accessible to spectroscopic analysis.

To no one is the problem of making measurements in short times more apparent than to enzymologists. For enzymes are catalysts which carry reactions of exceptional sluggishness into the domain of "fast reactions." This book is therefore aimed at biochemists and chemists engaged in the study of reaction mechanisms, especially those of biological significance. Its use allows the research kineticist to establish a laboratory for studying solution kinetics, where several instruments are available for the investigation of complex reactions throughout the entire chemical time range.

Each chapter is devoted to a single type of instrument or a closely related group of instruments. The authors, who have successfully built and operated the equipment they describe, are themselves involved in research on chemical kinetics. They have presented complete, self-contained descriptions of the principles and methods of construction of fast reaction instrumentation. The capabilities and limitations (the latter known only too well to the authors) of this equipment have been clearly and frankly stated. The level of treatment should enable the chemist or biochemist, with an elementary knowledge of instrumentation, to plan and construct a fast reaction laboratory, by building appropriate instruments selected from those presented in this volume.

Methods of an essentially chemical nature are covered in the book, thereby omitting nuclear and electron resonance techniques, which perturb physical states of the system. The most versatile techniques are the different relaxation methods, which comprise the opening section. The next section, consisting of a chapter on polarographic methods, was so placed to bring out its close relationship to the relaxation techniques using electric fields. The last two sections are on rapid mixing and irradiation.

It is clear now that no single apparatus can encompass all reactions over all time ranges. A rapid mixing device, whether built into another apparatus, temperature-jump, for example, or simply available in the laboratory, is always useful. The need for combining equipment is apparent even when photochemical excitation is a part of the reaction sequence. In compiling this book, the assumption has therefore been made that a reader will want to build more than a single fast reaction apparatus, and the chapters have been written to encourage this idea.

Although skill and good judgment are involved in the design, construction, and application of these instruments, each section displays a different emphasis. For the polarographic technique of Section II, most of the equipment is either commercially available or of a simple nature, yet no other method has proved to be so difficult to apply successfully. The application of this powerful technique to fast reactions requires a severely critical examination of the data. As a result, the main emphasis in this chapter has been placed on the proper acquisition and treatment of the data. Other electrochemical methods may also provide some entree into the area of fast reactions. These techniques do not have the scope and sensitivity of polarography, which has therefore been featured.

For rapid mixing and chemical relaxation—Sections I and III—the main problem is how to obtain the data in the first place; its treatment, although frequently difficult, nevertheless necessitates only standard kinetic techniques of analysis. The attitude that has guided the presentation in these sections is that the best results are achieved when the investigator constructs his own equipment. To this end, the user of this volume is supplied with full construction details. More importantly, the strengths and weaknesses of the components themselves are frankly stated. This objectivity has two advantages. First, it confers upon the builder of the instrument a really useful knowledge of the boundaries of his instrument. Second, when equipment is still in the developmental stage (as, in a sense, almost everything is today since technology advances so rapidly), a basis is presented for determining the selection of new components and accommodating future trends.

Unlike the techniques of Sections I, II, and III, the irradiation methods of Section IV are more specific in nature. Not only is the perturbation— light energy—itself a part of the reaction scheme, but the molecules susceptible of study by this method are also unique. This section has therefore received a different treatment. It begins with a chapter built around the concept of dosimetry, or the measure of energy absorption, fundamental to all radiation methods. Moreover, as only few classes of compounds are truly important in this field, this chapter on photochemistry concentrates on a class of particular biochemical importance, namely, nucleic acids. The most versatile tools for following rapid, photoinduced processes are repetitive pulse and fluorescence methods. Repetitive pulse techniques, based on the principle of flash photolysis, utilize the increased sensitivity brought about by sampling techniques to suppress the effect of disturbing, but otherwise inescapable, noise levels.

In general, the approach that has been adopted by the editor in planning this book is akin to that which he observed while working with Manfred Eigen and Leo DeMaeyer in Göttingen. There, design of equipment is an integral part of research, where apparatus is conceived and constructed as a part of solving scientific problems. The device that first emerges emphasizes the essential physicochemical solution to the problem. Appearance and convenience are of secondary importance at this stage, but as the equipment is used these aspects of the design gain in importance. Thus, no *research* instrument is conceived of as being complete, or finished, in the usual sense. Each apparatus is constantly being improved in sensitivity, ease of handling, and scope of applicability. Consequently, it is the editor's hope that this book will not only be useful to all kineticists, but that a more adventurous spirit will also be imparted to those chemists and biochemists who are reluctant to have first-hand experience with the construction of the equipment they use.

The editor wishes to thank the authors of the chapters for their willingness to contribute to this volume and for their cooperation in doing so. The assistance of the staff of Academic Press is also appreciated.

*Waltham, Massachusetts*                          KENNETH KUSTIN
*September, 1969*

# Table of Contents

# METHODS IN ENZYMOLOGY

EDITED BY

## Sidney P. Colowick and Nathan O. Kaplan

VANDERBILT UNIVERSITY
SCHOOL OF MEDICINE
NASHVILLE, TENNESSEE

DEPARTMENT OF CHEMISTRY
UNIVERSITY OF CALIFORNIA
AT SAN DIEGO
LA JOLLA, CALIFORNIA

# METHODS IN ENZYMOLOGY

EDITORS-IN-CHIEF

## Sidney P. Colowick        Nathan O. Kaplan

*In Preparation (Continued)*

Proteolytic Enzymes
*Edited by* GERTRUDE E. PERLMANN AND LASZLO LORAND

Photosynthesis
*Edited by* A. SAN PIETRO

Enzyme Purification and Related Techniques
*Edited by* WILLIAM B. JAKOBY

# Section I

# Chemical Relaxation

# [1] The Temperature-Jump Method

## By THAYER C. FRENCH and GORDON G. HAMMES

## I. Introduction

This chapter will be concerned with the construction of equipment for the measurement of relaxation spectra by the temperature-jump method. Since the innovation of this method,[1] several different laboratories have participated in further development of the instrumentation. Most of the equipment described here has been constructed and tested in our laboratories; in a few cases literature descriptions of the apparatus have been relied upon. An effort has been made to give enough detail so that construction of a temperature-jump apparatus by an interested reader should be relatively simple. Only a brief treatment of the general theory of relaxation spectra will be given since this subject has been discussed in considerable detail elsewhere.[2] The analysis of the actual raw data will be considered for a few special cases, but a comprehensive survey of the application of the temperature-jump method to biochemical mechanisms will not be attempted. Such a review has been recently published.[3]

If a solution, initially at thermal equilibrium and containing substances coupled by chemical equilibria, is perturbed by a rapid increase in temperature (typically with a rise time of a few microseconds), the concentrations of the chemical species will change to new equilibrium values at the higher temperature. The magnitude of the concentration changes is dictated by the laws of thermodynamics: for each equilibrium

$$\left(\frac{\partial \ln K}{\partial T}\right)_P = \frac{\Delta H^0}{RT^2} \tag{1}$$

[1] G. Czerlinski and M. Eigen, Z. Elektrochem. **63**, 652 (1959).
[2] M. Eigen and L. De Maeyer in "Technique of Organic Chemistry" (S. L. Friess, E. S. Lewis, and A. Weissberger, eds.), Vol. VIII, Part II, pp. 895–1054. Wiley (Interscience), New York, 1963.
[3] G. G. Hammes, Advan. Protein Chem. **23**, 1 (1968).

where $K$ is the constant pressure equilibrium constant of the chemical reaction, $\Delta H^0$ is the standard enthalpy change of the reaction, $R$ is the gas constant, and $T$ is the absolute temperature. Clearly if a sequence of coupled reactions occurs, the system will be perturbed by a temperature jump if any one reaction is characterized by a nonzero enthalpy change. Even if the equilibria are insensitive to temperature, a perturbation of the system can be achieved by coupling the reactions of interest to a temperature-dependent reaction. The most common example of this is a "pH jump" which can be achieved by use of a buffer with temperature-dependent ionization constants; with this technique pH-dependent reactions can be readily perturbed. Other concentration jump methods can be devised if needed, but, in general, equilibrium constants are sufficiently dependent on temperature that this is unnecessary.

The rates of decay of the concentrations to their new equilibrium values are characterized by linear first-order differential equations, of which the solutions are linear combinations of exponentials. Each exponential term is associated with a relaxation time, $\tau$, which may be thought of as a reciprocal first-order rate constant. Thus a complex reaction mechanism is associated with a spectrum of relaxation times. The explicit evaluation of the relaxation spectrum has been described in detail elsewhere (*cf.* footnote 2). In general, the relaxation times will have a unique dependence on the equilibrium concentrations, and determination of the relaxation times at

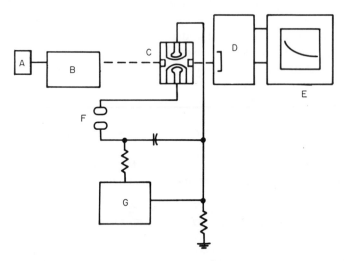

FIG. 1. Schematic diagram of a temperature-jump apparatus utilizing absorption spectrophotometry for the detection of concentration changes. A, Light source; B, monochromator; C, observation cell; D, photomultiplier emitter follower; E, oscilloscope; F, spark gap; G, high voltage.

various concentrations permits postulation of a reaction mechanism. The present time resolution of temperature-jump equipment is about $10^{-6}$ seconds, but in principle the method can be extended to even shorter times.

Thus far only the perturbation of a system at equilibrium has been considered. In principle, relaxation methods can be applied to any system where an overall net change in concentrations does not occur. Thus systems in a steady state can be perturbed and the relaxation spectrum determined if the steady state persists for a sufficiently long time. Application of steady state perturbation techniques to the elucidation of reaction mechanisms is still in a stage of infancy. However, construction of a temperature-jump apparatus directly coupled to a rapid mixing flow system will be considered. With this apparatus, perturbation of steady states with half-times as short as milliseconds can be accomplished. Thus the temperature-jump method can be applied to both reversible and essentially irreversible reactions. The temperature-jump method has a wide range of application to biochemical systems because of the sensitivity of such systems to temperature.

## II. Temperature-Jump Apparatus with Stationary Samples

### A. Temperature Pulse by Joule Heating

*1. General Principles.* A simplified diagram of a temperature-jump apparatus that uses joule heating is shown in Fig. 1. The temperature pulse is applied to an aqueous solution of electrolyte of low resistance by discharge of a high-voltage capacitor through the solution in a suitable electrode cell. The condenser is discharged when the breakdown voltage of a spark gap is reached. The values of the capacitor, the cell resistance, the voltage at discharge, and the volume of the cell are chosen to provide a rapid change in the temperature of the contents of the cell between the electrodes of several degrees. Detection of the concentration changes can be accomplished by use of absorption spectrophotometry, fluorimetry, polarimetry, and other suitable optical techniques. The observation windows are perpendicular to the path of the discharge. When the entire cell is placed in a brass thermostat that is in thermal contact with the solution by way of the ground electrode of the cell, the temperature of the sample solution returns to its initial value within a few minutes after the pulse. Temporal changes in the spectral properties of the cell contents are detected with a photomultiplier coupled to an emitter follower, and the event is recorded with an oscilloscope.

Convection currents, which disturb spectral measurements, commence within about 1 second after the discharge. The upper limit of relaxation times that may be conveniently measured is therefore of the order of a few hundred milliseconds, which corresponds to first-order rate constants of approximately 1–10 sec$^{-1}$. The lower limit for measurement of relaxation times depends on the rise time of the temperature pulse. If the inductance, $L$, of the high-voltage capacitor is low, the temperature will rise exponentially to a limiting value. With such capacitors, the shortest heating times possible thus far are of the order of $10^{-6}$ to $10^{-7}$ seconds.

2. *Heating.* An equation for the total temperature rise and its associated time constant is derived as follows: let the current delivered during discharge have the overdamped form

$$i = (V_0/R)e^{-t/RC} \tag{2}$$

where $RC/2 \gg 2L/R$, $V_0$ is the initial value of the voltage across the capacitor, $R$ is the resistance of the solution between the electrodes, and $C$ is the discharge capacitance in farads. If all other impedance losses in the circuit are small compared to that occurring in the solution, the rate of temperature rise is

$$\frac{dT}{dt} = \frac{i^2 R}{4.18 c_P \rho V} \tag{3}$$

where $c_P$ and $\rho$ are the specific heat capacity and density of the solution, and $V$ is the volume of the solution between electrodes that is heated. Substitution of Eq. (2) into Eq. (3) and integration from $t = 0$ to $t = t$ (with constant $c_P$, $\rho$, $V$, and $R$) yields the relationship,

$$\delta T(t) = \frac{CV_0^2}{8.36 c_P \rho V} [1 - e^{-2t/RC}] = \delta T_\infty [1 - e^{-2t/RC}] \tag{4}$$

values of $RC/2$ and $\delta T_\infty$ for the three cells that are described in a subsequent section are included in Table I.

Typical values of the time constant, $RC/2$, and the total temperature rise, $\delta T_\infty$, are 1–10 $\mu$sec and 8°. The actual time resolution of the apparatus

TABLE I
CHARACTERISTICS OF SOME TEMPERATURE-JUMP CELLS

| Cell[c] | $R^a$ (ohms) | $C$ ($\mu$F) | $V_0{}^b$ (kV) | Volume (ml) | $RC/2^a$ ($\mu$sec) | $\delta T_\infty$ (°C) |
|---|---|---|---|---|---|---|
| Standard (1) | 140 | 0.1 | 25 | 0.93 | 7.0 | 8 |
| Stopped flow-temperature jump (1) | 240 | 0.1 | 10 | 0.16 | 12 | 7.5 |
| Ten-cm (10) | 16 | 0.1 | 50 | 4.0 | 0.8 | 8 |

[a] Resistances when 0.1 $M$ KNO$_3$ is in cell at room temperature.

[b] Voltage across $C$ at the time of the spark gap discharge.

[c] In parentheses, path length in centimeters.

is a few times $RC/2$. Values of $RC/2$ shorter than a few microseconds lead to an increase in the intensity of the shock wave produced by the temperature rise. Cavitation of aqueous solutions results unless measurements are made near 4°, where the thermal expansion coefficient of water is zero. The resistance of the cell can be varied by use of different concentrations of inert electrolytes in the cell. In general any inert electrolyte, such as $KNO_3$, at a concentration of about 0.1 $M$, is used to achieve a sufficiently rapid and homogeneous temperature rise. For liquids the distribution of the energy of a temperature pulse among translational and rotational degrees of freedom is much more rapid than $RC/2$; it is complete in picoseconds.

When the temperature pulse is to be obtained by a high-voltage discharge through the sample solution, one convenient means of switching on the charged high-voltage capacitor is by means of a spark gap (Fig. 1,$F$). The capacitor is charged relatively slowly ($\sim$3 minutes) by a low-current, high-voltage power supply in series with a resistance, $R_c$, that is large enough to give a charging time constant ($R_cC$) of about 30 seconds. The time for charging a condenser with a voltage equal to 97% of the applied voltage is approximately $3R_cC$. The distance between electrodes may be adjusted so that discharge will occur automatically when the capacitor has been charged to the desired voltage. The entire spark gap assembly, with condenser, must be insulated and, for the higher voltages, surrounded by two electrostatic shields that are separated by an insulating medium. The aluminum outer shield is grounded. The copper inner shield is connected to the low-voltage side of the capacitor and also to the low voltage cell electrode by means of the shield of a high-voltage cable. The low-voltage side must be connected to the grounded shield by a resistance that would ensure safe dissipation of the energy of the discharge if it were delivered directly to the copper shield rather than to the cell.

The design of a shielded spark gap assembly that will deliver about 100 joules at 50 kV is illustrated in Fig. 2. This apparatus is suitable for use with either of the two larger cells that are subsequently described. A special polyethylene "disconnect"[4] may be used to link this compact unit with a suitable power supply.[5] The necessary electrical components are listed in Table II. The discharge resistor with a plastic insulator may be conveniently mounted in an upper corner at the low-voltage end and connected between the two shields by lugs. A trigger antenna (a few turns of insulated wire) and jack should be mounted near the window through which the spark gap may be viewed.

[4] Alden Products Co., Brockton, Massachusetts; No. 8111 series for 50 kV molded to M 1 No. RG-8/U cable.

[5] Neutronics Associates, 4 Hawthorne Street, Farmingdale, Long Island, New York; Model 24MR.

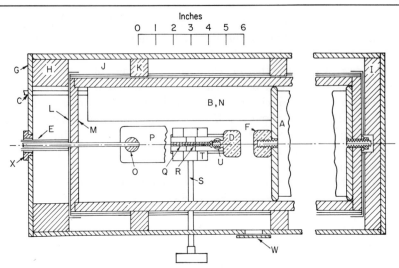

Fig. 2. Top view of horizontal cross section through the axis of the electrodes of a shielded spark gap assembly drawn to the scale indicated. Electrical components: *A*, high-voltage condenser; *B*, charging resistor; *C*, cable from high-voltage supply; *D*, spark gap electrode (low-voltage side); *E*, high-voltage cable (connected to adjustable electrode with tinned braid); *F*, spark-gap electrode (high-voltage side). Parts: *G*, outer shield; insulating supports *H*, *I*, *J*, *K*, high- and low-voltage ends, bottom and side pieces, respectively; *L*, inner shield; *M*, insulating box; *N*, insulator for charging resistor; *O*, centerpost; *P*, insulator for flexible electrode links (pictured cut away to expose rack); *Q*, rack; *R*, spur gear; *S*, shaft; *T*, shaft support; *U*, support for negative electrode; *W*, window shade; *X*, insulating bushing.

A very much simpler spark gap assembly suffices for use with the stopped flow-temperature jump cell, which can also be used as a micro cell (see subsequent sections) although the basic design is identical. Since less than 10 kV is required, elaborate precautions with shielding and insulation

TABLE II
Electrical Components of Spark Gap Assembly

| Item | Description |
| --- | --- |
| High-voltage condenser[a] | 0.1 $\mu$F, 50 kV DCW, $L < 0.1$ $\mu$H, $4\frac{3}{4} \times 6 \times 9\frac{1}{4}$, terminals at opposite ends on center |
| Charging resistor[b] | 300 M$\Omega$, 50 kV |
| Discharge resistor[b] | 10 M$\Omega$, 50 kV |
| Spark gap electrodes | Copper-tungsten Elkonite,[c] $1\frac{1}{2}$ D $\times l$, corners $\frac{1}{4}$ R |
| High-voltage cable | MIL. No. RG/8U |
| Flexible electrode links | Tinned braid |

[a] Plastic Capacitors, Inc., Chicago, Illinois, No. OP500-104NDA.
[b] International Resistance Co., Boone, North Carolina, No. MVO-16.
[c] P. R. Mallory and Co., No. 10W3.

are unnecessary, and the entire assembly is mounted on an ordinary chassis. Stainless steel spark gap electrodes about ¾ inch in diameter are surrounded by a plastic cylinder, and one of them is positioned by means of a brass rack and spur gear. A high-voltage pulse capacitor (0.1 mF) with low inductance is connected directly to the spark gap, mounted on the chassis, and covered with a plastic shield. A charging resistor of 40 MΩ is sufficiently large if a triggering spark is used to initiate discharge of the main capacitors. A circuit that is suitable for triggering the spark gap of such a temperature-jump apparatus is shown in the lower part of Fig. 12. Triggered spark gap assemblies are also available commercially (E G and G, Boston, Massachusetts). These are quite suitable and convenient for construction of the discharge circuit.

For voltages of 25 kV or less, the entire spark gap assembly can be conveniently replaced by a thyratron triggering circuit. Such an apparatus has been described by Kresheck et al.,[6] and a detailed circuit diagram is available from the American Documentation Institute, Auxiliary Publications Project, Photoduplication Service, Library of Congress, Washington 25, D. C. by ordering Document 8584 and remitting $1.25 for microfilms or $1.25 for photoprints.

3. *Detection of Concentration Changes.* Concentration changes may be detected by several optical methods. Presently absorption spectrophotometry is most commonly employed, although fluorimetry and polarimetry are also used. The total concentration change, $\delta c_{io}$, of a single component of a one-step mechanism when the temperature jump, $\delta T$, is small is:

$$\delta c_{io} = \delta T \frac{\partial \ln c_i}{\partial \ln K} \frac{\Delta H^0}{RT^2} \tag{5}$$

where activity coefficients have been neglected. For a single relaxation process

$$\delta c_i = \delta c_{io}(1 - e^{-t/\tau}) \tag{6}$$

while for multiple relaxation processes

$$\delta c_i = \sum_j A_j e^{-t/\tau_j} \tag{7}$$

where the $A_j$'s are constants that can be evaluated for a given mechanism.[2] Therefore any optical property which is a linear function of the concentration (for small concentration changes) can be used for the measurement of relaxation times.

For absorption spectrophotometry the equation relating concentration and light intensity, $I$, transmitted through the solution is

[6] G. C. Kresheck, E. Hamori, G. Davenport, and H. A. Scheraga, *J. Am. Chem. Soc.* **88,** 246 (1966).

$$I = I_0 e^{-\epsilon_i c_i l} \tag{8}$$

where $\epsilon_i$ is the molar extinction coefficient of the $i$th species (the decadic molar extinction coefficient usually cited is $\epsilon_i/2.303$), $l$ is the path length of the observation cell in centimeters, $c_i$ is the concentration in moles/liter, and $I_0$ is the light intensity before passage through the cell. For small concentration changes

$$\delta I/I_0 = -\epsilon_i l \delta c_i \tag{9}$$

so that the change in light intensity is directly proportional to the concentration change.

Fluorescent intensity is related to concentration by the equation

$$I_f = \text{const. } I_0[1 - e^{-(\epsilon_{\lambda e} + \epsilon_{\lambda f})c_i l}] \tag{10}$$

(neglecting the angular dependence of the fluorescence) where $\epsilon_{\lambda e}$ and $\epsilon_{\lambda f}$ are the extinction coefficients at the wavelengths of the exciting and fluoresced light, respectively. In this case for small concentration changes

$$\delta I_f/I_f = \frac{l(\epsilon_{\lambda e} + \epsilon_{\lambda f})e^{-(\epsilon_{\lambda e} + \epsilon_{\lambda f})c_i l}}{1 - e^{-(\epsilon_{\lambda e} + \epsilon_{\lambda f})c_i l}} \delta c_i \tag{11}$$

and the change in fluorescence intensity is again directly proportional to the concentration change. In the case of fluorescence the temperature-jump cell must be specially designed so as to eliminate interference between the exciting and fluorescent light. This will be discussed further in the section on cell designs.

Changes in optical rotation can also be related to concentration:

$$I = \text{const. } I_0 \sin^2 (\sigma + 10^{-2}[\phi]lc_i) \tag{12}$$

where $\sigma$ is the angle of rotation between the polarizer and analyzer, and $[\phi]$ is the molar rotation of the optically active species. Again for small concentration changes

$$\delta I/I = 2[\cot (\sigma + 10^{-2}[\phi]lc_i)]10^{-2}[\phi]l\delta c_i \tag{13}$$

For measuring changes in optical rotation a polarizer is placed in the incident beam and an analyzer is placed in the beam as it emerges from the cell. Any type of polarizer and analyzer may be used: Glan-Thompson prisms are probably most efficient, while Polaroid polarizers are most economical. It is convenient to fix one of these elements, while the other is allowed to rotate so as to vary the angle $\sigma$. A cell of long path length is necessary to produce measurable changes in rotation; a special cell designed for optical rotation measurements will be described in a subsequent section.

Changes in light intensity can be conveniently followed with a photomultiplier. A simple photomultiplier and emitter follower circuit is shown in Fig. 3. This circuit is adequate for measurement of relaxation times longer than 1 μsec when the minimum load resistor is used. The components are mounted in a single compact metal housing in order to minimize capacitance between stages and interference with the signal by external fields. Additional shielding with mu metal is sometimes necessary. The cathode of the photomultiplier is connected to a well-regulated power supply.[7]

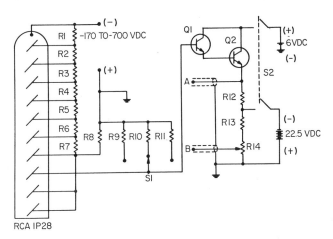

Fig. 3. Schematic diagram of a photomultiplier emitter follower. *Q1*, *Q2*, 2N336. *R1*, 150 KΩ; *R2–R9*, *R13*, 100 KΩ; *R10*, 300 KΩ; *R11*, 1 MΩ; *R12*, 5 KΩ; all resistors are 0.5 W and ±5%. *R14*, 100 KΩ potentiometer, 2 W, linear taper. *S1*, 1-pole, 3-position ceramic rotary switch; *S2*, DPST toggle switch.

The A (signal) and B (reference) outputs are connected to the corresponding inputs of a differential preamplifier (Tektronix Type 1A7) in a Tektronix Type 549 oscilloscope. Any similar oscilloscope-preamplifier combination would be suitable. The external trigger input is connected to the trigger antenna of the spark gap assembly. If the initial vertical position is properly set, a single trace will appear on the screen at the instant of triggering. To record an event, the shutter of the camera (Type C-12) that is mounted on the oscilloscope is opened, the spark is triggered manually, and the shutter is closed. The single trace that appears may be viewed directly on the screen and/or may be recorded on film. If a "memory" oscilloscope is used, the photograph can be taken after the event is viewed.

Although the detection circuit described employs a single beam, it is

[7] John Fluke Co., Inc., Seattle, Washington, Catalog No. 409A, 170–1530 V, 0–3 mA, ±0.1% regulation.

still a differential method since the initial photomultiplier signal is nulled by an adjustable potentiometer (output B). Changes in absorbance as small as $10^{-3}$ can be detected with this arrangement. In some cases, double-beam instrumentation may be advantageous; this is usually true if the lamp output is not stable. In this case the beam of light can be split by a prism or half-silvered mirror, one of the beams going through the cell to a photomultiplier, the other going through air to a second photomultiplier. The outputs of the two photomultipliers are balanced against each other via the oscilloscope preamplifier. An exact null is obtained through use of diaphragms that can be used to vary the amount of light incident on each photomultiplier. Single-beam operation is preferred because of the higher light intensity which passes through the cell (see below).

The RCA 1P28 tube that is shown in the circuit diagram of Fig. 3 contains a highly sensitive photocathode with a maximum in its spectral sensitivity at $340 \pm 50$ m$\mu$. The last two dynodes are held at the same potential as the seventh one so that the photomultiplier may be operated at a high light intensity without exceeding the limits of safety prescribed for the anode current. In some cases, it may be desirable to eliminate even more of the dynode stages. Any other photomultiplier with desirable spectral characteristics may be used. The voltage applied to the photomultiplier should be as large as is consistent with a linear response of the photomultiplier and the tube specifications, but not so great that the photocathode or anode become saturated.

The most important aspect of the design of a detection circuit is maximization of the signal to noise ratio $(S/N)$, which is given approximately by the equation

$$\frac{S}{N} = 1.4 \times 10^9 \left(\frac{\delta I}{I}\right)\left(\frac{I_0 A_g S_\lambda}{\Delta f}\right)^{1/2} e^{-\epsilon_i c_i l/2} \tag{14}$$

where $\delta I/I$ is the relative change in light intensity, $I_0 A_g$ is the available light intensity, $S_\lambda$ is the sensitivity of the photocathode at a particular wavelength, $\Delta f$ is the bandwidth; other symbols have been previously defined. Equation (14) is valid at room temperature and above, where noise is mainly provided by the "shot effect." Under these conditions the optimum signal-to-noise ratio for a single light-absorbing component is obtained at an absorbancy of 0.868. (This value is obtained by differentiating $S/N$ with respect to $\epsilon_i l$ and setting the derivative equal to zero.)

Clearly, the light intensity at the photomultiplier cathode should be as high as possible, within the region where the response of the photomultiplier to changes in light intensity is linear, and the bandwidth of the detection circuit should be as narrow as possible. With regard to the first requirement, it is the surface brightness, rather than the total energy output of the lamp, that is important. The lamp should provide as nearly as possible a point source of light and constant light intensity over the time range 1 $\mu$sec to 1 sec within an appropriate spectral range. The optical arrangement should be as efficient as possible. Generally this involves use of an efficient monochromator and use of lenses that either pass the light through the cell as a parallel beam or focus the light beam at the center of the cell. The light intensity should be uniformly distributed on the surface of the photomultiplier. Also the light beam must be sufficiently shielded from the electrodes so that transient changes at the electrode surface do not interfere with detection of the concentration changes. Use of recently developed photodiodes instead of photomultipliers may be advantageous.[8]

Quartz lamps with a tungsten-iodine cycle are operated at color temperatures of 2600°–3000°K, where light intensity is proportional to the third or fourth power of voltage. The low-voltage, direct-current power supply for such lamps must therefore be well regulated. Suitable commercial power supplies are available,[9] or a bank of storage batteries can be used. A 100-W quartz lamp with a short filament and an optimum color temperature of 2950°K is available commercially.[10] This lamp may be operated somewhat over the specified current ($\sim$10%) to increase its intensity. The lower limit of its useful spectral range is about 300 m$\mu$. It may be mounted vertically in a commercial illuminator or secured to an optical bench that holds a monochromator. Although high-pressure xenon, mercury, and deuterium lamps can provide light of high and constant intensity from a point source over a wide spectral range, random oscillations in the geometrical position of the arc lead to temporal variations in the intensity of light that is received at the photomultiplier. This problem was particularly noticeable with the Osram lamps (XBO 150 W, HBO 100, and HBO 200) that were employed in earlier investigations, although the difficulty could be mitigated by careful alignment of the optics and use of a double beam arrangement. This problem is somewhat less serious in capillary arc lamps. High-pressure lamps of improved design that are now available[11] are recommended for irradiation with ultraviolet light. Again a

---

[8] E G and G, Electronic Products Division, 160 Brookline Ave., Boston, Massachusetts; Silicon Photodiode SGD-444.

[9] Kepco, Inc., 131–38 Sanford Ave., Flushing, Long Island, New York 11352. Model No. SM75-8M or other similar models.

[10] Large Lamp Department of the Lamp Division of General Electric Co., Nela Park, Cleveland, Ohio. Catalog No. Q6 6A/T3/1CL, 45 or 100 W.

[11] For example, Large Lamp Department, General Electric Co., Nela Park, Cleveland, Ohio: 85-W mercury lamp, Catalog No. H85A31UU, and 500-W Xenon lamp, Catalog No. XE500T14; or Hanovia Lamp Division, 100 Chestnut St., Newark, New Jersey has a general line of useful arc lamps.

well-regulated power supply is needed. Either grating or prism mono-chromators may be employed where optical efficiency is appropriate to the spectral range that is required. Interference filters can also be used although they are not as convenient as a monochromator. A prism monochromator, such as the Zeiss M4QIII, has 35–40% transmittance in the range 250–700 m$\mu$, whereas a grating monochromator, such as Bausch and Lomb No. 33-86-26-07, has 30–55% transmittance in the same range.

As mentioned above, the bandwidth of the detection circuit should be as narrow as possible, consistent with the time constants being measured. Variation of the bandwidth can be accomplished by changing the load resistance of the photomultiplier. In the circuit described, the load resist-

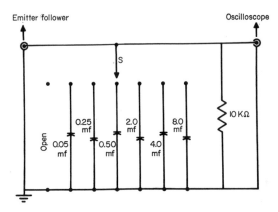

Fig. 4. A simple RC filter circuit for reducing high-frequency noise. (Courtesy of Dr. K. Kustin.)

ance can be varied from about 1 kohm to 1 Mohm. The variation in reso-lution time of the detection circuit is typically from less than 1 $\mu$sec to 150 $\mu$sec, whereas $S/N$ varies from about 300 to 3000. These numbers are intended only as approximate indications of the possible variations in $S/N$. Another simple way to increase the signal to noise ratio is to remove the high frequency noise by passage of the signal from the emitter follower through a simple RC filter such as that shown in Fig. 4. This filter circuit, when used with the emitter follower previously described, permits the apparatus time constant to be varied in steps from about 1 $\mu$sec to 2 msec, depending on the capacitance that is selected. In general a safe rule is that the resolution time of the apparatus be less than one-tenth of the relaxa-tion time being measured.

If several relaxation processes occur, improved precision in the measure-ment of the individual relaxation times can be obtained by eliminating all signal changes from the oscilloscope input which are much faster than the relaxation process of interest. This can be accomplished by use of a fast switching circuit which grounds out the photomultiplier signal until a specified time after the temperature jump. The basic components of a transistor switching circuit designed by Dr. T. B. Lewis are shown in the upper part of Fig. 5. The output of the photomultiplier tube is sent to an emitter follower with an output impedance less than 100 ohms and then to

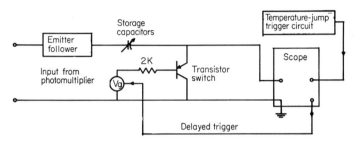

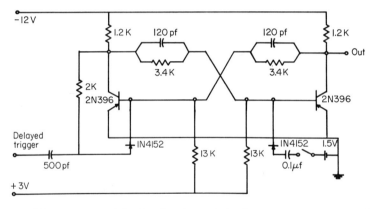

FIG. 5. The upper part of the drawing shows the basic components of a fast switching grounding circuit for a temperature-jump apparatus; the lower half of the drawing is the saturated flip-flop circuit used as a gating source ($V_g$) for the transistor switch. (Designed by Dr. T. B. Lewis.)

a bank of storage capacitors. When conducting, the transistor is a switch in the closed position and the oscilloscope is grounded, while with the transistor in a nonconducting state the switch is open and the oscilloscope is in series with the storage capacitors. The switch is maintained in the closed position until after the temperature jump has been produced, and opened at a predetermined time to observe the signal of the relaxation effect.

The transistor used for switching is a silicon chopper transistor, 2N 3319. The transistor is located in shunt across the output and operated in the "inverted" mode. The gating voltage, $V_g$, applied between base and collector maintains the switch closed by supplying a base current adequate to keep the transistor in saturation. When the gating source is triggered by a delayed pulse from a Tektronix oscilloscope, the voltage at the base changes in a microsecond to a value adequate to drive the transistor beyond cutoff, thus opening the switch. When conducting the chopper transistor has an offset voltage of less than 2 mV and very low emitter to collector resistance ($<10$ ohms). In the nonconducting state the resistance across the transistor is much greater than the impedance of the oscilloscope ($10^6$ ohms). The gating source is a saturated flip-flop circuit employing 2N396 transistors, as indicated in the lower part of Fig. 5. Two conditions must be satisfied for the observed trace on the oscilloscope to correspond only to the relaxation effect. First, when the switch is in the open position the charging time of the storage capacitor in series with the oscilloscope must be much longer than the observation time across the screen of the oscilloscope. It is necessary that only a negligible amount of the signal from the relaxation process be across the capacitor and the entire signal appear on the screen of the oscilloscope. Capacitors were selected for each time range (20 $\mu$sec–50 msec) so that less than 2% of the amplitude of the relaxation effect would have decayed onto the capacitor. All the capacitors were mylar (Balco Capacitor Co.), and the range of capacitance was 0.008 $\mu$F to 24 $\mu$F. The second condition involves the charging time of the storage capacitor and the series resistance in the closed position. A sufficient time must be allowed for the very fast signal change corresponding to the temperature jump from the initial equilibrium state to the higher temperature to appear completely across the capacitor. The switch is then opened to observe the relaxation process (from the higher temperature to the final equilibrium state) on the oscilloscope. The opening of the switch is accomplished by utilizing the horizontal display on the Tektronix 549 oscilloscope in the "A delayed by B" position. The scope is initially triggered at the start of the temperature jump and the delayed pulse from the oscilloscope triggers the gating voltage and initiates the delayed sweep on the oscilloscope. The series resistance with the switch closed is sufficiently small that the time for the start of the delayed pulse can be set at approximately 20% of the time/division on the observation time scale. This switch is an integral part of the stopped flow temperature-jump apparatus to be discussed later.

4. *Cells.* The most critical part of the temperature-jump apparatus from the aspect of design is the cell that holds the reaction mixture. Many different designs have been tested. We will describe here four of the cells in current use.

The following conventions are observed unless otherwise stated.

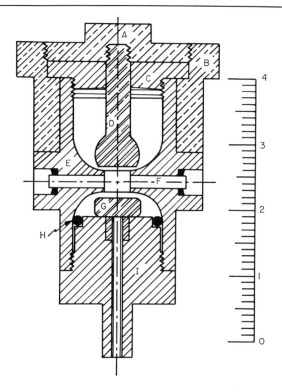

Fig. 6. Standard temperature-jump cell. A cross section in the plane of the cylindrical axis and light path is shown. *A*, Thermostat cap, brass 1-inch hex head and $2\frac{1}{4}$-24 thread. *B*, Thermostat, brass $3\frac{3}{32}$ outer diameter (o.d.) max $\times$ $1\frac{5}{8}$ high. *C*, Cell cap, plastic, $1\frac{3}{16}$ o.d. max $\times$ $\frac{7}{16}$ high. *D*, Negative electrode, stainless steel, $\frac{3}{4}$ o.d. max, $\frac{9}{16}$ face. *E*, Centerpiece, plastic, $2\frac{19}{32}$ o.d. max $\times$ $2\frac{39}{64}$ high; 24 thread; rectangular chamber between electrodes $0.394 \times 0.33 \times 0.312$ deep, corners $\frac{3}{64}$, 0.201 drill-through for windows. *F*, Window, quartz rod, high grade, 5.0 mm D $\times$ 22 mm long; optically polished and plane parallel faces to 1 wavelength (A. D. Jones Optical Co., Burlington, Massachusetts); cemented with epoxy resin. *G*, Positive electrode, stainless steel, $\frac{3}{4}$ o.d. max $\times$ $2\frac{1}{32}$ high, with tubulation $\frac{3}{16}$ o.d., light drive fit to part *I*. *H*, Gasket, Teflon, $1\frac{3}{8}$ o.d. $\times$ $\frac{1}{8}$ D. *I*, Bottom, plastic, $2\frac{1}{8}$ high.

Dimensions are in the English system; drawings are to scale; "plastic" refers to poly (methyl methacrylate) (Plexiglas, Lucite, Perspex, etc.), and all items of this material are thoroughly polished with an aqueous suspension of aluminum oxide and then stress-relieved at an elevated temperature; nonmagnetic types of stainless steel are used.

The design that is shown in Fig. 6 is the type most commonly used. It is patterned after a cell first described by Diebler.[12] This cell requires 8 ml

---

[12] H. Diebler, Doctoral Dissertation, Göttingen, Germany, 1960.

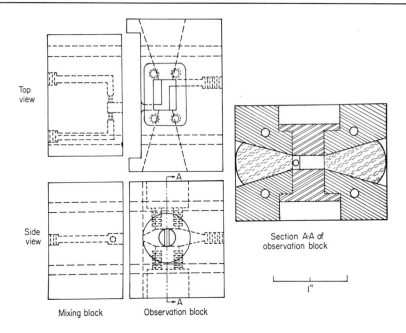

Top
view

Side
view

Mixing block        Observation block

Section A-A of
observation block

1"

FIG. 7. Construction of a stopped flow-temperature jump cell or micro temperature-jump cell.

*Kel-F observation block:* Wells that receive the flanges of the electrodes, 0.813 × 0.563 × 0.281 inch deep, are milled in the top and bottom of a block 1.25 × 1.5 × 2 inches, and the wells are extended (0.320 × 0.394 × 0.369 deep) to contain the main body of each electrode. A channel 0.080 inch wide keeps the electrodes at a fixed distance. As shown in the top view, this channel begins at the inlet corner of the well for the main body of an electrode, makes a right-angle turn at the centerline of the well, continues for 1 cm to form the observation channel, and exits in a right turn. To complete the channel, holes (0.080 in diameter) are drilled into the centers of the inlet and outlet legs, and the exit channel is threaded to receive a stainless steel connector. Optical grade quartz is used for the conical lenses, which are 0.787 inch long, with a polished plane face 0.197 inch in diameter and a polished spherical face of 0.316 inch radius of curvature. The tapered portion, which has a half-angle of 15°, is not polished.

*Plexiglas mixing block:* An exit channel, ⅛ inch in diameter is counterbored to a depth of 0.200 inch in a position that matches the inlet of the observation block. Inlet channels are drilled (No. 47) in the mixing block (1 × 1.5 × 1.563 inches) ¾ inch apart on center to a depth of about ⅞ inch; the right-angle turns are made by drilling and counterboring 0.180 inch from the exit face of the block to within 0.080 inch of the exit channel; each inlet channel is connected to the exit channel by boring two sets of jets, 0.020 inch in diameter offset from the centerline of the exit channel; plugs are inserted to complete the inlet channels.

*Assembly:* Each electrode is mounted to the Kel-F body with four machine screws. The exit hole of the mixing block is aligned with the inlet of the observation block, and four holes are drilled through the two pieces. The entire cell is then reassembled with appropriate Teflon gaskets, and the lenses are held in place with epoxy cement [brass of solution, and the length of the light path between quartz windows is 1 cm. When the faces of the electrodes are 0.47 inch apart (as shown) the volume that is heated by the electrical discharge is nearly 1 ml. When the solution between electrodes is 0.1 $M$ KNO$_3$, the resistance, $R$, of the cell is about 140 ohms (Table I). Discharge of an 0.1 mF capacitor at 25 kV will produce 8° rise in temperature with a time constant of 7 $\mu$sec. With this cell, repeated temperature jumps can be readily applied to the solution since thermal equilibrium is rapidly reestablished by convective mixing with the unheated solution.

Variations in these suggested operating conditions may be in the direction of higher values for applied voltage and lower values of capacitance. Care must be taken that no sharp edges are in the neighborhood of the electrode, or sparking may occur. The faces of the quartz rods must be flush with the walls of the chamber. It is quite important that the Teflon gasket and the threads of the bottom piece be coated lightly with silicone grease in order to prevent leakage of the solution. After the cell has been filled, the top, which consists of the thermostat and negative electrode, is screwed on, and any bubbles that have been trapped in the lower chamber or beneath the negative electrode are removed. The entire assembly may then be placed in a cylindrical thermostatted support that is mounted in the light path.

In Fig. 7, the temperature jump portion of a stopped-flow temperature apparatus is depicted. This can also be used as a microcell for an ordinary temperature-jump apparatus. The plastic used was Kel-F rather than Plexiglas since this material can be conveniently cleaned with nonaqueous solvents. This cleaning eliminates the formation of small bubbles on the walls that often interfere with absorbancy measurements. The total volume of this cell is about 0.16 ml and an 8° temperature jump can be obtained with a 0.1 $\mu$F capacitor and 10,000 V. The heating time constant is about 12 $\mu$sec with 0.1 $M$ KNO$_3$ in the cell. A thermostatted brass plate can be mounted on top of the cell in thermal contact with the ground electrode for temperature regulation. Conical quartz lenses focus the beam of light in the middle of the cell. Such an arrangement considerably enhances the signal to noise ratio. This cell is advantageous because of the small volumes required, but it has the disadvantage that repetitive temperature jumps may be inconvenient to apply.

---

retaining rings and "O" rings may also be mounted on the observation block at the end of each lens (not shown)]. The lower electrode is connected to the high-voltage source with a lug surrounded by an insulating sleeve, and the upper electrode is connected through the thermostat to the negative terminal of the high-voltage source.

A temperature-jump cell suitable for the detection of changes in optical rotation is shown in Fig. 8. This type of cell was first used by Lumry and his co-workers.[13] This cell, which has a 10-cm path length, requires 4 ml of solution and an applied voltage of 30–50 kV in order to get a sufficiently large temperature jump. With such large voltages, the electrodes must be very well polished. The windows, which are just quartz plates, about $\frac{1}{32}$ inch thick, can be mounted with stopcock grease or a glue. A portion of the solution in the light path is not heated and serves to dampen the shock wave accompanying the temperature jump. For most

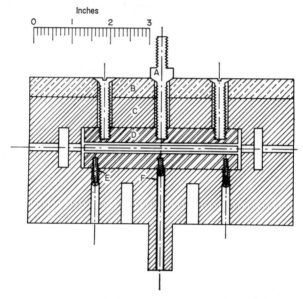

Fig. 8. Temperature-jump cell of 10-cm path length. Cross section through the axes of light path and electrodes, drawn to scale as indicated. $A$, Low-voltage terminal, $\frac{5}{8}$ o.d. $\times$ $2\frac{1}{2}$ long, threads 12-24 $\times$ $1\frac{3}{8}$ long and $\frac{1}{4}$–20 $\times$ $\frac{3}{4}$ long. $B$, Brass thermostat, $6\frac{13}{16} \times \frac{7}{12} \times \frac{1}{2}$. $C$, Plastic body, $6\frac{13}{16} \times 3\frac{3}{16} \times 2$, clearance for negative terminal on center and for 2 brass retaining screws on negative side and 2 nylon retaining screws on positive side; maximum dimensions of open chamber 1.250 deep $\times$ $4\frac{3}{16} \times 1\frac{1}{16}$; ridge 4 $\times$ $\frac{3}{16} \times 0.12$ high; upper section of stepped well (not shown), $1\frac{1}{16} \times \frac{1}{2} \times \frac{3}{4}$ deep and $\frac{1}{2}$ wall on outside edge, lower section of well (shown in cross section), $1\frac{1}{16} \times \frac{7}{32} \times \frac{1}{2}$ deep with $2\frac{5}{32}$ wall on outside edge; light path $\frac{3}{16}$ drill through on center; plug $1\frac{7}{8}$ o.d. max with concentric channel $\frac{5}{16}$ wide $\times$ 1 deep and tip $\frac{5}{8}$ o.d. $\times$ $1\frac{1}{8}$ long. $D$, Stainless steel electrode, 4 $\times$ $\frac{3}{8} \times \frac{3}{8}$, all exposed edges rounded $\frac{1}{16}$ R, polished all over. $E$, Clearance for 6-32 nylon screw, $c'$ bore 0.260 $\times$ 1 deep. $F$, High-voltage terminal stainless steel tube, $\frac{3}{16}$ o.d. $\times$ $2\frac{1}{2}$ long $\times$ $\frac{1}{32}$ wall, soldered to 6-32 stainless steel screw; sliding fit to plastic body. The plastic top, which is not shown, has a ridge which fits between the electrodes and is 0.25 from the ridge at the bottom of the cell.

[13] R. Lumry, private communication.

cases where absorption changes are measured, a 10-cm path length is not necessary and therefore this type of cell has not been widely used.

A cell suitable for fluorescence and absorption measurements has been described by Czerlinski.[14] A cross section of this cell is shown in Fig. 9. In the preferred optical arrangement for fluorescence measurements, two cells with sample and reference solutions are mounted equidistant and 90

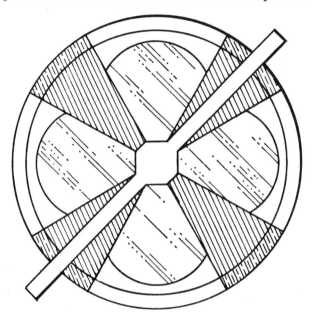

FIG. 9. Cross section of a fluoresence temperature-jump cell in the plane of the optical paths, after Czerlinski. Four conical light-gathering lenses of glass or quartz are mounted in a nylon or Teflon body of outer diameter 6 cm. Their plane faces, 5 mm in diameter, form four sides of an octagonal cavity that is 1 cm between these faces. The lenses are 24.2 mm long, with an included angle of 60° and a radius of the spherical surface of 14.2 mm. They are held in place with brass retaining rings. The cell is filled through the hole that passes through the center of the cavity. Thermostatted electrodes are mounted above and below the cross section of the cavity that is shown. The entire cell is surrounded by a brass tube that is grounded (see text footnote 14).

degrees apart with respect to the light source (e.g., General Electric high-pressure mercury arc type BH6). This cell was the first to have conical lenses that focus the light in the middle of the cell. The two photomultipliers are mounted equidistant and 90 degrees apart from the incident light beams to complete the W-shaped array. It is imperative that a very intense light source be used for fluorescence measurements in order to achieve a suitable

[14] G. Czerlinski, *Rev. Sci. Instr.* **33**, 1184 (1962).

signal to noise ratio. The original article should be consulted for further details.

### B. Temperature Pulse by Dielectric Relaxation[14a]

Ertl and Gerischer[15] have described a temperature-jump method that employs a microwave impulse generator to heat small volumes of solution. The absorption of microwave radiation by water is maximal at a frequency of about $10^{10}$ sec$^{-1}$ (3 cm). At this frequency about half of the energy will be absorbed after passing through 2 mm of water. Microwave generators are available that will produce about a 1° temperature jump in approximately 1 $\mu$sec. This method of heating is distinctly advantageous where

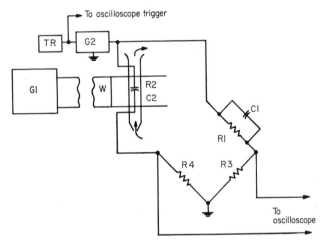

FIG. 10. Schematic diagram of the temperature-jump apparatus of Ertl and Gerischer (see text footnote 15). The microwave impulse generator, *G1*, transmits 3-cm radiation through a ferrite isolator and waveguide, *W*, to a cavity that contains a glass cell of about 3 mm inside diameter that is filled with an aqueous solution. The dimensions of the cavity, the thickness and orientation of the glass capillary, and the dielectric constant of the glass are chosen to provide maximum absorption of energy. Fluid may be passed through the cell continuously. The electrodes of the cell (*R2, C2*) are connected to a bridge to which a potential is applied when a trigger, *TR*, actuates a square-wave generator, *G2*. The trigger simultaneously actuates the sweep of an oscilloscope, which records the change in the conductance of the solution caused by the temperature jump.

water solutions of low ionic strength or nonaqueous solutions, whose microwave absorption regions are accessible, are to be investigated. However, the temperature jump is so small that thus far only a sensitive conductance method has been utilized to follow the concentration changes.

Of the components of the apparatus (Fig. 10), the microwave impulse

---

[14a] See also this volume [9].

[15] G. Ertl and H. Gerischer, *Z. Elektrochem.* **65**, 629 (1961).

generator is the most expensive.[16] Fortunately rebuilt generators, or parts of them, are often available from manufacturers of X-band heavy radars at a considerable saving. Once the generator has been chosen, it is necessary to compute optimal values for parameters that will eliminate reflections and will permit the maximum possible energy to be delivered to the sample. The original investigators irradiated their cell at a frequency of $9.6 \times 10^9$ sec$^{-1}$ and positioned the cell 0.78 cm from a node of the standing wave by use of a flexible waveguide of standard thickness (2.85 cm, MIL No. RG51/U). They could then calculate to a good approximation optimal values for the dielectric constant of the cell wall (6.92) and the cell wall thickness (2.6 mm). In this calculation the assumption was made that absorption by air and the cell wall is negligible. The cell was constructed from Jena Normalglas 16, which has a dielectric constant of 6.94 and a loss factor $tg\delta$, of 0.024 at a frequency of $10^{10}$ sec$^{-1}$. Use of a slightly smaller standard X-band wave guide, MIL No. RG52/U, changes the optimal dielectric constant to 6.76, a value that is close to the dielectric constant of domestic soda-lime glass (Corning No. 0080 with a dielectric constant of 6.71 and a loss factor of 0.017 at a frequency of $10^{10}$ sec$^{-1}$). Thus the parameters may be varied within somewhat narrow limits to facilitate the choice of components. The optimum inner diameter of the cell was 3 mm. Tubes of larger bore give rise to inhomogeneous heating as well as to a very small temperature jump. With tubes of smaller bore the standing wave tends to "overshoot" and accommodation of the electrodes is difficult.

Microwave heating has not been used very extensively, not only because it produces a much smaller temperature jump than is easily attained by Joule heating, but also because the apparatus is quite expensive.

## C. Temperature Pulse by Optical Heating[16a]

The temperature of a solution can also be raised rapidly by the absorption of light. The absorption of visible or ultraviolet light from a flash lamp has been used as a temperature-jump technique.[17] This method suffers from the drawbacks that the solution must have a high absorption coefficient at the wavelength of the light and that it is difficult to obtain high-power flash lamps with sufficiently short duration outputs. The use of lasers as heating devices is also possible. In principle, very short laser pulses (about $10^{-8}$ to $10^{-9}$ sec) can be used to cause temperature jumps. In practice, however, only small amounts of energy can be obtained in these

---

[16] Raytheon Co., Waltham, Massachusetts, manufactures a generator that is composed of a modulator, a magnetron, QKH 172; a circulator, CXH17; and a directional coupler. This apparatus delivers 400 kW with a duty cycle of 3.2 $\mu$sec. The pulse width may be varied between 1 and 7 $\mu$sec.

[16a] See also this volume [9].

[17] H. Strehlow and S. Kalarickal, Z. Elektrochem. 70, 139 (1966).

short pulses so that very small temperature jumps are obtained. Some success has been achieved in the use of a laser as a heating device,[18] but thus far none of the optical techniques appear to be as useful as resistive heating for generating temperature jumps.

## III. Stopped Flow-Temperature Jump Apparatus

The equilibrium temperature-jump method is limited to systems where the concentrations of reactants are comparable. If a steady state system is produced and some external parameter is varied, the approach of the system to a new steady state can be observed. This approach is again characterized by a spectrum of relaxation times. By use of rapid mixing techniques, steady states can be established within 1 msec. The feasibility of combining flow and temperature-jump methods has been often discussed.[19, 20] However, practical solutions to the many technical problems are still being solved. Two approaches are possible: one is to use a continuous flow of mixed reactants and to apply a temperature jump to the solution as it flows by an observation chamber; the other is to stop the flow rapidly and set up a *pseudo* steady state in which the concentration of some intermediate is essentially constant for a time longer than a few milliseconds. In practice a single apparatus can be used for both purposes, although optimal performance for both systems in a single apparatus is probably not possible. The apparatus proposed by Eigen and de Maeyer permits temperature jumps to be applied to solutions flowing through a tube of 2-mm bore. In our laboratory, we have built a stopped-flow temperature jump apparatus that has the advantage of small volume per experiment (less than 0.5 ml) with a 1-cm observation path. With stopped flow, electric field inhomogeneities due to flowing liquid and electric field breakdown due to cavitation of the flowing liquid do not occur. The continuous flow method is limited in the length of relaxation times that may be observed by the rate of flow, the volume of the observation chamber and the volume of the fluid heated. This is because all the liquid flowing by the observation point during the detection of the relaxation process must be heated.

The details of the stopped-flow temperature jump are available,[21] and

[18] W. Silfvast, J. Assay, and E. Eyring, *152nd Natl. Meeting, Am. Chem. Soc. New York, Sept., 1966*, Paper V-136; H. Hoffmann, E. Yeager, and J. Stuehr, *Rev. Sci. Instr.* **39**, 649 (1968).

[19] M. Eigen and L. De Maeyer *in* "Rapid Mixing and Sampling Techniques in Biochemistry" (B. Chance, R. H. Eisenhardt, Q. H. Gibson, and K. K. Lonberg-Holm, eds.), p. 175. Academic Press, New York, 1964.

[20] G. Czerlinski, see footnote 19, p. 183.

[21] J. E. Erman and G. G. Hammes, *Rev. Sci. Instr.* **37**, 746 (1966); E. J. Faeder, PhD Thesis, Cornell University, Ithaca, New York, 1969.

only the essential features of the apparatus will be presented here. A block diagram of the equipment is shown in Fig. 11. The flow apparatus is conventional in design (see this volume [6]). The four-jet mixer shown in Fig. 7 is coupled to the temperature-jump cell by a pressure fitting with a Teflon gasket. The mixing chamber is connected to 2-ml glass driving syringes by means of stainless steel connectors and short lengths of polyethylene tubing. The syringes are driven by a compressed air piston. The driving mechanism is mounted on a separate table from the flow and detection system to reduce the effect of mechanical vibrations on signal detection. After leaving the observation chamber, the fluid goes to a 5-ml stopping syringe coupled to the temperature-jump cell by a three-way Teflon-coated valve.[22] (Approximately 0.5 ml of each reactant is required for each mixing experiment.) The stopping syringe triggers the temperature jump through a microswitch.

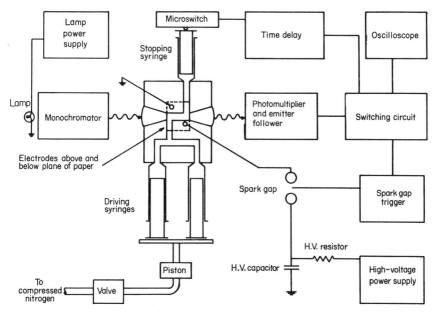

FIG. 11. Block diagram of a stopped flow-temperature jump apparatus (see text footnote 21).

The output of the microswitch goes to a time-delay circuit, which in turn triggers the spark gap. The temperature jump can be applied from 1 msec to several seconds after stopping the flow. The actual discharge circuit and time delay circuit are shown in Fig. 12. The detection system

---

[22] Three-way Hamilton Valve, Hamilton Co., Whittier, California.

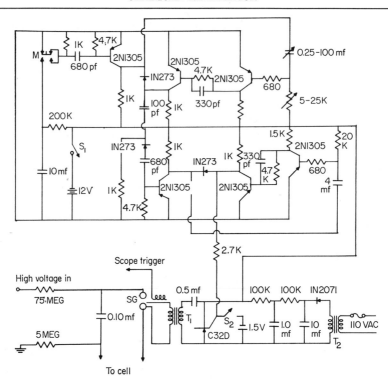

Fig. 12. Time-delay (upper) and trigger (lower) circuits for the stopped-flow temperature jump. $M$ is the stopping syringe microswitch, $T_1$ is a spark gap trigger transformer (TR 148A, E G and G, Boston, Massachusetts), $T_2$ is a 270-Volt output power transformer, and $SG$ is the spark gap (Triggered Spark Gap GP20-A, E G and G, Boston, Massachusetts).

is identical to that described in Section II,A,3. The output of the emitter follower goes to a switching circuit (Fig. 5) which grounds the signal until the temperature jump is applied to the solution. This is necessary because the concentration changes due to the temperature jump perturbation are imposed on a large concentration change caused by the overall reaction.

The dead time of the stopped-flow apparatus is about 10 msec with a driving pressure of about 1 atm. Since the temperature jump can be applied 1 msec after stopping the flow, reactions with overall half times of 15 msec or longer can be studied. The time constant for heating is about 10 $\mu$sec in 0.1 $M$ $KNO_3$. This apparatus should be of great utility for the study of enzymatic reactions that are essentially irreversible.

## IV. Experimental Procedure

In this section a brief description of the experimental procedure for carrying out a temperature-jump experiment with joule heating will be given. A useful test solution is a pH indicator in a buffered medium. If the buffer concentration is sufficiently high, the protolytic equilibria will be adjusted so rapidly that the observed rate of change in pH will be limited by the resolution time of the apparatus rather than by the rates of the chemical reactions that are perturbed. Possible test solutions are $2 \times 10^{-5} M$ phenol red in $0.2 M$ phosphate buffer, pH 7.6, or $2 \times 10^{-5} M$ phenolphthalein in $0.2 M$ glycine buffer, pH 9.2. After the test solution is put into the apparatus, the electrical connections checked, and the optical elements adjusted, the linearity of the photomultiplier should be examined. The desired load resistor is selected and the output of the photomultiplier is located on the screen by adjustment of the potentiometer that is connected to one channel of the differential preamplifier. The total signal change between light and dark, which is generally of the order of 1 V, is measured by blocking the light beam; a neutral density filter of known transmittance (typically about 90%) is then inserted in the light path, and the fraction of the signal lost should correspond to the known transmittance of the neutral density filter. A handy neutral density filter above 300 m$\mu$ is a thin piece of Plexiglas ($\sim\frac{1}{8}$ inch thick) that has been calibrated with an ordinary spectrophotometer. If the response of the photomultiplier is nonlinear, the voltage applied to the photomultiplier must be reduced. This voltage should be as high as is consistent with a linear response. A useful measure of the sensitivity of the detection system is the signal to noise ratio: this can be determined by measuring the peak-to-peak amplitude of the oscilloscope spot or trace (typically of the order of a millivolt) and dividing the total signal change between light and dark by this quantity. The value of the signal to noise ratio depends on the load resistor and filter network used, but it is typically of the order of $10^2$ to $10^3$.

A high-voltage pulse can now be applied to the solution. The sensitivity of the oscilloscope should be as high as is consistent with the noise level (typically about 5 mV/cm), and the spark gap should trigger a single sweep of the screen. The observed signal should display no vertical changes except at very short times, where the heating time of the apparatus should be measurable, or at very long times (greater than a few hundred milliseconds), where convection occurs.

Typical problems encountered at this point are (1) a large amount of 60-cycle noise and (2) the occurrence of cavitation due to the pressure wave accompanying the temperature jump. The 60-cycle noise is almost always caused by improper grounding. Care must be taken to avoid ground loops and to have the ground be as independent of other equipment as possible. Cavitation generally occurs in the first 100 $\mu$sec after the pulse and can be recognized as a large irregular decrease in light intensity. One can prevent this by careful degassing of the solutions and/or by lowering the temperature to the vicinity of 4°.

## V. Analysis of the Data

Very often a reaction mechanism is sufficiently simple or the relaxation times are sufficiently separated on the time axis so that by selection of an appropriate time base a single discrete relaxation time is observed. A typical example is shown in Fig. 13, where a relaxation effect characteristic

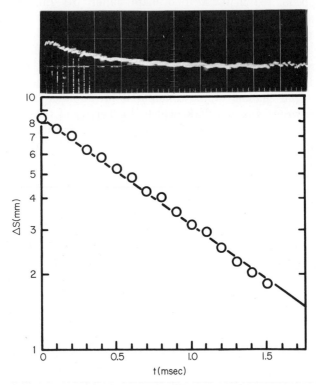

Fig. 13. Oscilloscope trace of the relaxation effect characteristic of ribonuclease isomerization and plot of logarithm of the signal amplitude versus time. $9.65 \times 10^{-5}\,M$ ribonuclease, $2 \times 10^{-5}\,M$ phenol red, $0.15\,M$ KNO$_3$, pH 7.0, 25°. Time scale on trace is 500 $\mu$sec per large horizontal division, $\tau = 1.03$ msec (see text footnote 23).

of the isomerization of ribonuclease is shown.[23] The progress of the reaction is observed by detection of the concomitant pH changes with phenol red. Experimental conditions are specified in the figure legend. For a single relaxation time

$$\Delta S = \Delta S_0 e^{-t/\tau} \tag{15}$$

where $\Delta S$ is the difference between the signal at $t = t$ and $t = \infty$ and $\Delta S_0$

[23] T. C. French and G. G. Hammes, *J. Am. Chem. Soc.* **87**, 4669 (1965).

is the difference between the signal at $t = 0$ and $t = \infty$. Therefore a plot of log $\Delta S$ versus $t$ should be a straight line with a slope of $-1/(2.303\tau)$. Such a plot is included in Fig. 13, and the linearity of the plot attests to the validity of the assumption of a single relaxation process. Another reaction mixture that can be conveniently used for test purposes is the following: 0.15 $M$ KNO$_3$, $2 \times 10^{-5} M$ phenol red, $1.11 \times 10^{-3} M$ Co$^{++}$, $5.34 \times 10^{-4} M$ glycylglycine, pH 7.7. The reaction observed is complex formation and dissociation between Co$^{++}$ and glycylglycine, and the relaxation time is 1.1 msec at 25°.[24] The experimental precision of the relaxation times depends on the amplitude of the observed effect. For the cases under consideration here $\pm 10\%$ is considered reasonable.

In some cases the characteristic relaxation times of a mechanism overlap so much that resolution of the relaxation spectrum into individual relaxation times is extremely difficult. One useful procedure is as follows. The signal amplitude, $\Delta S$, is in general given by

$$\Delta S = \sum_{i}^{n} \Delta S_{io} e^{-t/\tau_i} \tag{16}$$

A plot of log $\Delta S$ versus $t$ will be curved if $i > 1$. However, at sufficiently long times all the terms of the series except that characterized by the longest relaxation time, say the $n$th, will go to zero. Therefore, a straight line can be drawn through the curve at long times ($t \to \infty$) and the relaxation time and $\Delta S_{n0}$ determined. This term can now be subtracted from the series (i.e., the straight line subtracted from the curve) and the resultant series (or curve) is

$$\Delta S = \sum_{i}^{n-1} \Delta S_{io} e^{-t/\tau_i} \tag{17}$$

Repetition of this procedure will yield the $n$ relaxation times. In practice, if the relaxation times do not differ by at least a factor of two, resolution of the relaxation spectrum is not possible. In Fig. 14, an oscilloscope trace of a relaxation process that can be described by two relaxation times is given. The process is the stacking of acridine orange on poly-L-glutamic acid chains.[25] The appropriate logarithmic plot and resolution into two straight lines is also shown. In this case, $\tau_1 = 30$ $\mu$sec and $\tau_2 = 338$ $\mu$sec. Other experimental details are given in the figure legend. The precision in the relaxation times is about $\pm 15\%$. Many alternative procedures can be devised for the resolution of relaxation spectra, but the above method has proved most useful in our hands.

[24] G. G. Hammes and J. I. Steinfeld, *J. Am. Chem. Soc.* **84,** 4639 (1962).
[25] G. G. Hammes and C. D. Hubbard, *J. Phys. Chem.* **70,** 1615 (1966).

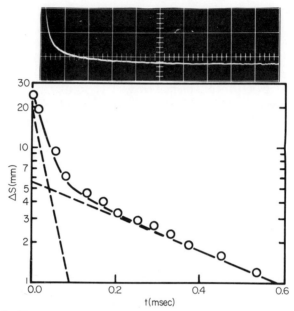

Fig. 14. Oscilloscope trace of the relaxation effect characteristic of the polyglutamic acid-acridine orange interaction and plot of logarithm of the signal amplitude versus time; $5.34 \times 10^{-3} M$ polyglutamic acid (in terms of monomer concentration), $5.0 \times 10^{-5} M$ acridine orange, $0.1 M$ Tris-acetate buffer, pH 7.5, 25°. Time scale on trace is 200 $\mu$sec per large division on the horizontal axis, $\tau_1 = 30$ $\mu$sec, $\tau_2 = 338$ $\mu$sec. The straight lines are those characteristic of the single relaxation times, while the curve is that calculated from the equation $\Delta S$ (mm) $= 18.8e^{-t/\tau_1} + 5.6e^{-t/\tau_2}$ (see text footnote 25).

The problem of interpreting the measured relaxation times in terms of mechanism is beyond the scope of this chapter. All the general pitfalls of any kinetic study are present, and each system investigated poses a unique problem. The recent literature should be consulted for specific examples.

### Acknowledgment

The authors are indebted to their many colleagues who have been actively engaged in the development of the temperature-jump technique. G. G. H. is grateful to the National Institutes of Health for generous support of his work.

# [2] The Pressure-Jump Method

*By* M. T. Takahashi and R. A. Alberty

## I. The Pressure-Jump

### A. Introduction

The pressure-jump technique utilizes a sudden perturbation of chemical equilibria induced by a rapid increase or decrease in pressure; the approach to the new equilibrium concentrations of the chemical species is followed by sensitive conductimetric measurements. Like sound absorption methods, a volume change in the overall reaction being studied is required; unlike sound absorption, however, where stationary techniques are utilized, a single-step function decrease or increase in pressure is utilized. The pressure-jump technique is limited to study of relaxation times of greater than $5 \times 10^{-5}$ second and may be used for reactions that are too slow to be studied by the sound absorption technique. In general, smaller volumes of reactants can be used, and the simplicity of the technique makes it a convenient method for studying aqueous solution reactions whose equilibrium positions can be perturbed by changes in pressure.

The first pressure-jump apparatus was developed by Ljunggren and Lamm.[1] The pressure rise was achieved by opening a valve to a nitrogen tank, and in this manner the pressure could be raised 150 atmospheres in 0.05 second. The course of the adjustment to the new equilibrium condition at 150 atmospheres was followed using conductivity measurements with an alternating current Wheatstone bridge. The rectified off-balance voltage was recorded on an oscillograph. The system $CO_2$ in water was studied in this apparatus.

[1] S. Ljunggren and O. Lamm, *Acta Chem. Scand.* **12,** 1834 (1958).

$$CO_2 + H_2O \rightleftharpoons H_2CO_3 \rightleftharpoons H^+ + HCO_3^-$$

Strehlow and Becker[2] devised an improved apparatus that incorporated a rupture disk into the pressure bomb containing the conductivity cells. Rupturing the disk permitted pressure drops of the order of 50–60 atmospheres within 60 $\mu$sec. A pressure drop to equilibrium conditions at 1 atmosphere does have the advantage that equilibrium data and rate constants obtained are applicable for the conditions employed in other kinetic studies.

Another important improvement by Strehlow and Becker was the addition of a second, reference cell in the pressure bomb. The reference cell provides compensation for the temperature drop associated with the adiabatic expansion of the solution and of the pressure transmitting liquid. Since conductivity is so strongly temperature dependent and may be measured with a very high precision, this temperature drop has a very large effect on the observed off-balance voltage at the end of an experiment. The magnitude of the temperature change accompanying the adiabatic expansion in water at 20°, with a pressure change of 60 atmospheres, is given by

$$\Delta T = \frac{TV\alpha}{C_P} \Delta P \tag{1}$$

where $V$ = molar volume; $C_P$ = molar heat capacity; and $\alpha = (\partial V/\partial T)_P(1/V)$

$$\Delta T = -0.09°$$

For paraffin oil as the pressure-transmitting fluid, the corresponding temperature drop is

$$\Delta T = -0.5°$$

Since the changes in conductivity resulting from a change in pressure are due not only to perturbation of chemical equilibrium, but result, as well, from changes in the concentration of ions due to compression of the solution into a smaller volume and the effects of pressure on the mobilities of the ions, the reference cell provides a way of compensating for these physical changes. The reference cell is customarily filled with a salt solution that has a very low change of conductivity with pressure, such as $KNO_3$ or $KCl$. The relaxation times for the physical changes cited above are much shorter than that for the pressure relaxation, but in lieu of reference cell compensation, their presence would cause an off-balance voltage that might mistakenly be attributed to perturbation of chemical equilibria with very fast relaxation times.

[2] H. Strehlow and M. Becker, *Z. Elektrochem.* **63**, 457 (1959).

Reactions with relaxation times in the range from $10^{-4}$ to 5 seconds can be conveniently studied with the Strehlow apparatus. The upper limit arises from conductivity changes due to thermal equilibration effects that occur following the pressure drop. A thermal relaxation time of the order of 5 seconds can be observed under conditions of high sensitivity.

## B. The Effect of Pressure on Chemical Equilibrium

The pressure-jump method depends upon the existence of a volume change in the reaction under study. Such a volume change makes it possible to displace reactant concentrations from their equilibrium concentrations at one atmosphere, since

$$\left(\frac{d \ln K}{dP}\right)_T = -\frac{\Delta V^0}{RT} \tag{2}$$

where $\Delta V^0$ = standard volume change for the reaction, and $K$ is the molal equilibrium constant.

Assuming a $\Delta V^0$ of $-10$ ml per mole, a reasonable value for a charge neutralization reaction occurring in aqueous solution, and a $\Delta P$ of 65 atmospheres at 25°

$$\frac{\Delta K}{K} = \frac{(10)(65)}{(82)(298)} = 2.7 \times 10^{-2}$$

Thus a 2.7% change in the equilibrium constant results from a 65-atmosphere pressure change. That this is easily within the sensitivity of the detection circuit is seen by the following calculation. Consider the reaction

$$A^+ + B^- \rightleftharpoons C$$

Let

$$\alpha = \frac{m_A}{m_0} = \frac{m_B}{m_0}$$

$$1 - \alpha = \frac{m_C}{m_0}$$

and from Eq. (2)

$$\left(\frac{\partial K}{\partial P}\right)_T = -K \frac{\Delta V^0}{RT} \tag{3}$$

$$\left(\frac{\partial \alpha}{\partial P}\right)_T = \frac{\left(\frac{\partial K}{\partial P}\right)_T}{\left(\frac{\partial K}{\partial \alpha}\right)_T} \tag{4}$$

$$\left(\frac{\partial K}{\partial \alpha}\right)_T = \frac{\alpha - 2}{\alpha^3 m_0} \tag{5}$$

$$\left(\frac{\partial \alpha}{\partial P}\right)_T = \frac{\alpha(1 - \alpha)}{2 - \alpha} \frac{\Delta V^0}{RT} \tag{6}$$

$$\frac{\Delta \alpha}{\alpha} = \frac{(1 - \alpha)}{(2 - \alpha)} \frac{\Delta V^0}{RT} \cdot \Delta P \tag{7}$$

The maximum change $\Delta \alpha$ occurs at $\alpha = 0.586$, where $m_A = m_C$. For $\alpha = 0.586$ we get

$$\frac{\Delta \alpha}{\alpha} = 0.29 \frac{(10)}{(82)(298)} 65$$

$$= 7.8 \times 10^{-3}$$

Thus an 0.8% change in the degree of dissociation results from the pressure drop of 65 atmospheres, and this is easily detectable using the apparatus described below.

The pressure-jump method has been used primarily to study inorganic reactions and simple organic reactions. Such studies have included the reactions involved in the formation of $AlSO_4^+$,[3] $VOSO_4$,[4] $FeCl^{2-}$,[5] and the hydration of pyruvic acid[2] and ninhydrin.[6] Conformational changes in proteins, which are often associated with a change in the number of hydrogen ions or other bound ions, have also been studied with the pressure-jump technique.

## II. The Pressure-Jump Apparatus

### A. Introduction

The pressure-jump apparatus described is designed following the general plan of Strehlow and Becker.[2] While the basic plan follows their apparatus quite closely, some changes have been made for greater convenience in use and to improve the design where possible. The apparatus can be divided into two major components: the pressure bomb or autoclave with its associated parts, and the monitoring and recording system consisting of an alternating current Wheatstone bridge and oscilloscope.

### B. The Pressure Bomb

The pressure bomb containing the conductivity cells is of the design shown in Fig. 1. The autoclave is constructed of stainless steel. The thickness of the stainless steel employed makes it sufficiently strong so that it is

[3] B. Behr and H. Wendt, Z. Elektrochem. **66**, 223 (1962).
[4] H. Strehlow and H. Wendt, Inorg. Chem. **2**, 6 (1963).
[5] H. Wendt and H. Strehlow, Z. Elektrochem. **66**, 228 (1962).
[6] W. Knoche, H. Wendt, M. L. Ahrens, and H. Strehlow, Collect. Czech. Chem. Commun. **31**, 388 (1966).

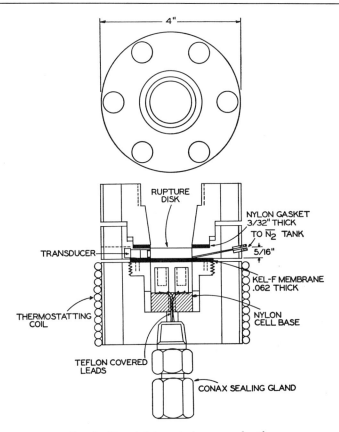

Fig. 1. The stainless-steel pressure bomb.

not the weakest element in the pressurized system, and in addition the rupture disk provides a built-in safety blowout feature.

The most important design consideration in construction of the autoclave is to minimize the apparatus relaxation time $\tau_p$. This is achieved by minimizing both the height of the pressurized gas layer and the distance between the bottom of the conductivity cells and the Kel-F membrane. This requirement arises from the fact that the sharpness of the rarefaction wave which moves down the gas layer decreases as it moves down the unexpanded gas column. Unlike shock waves, which get progressively sharper as they propagate down a shock tube, the rarefaction wave gets more and more diffuse,[7] and thus the pressure relaxation time becomes longer as one increases the height of the layer of compressed gas. Theo-

[7] J. N. Bradley, "Shock Waves in Chemistry and Physics," Chap. 2. Wiley, New York, 1962.

retically the lower limit to the height of the compressed gas layer is dictated by the amount of PV work required to rupture the rupture disk. The disk should rupture catastrophically so that the time of rupture does not determine the pressure relaxation time. Because of the many variables involved, the minimal height of the gas layer required to do this must be determined experimentally. In our apparatus the lower limit is set by the height of the pressure transducer used to trigger the oscilloscope. The Kistler pressure transducer is $\frac{1}{4}$ inch in diameter, and therefore the height of the pressure chamber was made $\frac{5}{16}$ inch.

Strehlow has reported shortening of the apparatus relaxation time by use of thinner layers of pressurizing gas and also by use of liquids in place of gas in the pressurizing chamber.[6] Apparatus relaxation times of 40 $\mu$sec have been achieved using kerosene in the gas chamber.

Some important considerations in the design and operation of the autoclave are the following:

1. The $\frac{1}{16}$ inch bore leading into the pressure chamber from the nitrogen tank is reduced further by welding in $\frac{1}{32}$ inch i.d. capillary tubing. This is done to restrict the flow of gas upon rupture of the rupture disk and thereby reduce the sudden decompressive shocks to the pressure regulator. Such sudden shocks can seriously damage the gauge.

2. The rupture disk is retained by a nylon gasket constructed so that fully 1 inch of the diameter of the disk is secured down. This prevents slippage of the disk under pressure, gives more reproducible breaks, and allows containment of higher pressures. Nylon gaskets $\frac{3}{32}$ inch thick were found to be satisfactory despite a slight tendency to cold flow and swelling in humid weather.

3. A Kel-F membrane (0.062 inch thickness) is used to contain the mineral oil in the cavity. This is used in place of the polyethylene membrane recommended by Strehlow since that membrane tended to buckle and deform during the tightening down of the bomb. Such deformation leads to an extremely distorted surface being presented to the pressure chamber, which in turn could lead to a nonuniform rarefaction wave moving into the cavity. The Kel-F is much more resistant to deformation under pressure and is more transparent. This transparency allows easier visual checking for the presence of unwanted bubbles trapped in the mineral oil. Such bubbles lead to rapid oscillation of the oscilloscope trace after the pressure drop.

4. Mounting of the conductivity cell base in the cavity is a critical part of the autoclave design. Tight anchorage of the cell base to the pressure bomb is highly unsatisfactory because vibration induced by the rarefaction wave causes large oscillations in the oscilloscope trace after approximately 100 $\mu$sec. Presumedly these arise because of changes in capacitive couplings

of the cells to the autoclave; these can be quite large effects, sometimes causing off-scale oscillations of the trace.

The mounting shown in Fig. 1 proved to be satisfactory. The leads of the Teflon-covered wire going through the Conax sealing gland are brought up through a small bore in the center of the cylindrical cell base. The outside diameter of the cell base is made only slightly smaller than the inside diameter of the autoclave. The sole support for the cell base is thus supplied by the Teflon-covered wires. While this would seemingly be an extremely insecure mounting, it eliminates the oscillation of the oscilloscope trace due to vibration.

5. Pressures in excess of the 55 atmospheres cited by Strehlow were tested as a means of increasing the sensitivity of the apparatus. Rupture pressures of 65 atmospheres are now routinely achieved using 0.040-inch stainless steel disks (purchased from Precision Steel Warehouse, Franklin Park, Illinois). The search for a suitable rupture disk material, one that holds to the requisite pressure and yet ruptures catastrophically, is extremely time consuming. There are so many variables involved that determine rupture pressure, including metal thickness, metal temper, tendency to work-hardening, that the only way to decide the suitability of a given material is to try it in the apparatus. Of a variety of materials tried, including 0.004- to 0.007-inch thicknesses of steel, phosphor-bronze, Mylar, copper, stainless steel and brass, only 0.005-inch brass and 0.004-inch stainless steel shim stock were found to be suitable. The former has a spontaneous rupture pressure of 890 psi and the latter 975 psi.

The stainless steel rupture disks cleave cleanly along the circumference and also tend to bulge less under pressure than do materials with less tensile strength. This latter characteristic serves to reduce the height of the gas layer in the compression chamber. Spontaneous rupture pressures can be reproduced to ±20 psi at approximately 975 psi. In order to get reproducible spontaneous rupture pressures, care has to be taken to raise the pressure in approximately the same manner every time. Too rapid raising of the pressure results in small pinholes being made in the disk by the rupturing needle instead of catastrophic rupture. Also the puncturing of the disk should be done within 10 seconds of reaching the desired pressure, since otherwise work-hardening occurs that can also result in pinholes instead of catastrophic rupture. Strehlow[4] has found that, at pressures near 90 atmospheres, cavitation in the conductivity cells becomes a very serious problem. Cavitation effects can be reduced by careful degassing of the solutions before pressurization.

Rupturing the stainless steel disk produces a substantial report. In order to reduce this noise and to prevent the fragments of stainless steel from flying out at high velocity, a muffler is mounted on top of the pressure

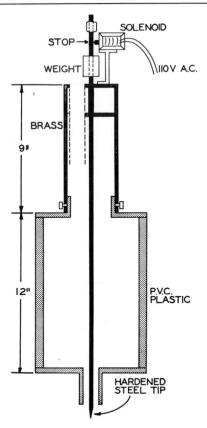

Fig. 2. The muffler and the solenoid used to retain the rupturing needle.

bomb. The muffler is constructed of polyvinyl chloride plastic pipe approximately 17 cm in diameter with brass end flanges. The lower flange is threaded to allow screwing it into the top of the autoclave. The muffler volume must be made sufficiently large so as to reduce the possibility of a relaxation time due to time of air escape.[4] The upper part of the muffler is baffled, and is used to mount the solenoid and puncturing rod. Lining of both top and bottom sections of the muffler with 1½-inch Corning acoustical glass wool matting (absorption coefficient approximately 0.8) reduces the noise level to a comfortable level and in addition gets rid of microphonics. The glass wool also serves to reduce the possibility of reflection of shock waves back into the pressure chamber. Figure 2 shows the final form of the muffler used. After about 100 experiments the glass acoustical wool must be renewed because the metal fragments destroy the mat and microphonics can be detected again.

As shown in Fig. 2, a weighted brass rod with a hardened steel tip is held suspended by a stop. When the desired pressure has been reached this stop can be pulled out suddenly by activating a solenoid (Guardian Electric 115 V, continuous duty) to initiate the pressure drop.

The special miniature conductivity cells were made by Research Apparatus, Inc., Franklin Park, Illinois, following the design of Fig. 3. Total volume contained is 0.3 ml with cell constants of 0.8–1.0 cm$^{-1}$. The cells are made with bright platinum rectangular electrodes sealed flush into the glass, care being taken to minimize any sharp edges in the cells. Such

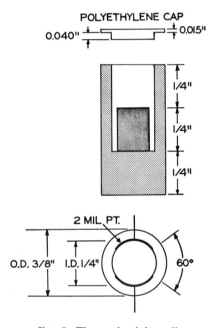

FIG. 3. The conductivity cells.

precautions are necessary since cavitation effects are aggravated by any such rough surfaces. Cell caps made of polyethylene can be used to cap the cells in instances where substances soluble in mineral oil are being investigated. The cell base upon which the two cells are mounted is shown in Fig. 4. The cells are sealed to the nylon base with epoxy resin.

Electrical leads are brought into the pressurized cavity by means of a Conax thermocouple sealing gland (Model TG-14-A4). This sealing gland when equipped with Teflon sealant is nominally rated to hold at 1500 psi.

Thermostatting of the autoclave is provided by a coil of ¼-inch copper tubing wound around the base of the autoclave. Water is circulated and temperature maintained constant with a Haake "F" thermostat. This unit

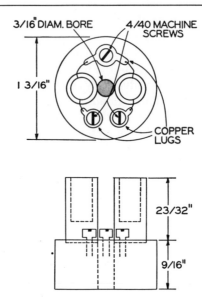

FIG. 4. The conductivity cells showing cell mounting and connectors.

is nominally rated to control to ±0.02°. Better thermal contact between the circulating water and the autoclave is achieved by soldering the coil to the autoclave.

## C. The Monitoring and Recording System

Changes in the concentrations of reactant species are followed using conductivity measurements at a frequency of 100 kc. The two conductivity cells in the autoclave form two arms of an alternating current Wheatstone bridge whose off-balance voltage is read with an oscilloscope. A diagram of the system is given in Fig. 5.

In the construction of a Wheatstone bridge circuit to be operated at 50–100 kc with a sizable frequency band pass and high sensitivities, pains must be taken to eliminate stray capacitances, since stray capacitances to ground can introduce serious errors in bridge balance and, even more importantly, make bridge balance uncertain and dependent upon the immediate environment of the bridge. Thus, even the position of the operator may change the balance point in an unshielded bridge. In order to eliminate such variations in the balance of the bridge, a Campbell-Shackelton shielded ratio box is used as the nucleus of the bridge. With proper shielding of the other component arms of the bridge, capacitive effects can be limited to the four corners of the bridge, thus making capacitive shunts definite and independent of the environment. Such capacitive effects can be

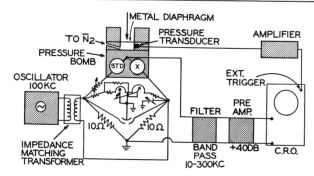

FIG. 5. Block diagram of the monitoring system.

subsequently balanced out in order to study the conductance of the electrolytes in the two variable arms. For the purpose of obtaining relaxation times from the data, absolute values of conductance are not required, but it is crucial to minimize extraneous effects during the course of the relaxation process under study.

The bridge described in detail below represents a wholly shielded bridge. Since the fractional resistance changes being observed are extremely small, $(\Delta R/R) \leq 0.001$, stable bridge balance must be achieved to better than 1 part in 10,000.

*1. Oscillator.* The oscillator is a Hewlett-Packard Model 200-T having a frequency range of 250–100,000 cps. The internal impedance is 600 ohms with an output of 20 V open circuit. The output is balanced to ground within 1% with attenuator at zero setting. Desirable and necessary features include the low hum and noise ($<0.03\%$ of rated output) and low distortion ($<0.5\%$ over entire frequency range). Harmonics must not exceed $0.01\%$ in order to achieve the balance to 1 part in 10,000 quoted above since bridge balance is frequency dependent.

*2. Campbell-Shackelton Ratio Box.* The ratio box has two doubly shielded matched 10-ohm resistors, a variable differential slide wire, and a small differential air capacitor which can be used in fine balancing of the bridge. Ten-ohm ratio arms are used to reduce the effect of capacitance to ground and thus do away with the need for more sophisticated bridge arrangements, for example, a Wagner grounding device. In fact, 10-ohm arms are rather an extreme value for the ratio arms, and 100-ohm arms are satisfactory. The 10-ohm arms do provide one advantage, however, and that is that since they represent usually 10 times smaller resistance than the cells, they determine the voltage applied to the bridge. The absolute magnitude of the off-balance signal from the bridge, in this case, is

$$\Delta E = \frac{\Delta R}{R}\frac{E}{2} \tag{8}$$

where $E$ = applied voltage, and $\Delta R/R$ = fractional change in resistance in the sample cell. This has the advantage that the off-balance signal is proportional to the magnitude of the displacement from equilibrium of the chemical system under investigation. A General Radio Type 578-C doubly shielded impedance matching transformer was substituted for the doubly shielded transformer in the ratio box. The transformer has a 60:240 turn ratio and is ±6 db between 2 kc and 500 kc. The impedance ranges are 20–2000 ohms for the low impedance winding and 0.4–40 kohms for the high impedance winding. When used between the Hewlett-Packard oscillator and the ratio box, a maximum voltage of 3 V peak-to-peak is applied across the bridge. One obvious way of increasing the sensitivity of the apparatus is to increase this voltage applied across the bridge. However, Strehlow[8] has found that at voltages exceeding 3 V (rms), a bridge of this sort cannot be balanced to 1 part in 10,000. This results from the fact that the bridge impedances cannot be adequately represented by simple resistive arms shunted by capacitances.

A General Radio 1432-U decade ac–dc resistance box is used in series with the reference cell. This box has a total resistance of 111.1 ohms with the smallest decade in 0.01-ohm multiples. Higher values of resistance are never used because the resistance of the two cells is kept as close as possible, usually within 20 ohms at the very least. Low zero inductance and shunt capacitance are required for balancing resistors.

*3. Amplifiers.* With a 3-V peak-to-peak voltage applied across the bridge, one can expect off-balance voltages of 1 mV or less according to Eq. (8).

$$\Delta E = \frac{(3)}{2}\,(10^{-3}) = 1.5 \times 10^{-3}\ \text{volt}$$

The value of $\Delta R/R = 10^{-3}$ is a typical value which can be expected for $\Delta P = 55–65$ atmospheres. This means that an additional gain of approximately 40 db is required in the amplifying system. A Scott 140 B Decade amplifier is inserted between the bridge and the oscilloscope to provide the needed 40-db amplification. Similar amplifiers that have fast risetimes and low equivalent input noise are satisfactory.

*4. Bandpass Filter.* Since the whole frequency band is unnecessary and contributes unwanted noise, a bandpass filter is inserted between the Scott amplifier and the oscilloscope. The 3-db points are at approximately 10 kc and 200 kc. At 60 cps and at 900 kc and higher frequencies the gain is down 40 db. This attenuation is required to get rid of 60-cycle hum and to keep down noise from the local radio stations. In order to study relaxation

---

[8] H. Strehlow, personal communication.

phenomena with half-lives of the order of 40 $\mu$sec, a bandpass of about 5–10 times the reciprocal of the relaxation time is required. The bandpass should therefore be about $5 \times 1/(40 \times 10^{-6}) = 125,000$ cycles at the carrier frequency.

Further reduction of the noise levels can be achieved by further decreasing the bandwidth of the amplifier. Increased signal-to-noise ratios can be obtained with a variable bandwidth amplifier;[9] smaller frequency bandwidths can be used where slower relaxation processes are being studied.

5. *Oscilloscope Recording.* A Tektronix 545 A oscilloscope with D plug-in preamplifier is used. This combination has a rise time of 0.18 $\mu$sec and a maximum sensitivity of 1 mV/cm from dc to 350 kc.

Recordings of the single-sweep traces on the oscilloscope are made with a Beattie-Coleman Oscillotron direct-view camera system. This system is equipped with a large Polaroid back, 75 mm f/1.9 lens camera (Model No. 12670), direct view camera mount, and viewing hood (Model No. 12365). The camera provides an object-to-image ratio of 1:0.9. Polaroid Land film type 44 (400 ASA equivalent) was used. An electric shutter actuator (Model No. 29560) is used to open the camera remotely just before the single trace appears on the screen. This shutter actuator is activated by the same switch which turns on the solenoid used for dropping the puncturing needle. The height of the tip of the rod above the rupture disk is arranged so that the camera shutter has sufficient time to open before the pressure decay occurs.

6. *Transducer Triggering.* A great deal of trouble was experienced with the method of triggering the oscilloscope described by Strehlow and Becker.[2] In their method, the tip of the puncturing needle making contact with the rupture disk completed a circuit including a battery attached to the external trigger circuit of the oscilloscope. This extremely simple trigger was found to be susceptible to outside effects, especially the field of the solenoid used to drop the puncturing needle. Premature triggering often resulted. However, the worst feature of this method of triggering is the uncertainty in the delay time required between the time of impact of the needle upon the rupture disk and the onset of the pressure decay in the cell cavity. It was therefore decided to use a transducer in the gas chamber itself to trigger the oscilloscope sweep. A Kistler No. 603 quartz transducer was used with a special low-noise coaxial cable (Kistler No. 575-5, 100 pf). The transducer was mounted flush to the bore of the compression chamber. This transducer has a risetime of 1 $\mu$sec (natural frequency 350 kc) and a theoretical sensitivity of 0.2 picocoulomb/psi. The theoretical voltage out ($E_{out}$) is therefore

[9] G. B. Miller, *Electronics* **31**, 79 (1958).

$$E_{out} = \frac{Q}{C} \quad \text{where } Q = \text{charge in coulombs}$$

$$C = \text{capacity of cable, farads}$$

$$= \frac{(840)(0.2) \times 10^{-12}}{100 \times 10^{-12}F} \text{ coulomb}$$

$$= 1.7 \text{ V out for 840 psi pressure drop}$$

At the end of the 100 pf cable one obtains 0.8 V out for a pressure drop of 840 psi. The external trigger on the oscilloscope requires at least 0.2 V to trigger reproducibly, but the transducer output does not reach this level fast enough, and one therefore loses the first 10–20 $\mu$sec of the pressure decay. The voltage output from the transducer must therefore be amplified so as to get greater than 0.2 V out within 20 $\mu$sec. A simple one-stage amplifier utilizing a 6 AK5 tube was inserted between the transducer and the external trigger on the oscilloscope. The amplifier has a gain of about 20 db. The amplified voltage reaches 0.2 V within 10 $\mu$sec. This amplified triggering voltage permits seeing the complete pressure decay.

The transducer has the advantage that it requires no delay and the sweep speed can be set at any time scale desired and the pressure relaxation followed without missing any part. The only disadvantage results from the fact that the diameter of the transducer determines a lower limit for the height of the pressure chamber. However, the greater convenience of this triggering arrangement compensates for the possible increase of a factor of 2 in the pressure relaxation time $\tau_p$, over the lowest value obtained thus far.

## D. Experimental Limitations

Chemical reactions are generally less sensitive to pressure changes as compared to temperature changes. Therefore, the highest possible sensitivity is needed to detect equilibrium displacements. This need for sensitivity has been the principal reason for use of conductimetric measurement of equilibrium perturbations in the pressure-jump apparatus. However, conductimetric methods, while extremely sensitive on an absolute basis, suffer from a lack of specificity. It is often desirable and necessary to provide adequate levels of buffering and ionic strength in solutions of biological materials. In view of the relatively small molar volume changes involved, the presence of even relatively low levels of inert electrolyte renders the conductance changes of interest undetectable when pressure perturbations of from 50–100 atmospheres are employed. In the experiments on bovine plasma albumin cited below, the protein itself was the predominant buffering species in the solution.

Optical methods of following concentration changes offer real advantages to the pressure-jump technique. Monitoring equilibrium displacements by ultraviolet or visible spectrophotometric measurements would permit use of convenient levels of buffering and wider ranges of concentration of reactants and products than are currently accessible with conductimetric techniques. Optical windows made of sapphire or quartz have been utilized for substantially higher hydrostatic pressures, and it should be practicable to utilize similar window construction for a pressure-jump cell. Such windows would have to be capable of withstanding repeated decompression shocks of 100 atmospheres. However, it is anticipated that devising suitable optical windows capable of withstanding repeated stresses and compensation for cell vibration can be accomplished, and that optical monitoring is well worth attempting.

The development of optical monitoring techniques for pressure-jump, permitting observation of changes in concentration of specific chemical species, should substantially increase the versatility of the method. Increased versatility would result also from use of coupled buffer systems with substantial molar volume changes for ionization. Such buffers could be used to induce pH changes in systems that are more sensitive to changes in pH than to changes in pressure.

Suggestions were made by Eigen and De Maeyer that shock waves might be utilized to provide rapid and large pressure rises.[10] (A device based on this principle has been constructed and will be described below.) However, in view of the scarcity of equilibrium data for reactions at elevated pressure, and for biological reactions in particular, kinetic data at high pressures are less useful. Since pressure rises up to atmospheric pressure are limited to one atmosphere, the pressure drop method seems to be inherently the most useful technique at present.

### E. Closely Related Pressure Perturbation Techniques

The discussion in this chapter is concerned primarily with the pressure-jump technique originated by Strehlow and Becker. However, a number of closely related techniques which utilize pressure perturbation to study chemical relaxation in solution have been reported in recent years. These techniques were designed to extend the time resolution of the pressure-jump method to shorter relaxation times. Generation of pressure alterations using shock waves,[10a] standing sound waves,[10b] and electromechanical techniques[10c] have been reported.

---

[10] M. Eigen and L. De Maeyer, in "Technique of Organic Chemistry" (S. L. Friess, E. S. Lewis, and A. Weissberger, eds.), Vol. VIII, pt. II, p. 895. Wiley (Interscience), New York, 1963.
[10a] A. Jost, Ber. Bunsenges Physik. Chem. 70, 1057 (1966).
[10b] H. Wendt, Ber. Bunsenges Physik. Chem. 70, 556 (1966).
[10c] H. Hoffman and K. Pauli, Ber. Bunsenges Physik. Chem. 70, 1052 (1966).

Pressure-jumps arising from reflected shock waves generated in a water medium have been used by Jost[10a] to study fast reactions in aqueous solution. A rupture disk separating the high and low pressure compartments in a shock tube is broken at approximately 1000 atmospheres, causing a 500 atmosphere shock wave to propagate down to the measuring cell. Optical measurement of changes in reactant concentrations were made behind the reflected shock wave ($\sim$1000 atmospheres amplitude). The duration of the high-pressure pulse is determined by the geometry of the shock tube. The lower limit of the time resolution is set by the slit width of the optical monitoring system. Relaxation times from 2 msec to 2 $\mu$sec can be studied using the apparatus described. The reactions of the indicator Tropaeolin O with ammonia and water was studied with this system and the pH dependence of rate constants for proton transfer were determined.

The feasibility of optical monitoring techniques for a pressure-jump device of Strehlow's basic design is certainly supported by the successful use of sapphire optics in the shock wave technique. Sufficiently large extinction coefficients of reactants or products, would of course be required, or, alternatively, coupled dye systems could be utilized. Since the magnitude of the pressure change is smaller, the compression (decompression) shock to the window should be considerably smaller in a Strehlow device utilizing optical monitoring.

Wendt[10b] has devised a method for measuring chemical relaxation using sinusoidal pressure changes developed in standing sound waves. Sensitive conductivity measurements at high frequency are used to determine relaxation times together with amplitude and phase angle measurements. Since the basic principle is the same as that used in pressure-jump techniques, this technique can be used to extend pressure-jump relaxation measurements down to 5 $\mu$sec. Theoretically, the technique can be extended to give resolution down to 0.5 $\mu$sec. The system, as presently constituted, has been used to study chemical relaxation processes occurring in cobalt (II)-malonate and cobalt (II)-tartrate solutions.

Hoffman and Pauli[10c] have developed an electromechanical technique for studying chemical relaxation via pressure perturbation. Using the method of Eisenmenger,[10d] pressure waves of 25 atmospheres and 12 $\mu$sec duration can be generated and the equilibrium displacements followed by conductivity measurements. The rectangular pressure pulses are generated by discharge of a high-voltage capacitor through a flat coil in front of a copper membrane. The displacement of the copper membrane results in the propogation of a pressure wave through a steel tube into a measuring cell.

[10d] W. Eisenmenger, *Acustica* **12**, 185 (1962).

Since rupture of a membrane is not required, multiple kinetic measurements can be conveniently made on the same solution and the signal to noise ratio thereby reduced. The duration of the pressure wave can be increased to 100 $\mu$sec by appropriate manipulation of the discharge condenser network. This technique has been used to study chemical relaxation resulting from the complexation reaction between cobalt (II) and malonate ions.

The pressure-jump technique has been used by Hurwitz and Atkinson[10e] to study uranyl ion equilibria in aqueous nitrate containing media. The apparatus used was constructed according to a design of Hoffman and co-workers.[10f]

The prime limitation of the more sophisticated techniques for producing pressure pulses lies in the time ranges accessible to convenient measurement. The Jost technique and the Hoffman and Pauli technique have a limited time of sustained pressure elevation which cannot be conveniently extended to include study of relatively slower relaxation processes. The use of standing sound waves is at present suitable only for fairly large volumes of solution and low degrees of dissociation ($\alpha \simeq 0.1$). At present these more elegant techniques are useful adjuncts for extending the range of pressure-jump measurements down to the microsecond region. The Strehlow device appears to be the most versatile and simple pressure-jump technique at this time, since the accessible time range is the broadest and the apparatus is relatively easy to construct and operate.

### III. Applications

### A. Acquisition and Interpretation of Data

Kinetic data are obtained with the pressure-jump apparatus in the following manner. The solution to be studied is pipetted into the sample cell and then a KCl solution with only slightly higher conductance (adjusted so that the conductance difference is less than 10%) is pipetted into the reference cell. The cell cavity is then filled with paraffin oil or other suitable pressure transmitting fluid. The cell cavity is covered with the Kel-F membrane and the rest of the pressure bomb including the rupture disk is assembled and tightened down. The best possible balance of the bridge is achieved after thermal equilibration has occurred in the cells. The autoclave is then pressurized to produce an off-balance bridge voltage and the rupture disk is punctured.

The sudden release of the applied pressure of 60–65 atmospheres causes

[10e] P. A. Hurwitz and G. Atkinson, *J. Phys. Chem.* **71**, 4142 (1967).
[10f] H. Hoffman, J. Stuehr, and E. Yeager, ONR Contract Nonr 1439(04), Project NR 384-305, *Tech. Rept.* **27**, 12 (1964).

a return of the concentrations of various species in solution to their equilibrium values at atmospheric pressure. The adiabatic pressure change results in an equilibrium displacement given by:

$$\left(\frac{d \ln K}{dP}\right)_s = -\frac{\Delta V}{RT} + \frac{V \alpha_T}{C_P} \frac{\Delta H}{RT} + \beta_s \Delta \nu \tag{9}$$

where $K$ is the molar equilibrium constant, $\Delta V$ the molar volume change, $\Delta H$ the molar enthalpy change, $\alpha_T$ the thermal expansion coefficient, $C_P$ the heat capacity of the solution at atmospheric pressure, $\beta_s$ the isentropic compressibility, and $\Delta \nu$ the change in number of moles for the reaction.[10g]

The chemical species approach their equilibrium values at atmospheric pressure exponentially with characteristic relaxation time $\tau$. The first-order return of concentrations to their equilibrium values at atmospheric pressure is recorded as a decay of the off-balance voltage of the bridge circuit. This decay is seen as an exponential amplitude modulation of the carrier frequency of 100 kc. The voltage change is proportional to the relative changes in resistance of the solution in the sample cell. This relative change in resistance is given by Eq. (10) for the case of an exponential pressure decay.

$$\frac{\Delta R}{R \Delta P} = r_1 e^{-t/\tau_p} + r_2 \frac{\tau_p e^{-t/\tau_p} - \tau e^{-t/\tau}}{\tau_p - \tau} \tag{10}$$

where $\tau_p$ is the pressure relaxation time (apparatus constant), $\tau$ the chemical relaxation time, and $r_1$ and $r_2$ are amplitude factors, which may be either positive or negative.

Since for most practical application, $t \gg \tau_p$, the second, chemical, term of Eq. (10) predominates; for a system with a single relaxation time this results in

$$\frac{\Delta R}{R \Delta P} = r_2 e^{-t/\tau} \tag{11}$$

The appropriate amplitude term $r_2$ can be calculated for simple chemical systems as illustrated above in Eqs. 3 through 7. For example, for the simple bimolecular reaction

$$A^+ + B^- \underset{k_2}{\overset{k_1}{\rightleftharpoons}} C$$

the amplitude term for an adiabatic pressure change[4] is

$$r_2 = \frac{1 - \alpha}{2 - \alpha}\left(-\frac{\Delta V}{RT} + \frac{V \alpha_T}{C_p} \frac{\Delta H}{RT} + \beta_s \Delta \nu\right) \tag{12}$$

where $\alpha$ = degree of dissociation. A generalized theoretical treatment of

[10g] In general, in aqueous solution, the first term predominates (footnote 4).

relaxation spectra for multistep reaction systems is presented by Eigen and De Maeyer.[10] The use of transformation matrices for calculation of the amplitude factors is discussed for complex multistep reactions.

Linearization of rate expressions is possible when small perturbations from equilibrium are utilized. For pressure changes of 100 atmospheres or less, the magnitude of molar volume changes in aqueous solution are sufficiently small to ensure that this condition obtains. The chemical relaxation times can thus be related to rate constants and the equilibrium concentrations of chemical species in solution. In the example shown above:

$$\frac{1}{\tau} = k_2 + k_1(\bar{C}_A + \bar{C}_B)$$

where $\bar{C}_A$ and $\bar{C}_B$ represent equilibrium concentrations of A and B, respectively.

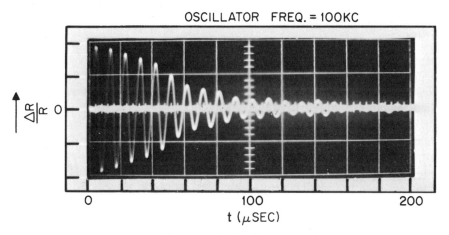

FIG. 6. The pressure relaxation determined using 0.05 $M$ MgSO$_4$ in the sample cell at 25°. The apparatus constant is $\tau_p = 50$ $\mu$sec.

The apparatus relaxation time $\tau_p$ is determined using MgSO$_4$. The dissociation of MgSO$_4$ is suitably pressure dependent, but the relaxation time for the system is considerably faster than the pressure relaxation. Thus, the decay of off-balance voltage is characteristic of the pressure drop in the conductivity cell. Figure 6 shows the pressure relaxation observed with 0.05 $M$ MgSO$_4$. The ordinate is proportional to $\Delta R/R$ while the major divisions of the abscissa each represent 20 $\mu$sec.

Relaxation times are derived from slopes of plots of log $[(\Delta R/R)_t -$ $(\Delta R/R)_\infty]$ versus time, $t$, where $(\Delta R/R)_\infty$ represents the limiting amplitude at long times. The pressure relaxation for this apparatus shows an exponential decline with a relaxation time of 50 $\mu$sec. Relaxation times for chemical reactions showing single relaxation times (where $\tau \gg \tau_p$) are determined in the same manner. Multistep reactions can exhibit a spectrum of relaxation times. In cases where the individual $\tau$'s are not widely separated in time, resolution of $\tau$'s can be achieved using mathematical procedures[10] or graphical procedures. The latter technique, based upon methods used for separation of half-lives in mixtures of radioactive isotopes, utilizes back-extrapolation of limiting slopes at long time intervals. This technique can also be used for studying reactions whose relaxation times are not too different from $\tau_p$. It was necessary to resolve $\tau$'s using graphical techniques in the study of bovine plasma albumin which is described below. In general, $\tau$'s must differ by at least a factor of 2 in order to detect more than one relaxation time—experimental limitations render resolution of separate relaxation times in such cases extremely difficult even under the best conditions.

## B. Systems of Biological Interest

To date, the pressure-jump technique has not been used to study enzyme-catalyzed reactions. However, use of pressure as the perturbing external parameter should be useful in cases where enthalpic changes are small or where heat-labile proteins are involved. Pressure-jump techniques, like temperature-jump techniques, should be useful for studying elementary steps of enzyme-catalyzed reactions, especially where charge neutralization is involved. Charge neutralization generally results in appreciable volume changes in aqueous solution. Such large volume changes are due to effects on water structure, such as electrostriction of solvent. A tabulation of values for volume changes for some reactions in aqueous solution is shown in the table.[11,12] The volume change for the second ionization of phosphoric acid is seen to be large. This result follows from the Nernst-Drude equation[13] for solvent electrostriction.

$$\Delta V_{\text{el}} = \frac{V \kappa_s Z^2 e^2}{2 r \epsilon^2} \cdot \frac{\partial \epsilon}{\partial V}$$

where $\Delta V_{\text{el}}$ is the volume change in a medium of dielectric constant $\epsilon$ surrounding a sphere of radius $r$ carrying a charge of $Ze$ with $e$ the electronic charge, $V$ the volume of the dielectric, and $\kappa_s$ the compressibility of the solvent. This equation suggests that reactions of biological interest involving charge neutralization of multiply charged chemical species of high charge density should be particularly suited for study with the pressure-jump technique.

[11] S. D. Hamann, in "High Pressure Physics and Chemistry" (R. S. Bradley, ed.), Vol. 2, p. 131. Academic Press, New York, 1963.

[12] W. Kauzmann, A. Bodanszky, and J. Rasper, J. Am. Chem. Soc. **84**, 1777 (1962).

[13] P. Drude and W. Nernst, Z. Physik. Chem. (Frankfurt) **15**, 80 (1894).

VOLUME CHANGES FOR IONIC REACTIONS IN AQUEOUS SOLUTION[a]

| Reaction | $\Delta V$ (cm$^3$/mole) |
|---|---|
| $CH_3CO_2H \rightleftharpoons CH_3CO_2^- + H^+$ | $-10.9$ |
| Pyruvic acid $\rightleftharpoons$ pyruvate $+ H^+$ | $-12$ |
| Succinic acid $\rightleftharpoons$ succinate $+ H^+$ | $-11.9$ |
| $H_3PO_4 \rightleftharpoons H_2PO_4^- + H^+$ | $-16$[b] |
| $H_2PO_4^- \rightleftharpoons HPO_4^{2-} + H^+$ | $-28.1$[b] |
| $NH_3 + CH_3CO_2H \rightleftharpoons NH_4^+ + CH_3CO_2^-$ | $-17.5$ |
| $CH_3NH_2 + H_2O \rightleftharpoons CH_3NH_3^+ + OH^-$ | $-24.9$ |
| $CH_3NH_2 + C_2H_5CO_2H \rightleftharpoons CH_3NH_3^+ + C_2H_5CO_2^-$ | $-18.0$[b] |
| $I_2 + I^- \rightleftharpoons I_3^-$ | $-5.4$[b] |
| $(Mg^{2+}SO_4^{2-})_{ion\ pairs} \rightleftharpoons Mg^{2+} + SO_4^{2-}$ | $-7.3$[b] |

[a] See text footnotes 11 and 12.
[b] Represent values obtained by extrapolation to infinite dilution.

The pressure-jump technique has been used to study rates of metal–ligand complexation reactions and to study rates of conformational changes in proteins. The volume change and conductivity changes that accompany the structural alterations in bovine plasma albumin make it possible to study these changes with the pressure-jump technique.

*Study of Conformational Changes in Bovine Plasma Albumin.* A great deal of work has been done on the unusual physicochemical properties of bovine plasma albumin (BPA) at low pH. Tanford's work[14] showed an anomalous steepening of the titration curve below pH 4.3, which was attributed to a molecular expansion caused by intramolecular electrostatic repulsions. Tanford[15] also presented viscosity data as evidence for a two-stage conformational change occurring below pH 4.3. The first stage, which is delineated at higher ionic strength, was termed a conversion to an "expandable" form, while the second stage was interpreted as an expansion of the "expandable" form. Subsequent electrophoretic and optical rotatory studies by Foster[16] showed the existence of an "N–F" isomerization in the pH range 3.5–4.5, the "F" designating an electrophoretic component which moves faster than the "N" normal form. Weber[17] has reported pepsin digestion studies on BPA that support Foster's subunit model. Foster's[18] model for BPA consists of four subunits connected by poly-peptide links. This model accounts satisfactorily for many of the unusual

[14] C. Tanford, S. A. Swanson, and W. S. Shore, *J. Am. Chem. Soc.* **77**, 6414 (1955).
[15] C. Tanford, J. G. Buzzell, D. G. Rands, and S. A. Swanson, *J. Am. Chem. Soc.* **77**, 6421 (1955).
[16] K. Aoki and J. F. Foster *J. Am. Chem. Soc.* **79**, 3385 (1957).
[17] G. Weber and L. B. Young, *J. Biol. Chem.* **239**, 1415 (1964).
[18] J. F. Foster, *in* "The Plasma Proteins" (F. W. Putnam, ed.), p. 179. Academic Press, New York, 1960.

properties of BPA at low pH. Kauzmann,[19] using dilatometric measurements, found abnormal volume changes accompanying acid dissociation in BPA below pH 4. An unexpected volume decrease was found superimposed upon the increase in volume associated with the titration of carboxylate groups.

Temperature-jump studies[20] on solutions of BPA and benzyl orange yielded a relaxation time of 20 msec at 25° and pH 3 which is independent of BPA concentrations. This same relaxation time was obtained in the absence of dye in the pH range 2–4, using a temperature-jump apparatus equipped for polarimetric detection. These results were tentatively ascribed to a conformational change in BPA. These results plus the volume changes reported in the same pH range by Kauzmann[19] led us to study BPA at low pH with the pressure-jump apparatus.[21]

Three relaxation times are observed with BPA in chloride media at acid pH. In the pH range 2.5–4.5 there is a relaxation time shorter than the apparatus relaxation time. This rapid reaction is ascribed to combination of protons with carboxyl groups, since the carboxyl group is the only group ionized appreciably below pH 5. The volume changes found by Kauzmann[19] in this pH range are primarily due to this group. Eigen[22] has shown that such protolytic reactions are quite rapid processes in small molecules. There are also two slower relaxation times that show a smaller conductivity change. Figure 7 shows the relaxation observed at pH 3.13 in 0.5% BPA. This relaxation process shows two relaxation times, $\tau_1$ and $\tau_2$, with $\tau_1$ designating the faster relaxation time. These relaxation times are found only in the pH range from 2.8 to 3.8. The upper limit corresponds well to the pH at which Kauzmann has found that abnormal volume changes attributed to conformational changes disappear. The two slower relaxation times occur in a pH range lower than that of the N-F transition (pH 3.5–4.5). They are observed in the pH range where there are changes in intrinsic viscosity, sedimentation coefficient, and specific rotation. These changes in physical properties have been attributed to expansion of the molecule.[15] Both $\tau_1$ and $\tau_2$ were found to be independent of albumin concentration over a 4-fold change in concentration. This indicates that the reactions observed are unimolecular; the presence of two successive relaxation times thus signifies two consecutive unimolecular steps.

Both $\tau_1$ and $\tau_2$ are strongly pH dependent, with $\tau_1$ showing a pH maximum at about pH 3.1. The effect of changing the net charge on the BPA

[19] W. Kauzmann, *Biochim. Biophys. Acta* **28**, 87 (1958).

[20] R. A. Alberty, A. Froese, and A. H. Sehon, *Abstr. 139th Meeting Am. Chem. Soc. Chicago, Sept., 1961*.

[21] M. T. Takahashi and R. A. Alberty, in preparation.

[22] M. Eigen, *Z. Elektrochem.* **64**, 115 (1960).

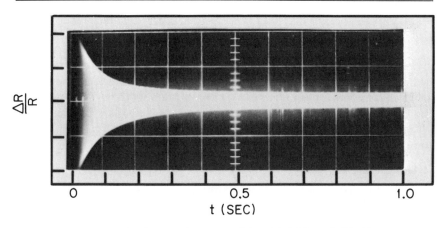

FIG. 7. The relaxation in 0.5% BPA containing 0.013 $M$ total chloride at pH 3.13. Values of the two relaxations times are: $\tau_1 = 40$ msec and $\tau_2 = 200$ msec.

molecule was studied by utilizing anions that are more strongly bound to BPA than is chloride ion. In perchlorate and thiocyanate media, the experiments could be extended down to pH 2, and two successive relaxation times independent of BPA concentration were observed. The same functional dependence on hydrogen ion concentration is observed for $\tau_1$ with the exception that the pH maximum is shifted to lower pH values.

The simplest mechanism consistent with the pH dependence of the relaxation times involves a stepwise conversion of an "expandable" form A, to an "expanded" form C via an intermediate form B, as formulated in the following mechanism.

$$\begin{array}{ccc}
A & \rightleftharpoons B & \rightleftharpoons C \\
\updownarrow & \updownarrow & \updownarrow \\
AH & BH & \rightleftharpoons CH \\
\updownarrow & \updownarrow & \updownarrow \\
AH_2 & \rightleftharpoons BH_2 & \rightleftharpoons CH_2
\end{array}$$

The relaxation behavior of BPA shows many similarities to the kinetics of the "reversible" denaturation of conalbumin[23] and hemoglobin.[24] All three proteins show a complex dependence of rates on pH, characterized by maxima in the rates at given pH values. The absence of large temperature effects on the BPA relaxation times at pH 3.2 parallels the low energies of activation observed with hemoglobin and conalbumin. In addition, conalbumin at pH 3.09 in the presence of nitrate ion was found to have

[23] A. Wishnia and R. C. Warner, *J. Am. Chem. Soc.* **83**, 2065 (1961).
[24] J. Steinhardt and E. M. Zaiser, *J. Am. Chem. Soc.* **75**, 1599 (1953).

two successive relaxation times of 45 and 235 msec in the pressure-jump apparatus.

*Other Systems.* Preliminary experiments on $\beta$-lactoglobulin and $\alpha$-chymotrypsin yielded relaxation times attributable to conformational changes. In lysozyme and ribonuclease, however, for which no conformational changes have been observed by equilibrium methods down to pH 2 at room temperature, no relaxation processes slower than the apparatus relaxation time were detected. The pressure-jump method thus seems to be useful for the study of conformational changes in protein molecules.

## C. Concluding Remarks

Steady-state studies of enzyme-catalyzed reactions give only limited information regarding the elementary steps in the mechanism. For the case of an extended Michaelis-Menten mechanism for single substrate, S, and product, P, with $n$-intermediates

$$E + S \rightleftharpoons X_1 \rightleftharpoons \cdots X_n \rightleftharpoons E + P \tag{13}$$

Peller and Alberty[25] have shown that individual rate constants cannot in general be determined using steady-state measurements; instead, one obtains lower limits for the rate constants from the maximum velocities and Michaelis constants. The study of the transient state using relaxation techniques offers a means for investigation of the individual reaction steps. The above mechanism is characterized by $n + 1$ relaxation times,[26] and investigation of the relaxation times as a function of enzyme and substrate concentration can, in principle, give a complete kinetic description of the elementary steps in the reaction. High concentrations of enzyme and substrate can be utilized since very rapid reactions are now accessible to study. Relaxation spectra for a variety of reaction mechanisms typical of those catalyzed by enzymes are treated by Alberty[27] and Eigen.[10]

Although the pressure-jump technique has not yet been applied to enzyme-catalyzed reactions, it is felt that this represents a potentially fruitful area for study. The development of optical monitoring techniques should permit studies of the transient state in enzyme catalysis, subject to the following limitations. First, the reaction must result in an appreciable volume change. Enzyme reactions involving charge neutralization are therefore logical candidates for study. Second, the overall reaction must be reversible. Perturbation of steady-state equilibria has been reported[28] using

[25] L. Peller and R. A. Alberty, *J. Am. Chem. Soc.* **81,** 5907 (1959).

[26] R. A. Alberty and G. G. Hammes, *Z. Elektrochem.* **64,** 124 (1960).

[27] R. A. Alberty, G. Yagil, W. F. Diven, and M. T. Takahashi, *Acta Chem. Scand.* **17,** S 34 (1963).

[28] H. Diebler, M. Eigen, and P. Z. Matthies, *Z. Naturforsch.* **16b,** 629 (1961). See also this volume [1], p. 24.

temperature jump in combination with a rapid flow device; however, this technique does not appear to be feasible for the pressure-jump method. Thus, reactions with extremely large or small equilibrium constants cannot be studied.

The following types of elementary steps in enzyme catalysis should be amenable to study with the pressure jump: (a) enzyme–substrate complex formation; (b) metal–enzyme and metal–substrate complexation reactions; (c) catalytically significant conformational changes in enzyme molecules. The pressure-jump apparatus using conductimetric monitoring is well suited for such study.

# [3] Ultrasonic Methods

## By FRIEDER EGGERS and KENNETH KUSTIN

## I. The Application of Ultrasonics to Chemical Relaxation

### A. Introduction

Ultrasound of a wide frequency range has been used as a measurement tool in chemical kinetics for only a few decades. A considerable spectrum of relaxation times can be covered by a variety of techniques, but owing to several technical difficulties ultrasonic methods have not yet become as common as electrical or optical research techniques. The conversion from electrical to sound energy and the reverse, which all ultrasonic measurements in the laboratory employ, is rather inefficient. Mechanical parts for ultrasonic test apparatus must be manufactured to a high degree of precision, and not much ultrasonic research equipment is so far on the market.

This chapter is limited to ultrasound in liquids only, and to compressional waves; viscosity effects and shear waves are neglected. The restricted treatment here can cover only the most essential facts for the application to kinetics. The reader will find the theoretical aspects of acoustic phenomena covered in many books.[1,2] Extensive discussions of the technical aspects of acoustics are likewise fully represented.[3-6] Articles pertaining specifically to chemical relaxation techniques have also recently appeared.[7-10]

The frequency range reached by acoustical measurements ranges at present from a few kHz to more than 2 GHz. This presentation deals with methods for a range from about 50 kHz to 100 MHz (i.e., time constants from 3 microseconds down to 1.5 nanoseconds), which is of special importance for biochemical investigations. The two parameters that determine ultrasonic wave propagation in liquids are the sound velocity $v$ and the sound absorption $\alpha$. Both values can depend on frequency. Therefore, it is necessary to measure both over a wide frequency range; that is, one larger than 10:1 for a single relaxation process. Because it is difficult to cover such a wide range in a single experimental technique, it is in most cases necessary to combine several different types of apparatus.

From the sound velocity and absorption as functions of frequency and temperature it is possible to calculate the complex compressibility $\chi$, the absorption per wavelength $\alpha\lambda$, the molar cross section $Q$, the molar volume change of a reaction $\Delta V$, and also the molar enthalpy change $\Delta H$ and the activation energy of a reaction. These values and their relations are discussed in many articles.[8-10]

## B. Generation and Detection of Ultrasound in Liquids

For performing ultrasonic measurements, suitable transducers must convert electrical into sound energy and reverse. These transducers can be

[1] S. R. De Groot and P. Mazur, "Non-Equilibrium Thermodynamics," North-Holland Publ., Amsterdam, 1962.

[2] K. F. Herzfeld and T. A. Litovitz, "Absorption and Dispersion of Ultrasonic Waves." Academic Press, New York, 1959.

[3] B. Carlin, Ed., "Ultrasonics," 2nd ed. McGraw-Hill, New York, 1960.

[4] T. F. Hueter and R. H. Bolt, "Sonics." Wiley, New York, 1955.

[5] H. J. McSkimin, in "Physical Acoustics" (W. P. Mason, ed.), Vol. I, Part A, p. 271. Academic Press, New York, 1964. See also, entire series.

[6] E. G. Richardson, "Ultrasonic Physics," 2nd ed. Elsevier, Amsterdam, 1962.

[7] G. S. Verma, Rev. Mod. Phys. **31,** 1052 (1959); a survey on different methods with more than 200 references summarizing the earlier literature.

[8] M. Eigen and L. De Maeyer, in "Technique of Organic Chemistry" (A. Weissberger, ed.), 2nd ed., Vol. VIII, Part 2, p. 895. Wiley (Interscience), New York, 1963.

[9a] R. O. Davies and J. Lamb, Quart. Rev. (London) **11,** 134 (1957).

[9b] J. Lamb, in "Dispersion and Absorption of Sound by Molecular Processes" (D. Sette, ed.), p. 101. Academic Press, New York, 1963.

[10] J. Stuehr and E. Yaeger, in "Physical Acoustics" (W. P. Mason, ed.), Volume II, Part A, p. 351. Academic Press, New York, 1965.

classified into several groups:[11a,b] (1) electrostatic; (2) magnetostrictive, e.g. Ni-wires; (3) piezoelectric, e.g., crystalline quartz, barium titanate and other ferroelectrics, and also evaporated or sputtered thin films; (4) optical effects—Debye-Sears, *for detection only*.[12a]

Electrostatic transducers for use in liquids have been described in the literature and used for some measurements,[12b] but they are not easily available and usable in the biochemical laboratory and will therefore not be discussed further. Magnetostrictive transducers find technical application in ultrasonic cleaners and other high-power equipment.[4] Because rather low power is sufficient for ultrasonic measurements, which avoids heating effects from the ultrasonic absorption, X-cut quartz plates (with the crystal X-direction perpendicular to the surface) are most commonly used in the laboratory. The physical properties of crystal quartz are extensively described in a monograph by Mason.[13] When an alternating electrical field is applied, the disk oscillates, thus radiating sound into the adjacent medium. This effect is reversible; the quartz can also act as a sound receiver, delivering a voltage upon incident sound.

The fundamental frequency $f_0$ of an X-cut quartz plate is determined by the fact that the thickness $d$ equals half of the corresponding sound wavelength, from which follows the relation

$$f_0 = v_{x\text{-quartz}}/2d \tag{1}$$

with

$$v_{x\text{-quartz}} = 5.72 \times 10^5 \text{ cm sec}^{-1}$$

Quartz plates have their optimum response at the fundamental frequency and can be manufactured with fundamental frequencies higher than 100 MHz, but they become extremely thin and difficult to handle. It is possible to have the plate resonate in odd harmonics $3 f_0$, $5 f_0$, etc. (with decreasing efficiency) and use it for a set of different frequencies (overtone operation). Recently a new technique has been developed to generate ultra-high frequency sound (hypersound) on the polished surface of a quartz block by applying an RF electrical field.[11a] Frequencies of more than 10 GHz have been attained this way, but the method has a rather low

[11a] A. Barone *in* "Encyclopedia of Physics" (S. Flügge, ed.), Vol. XI, Part 2, p. 74. Springer Verlag, Berlin, 1962.

[11b] L. Bergmann, "Der Ultraschall," 6th ed. Hirzel, Leipzig, 1954.

[12a] Brillouin scattering techniques for measuring acoustic attenuation in the microwave frequency range should also be mentioned.

[12b] H. Siegert, *Acustica* **8**, 48 (1963).

[13] W. P. Mason, "Piezoelectric Crystals and their Application to Ultrasonics." Van Nostrand, Princeton, New Jersey, 1950; cf. Ref. 3, Chap. 2; also Ref. 4, Chap. 4.

conversion efficiency; also room temperature attenuation in liquids increases rapidly at higher frequencies.[14]

Barium titanate and other ferroelectrics such as lead zirconate or lead titanate are polarized ceramics and have the advantage of a high electromechanical coupling coefficient, providing an efficient conversion. They are rather cheap and can be prepared in any size and shape desired. Relatively high dielectric constants prohibit their use at frequencies much higher than 5 MHz due to electrical matching problems. Further information can be found in the literature.[15] Recently, evaporation and sputtering techniques have been developed that deposit piezoelectric materials as CdS or ZnO on a solid substrate in a uniform and oriented thin film; this provides excellent transducers with fundamental frequencies from about 100 MHz to several GHz.[16]

Finally, an optical effect, discovered by Debye and Sears and independently by Lucas and Biquard, must be mentioned.[11a,b] It has been used widely to measure the energy density of a sound wave in a liquid. A traveling sound wave generates a density modulation in the liquid which produces a modulation of the optical refraction index, thus providing an optical diffraction grating. The light intensity diffracted into first order is proportional to the sound intensity, which is proportional to the square of the alternating pressure amplitude. The theoretical aspects are rather complicated and need not be treated here. The reader interested in this method will find an extensive presentation in the literature cited.

## C. The Plane Wave Sound Field and Its Parameters

Sound fields of small amplitude obey a differential equation, often called a wave equation, which in its one-dimensional form, is

$$\frac{\partial^2 p}{\partial t^2} = v^2 \frac{\partial^2 p}{\partial x^2} \tag{2}$$

Here $p$ is the alternating sound pressure (amplitude $p_0$) in the liquid, $v$ is the propagation velocity (phase velocity) of the wave, $x$ is the distance traveled, and $t$ is the time. A simple solution of this equation is an unattenuated sinusoidal plane wave, $p = p_0 e^{i\omega(t-x/v)}$, where $i = \sqrt{-1}$. The attenuation—which stems from several physical and chemical reasons—can be included by a decreasing exponential factor, $\exp(-\alpha x)$, resulting in the equation

[14] K. G. Plass, *Acustica* **19**, 236 (1967/68).

[15] D. A. Berlincourt, D. R. Curran, and H. Jaffe, *in* "Physical Acoustics" (W. P. Mason, ed.), Vol. I, Part A, p. 169. Academic Press, New York, 1964.

[16] J. de Klerk, *in* "Physical Acoustics" (W. P. Mason, ed.), Vol. IV, Part A, p. 195. Academic Press, New York, 1966.

$$p = p_0 \cdot e^{i\omega t} \cdot e^{-(\alpha + i\kappa)x} \tag{3}$$

with the angular frequency $\omega = 2\pi f$, the wave number $\kappa = \omega/v = 2\pi/\lambda$, the attenuation (or absorption) coefficient, $\alpha$, and the wavelength $\lambda$. Analogous equations exist for the incremental changes of density, particle velocity, and temperature in the sound wave. This wave travels in the positive $x$-direction. Linear superposition with another wave of the same frequency and amplitude traveling in the opposite direction results in the formation of standing waves. Standing waves exhibit nodal planes (with zero amplitudes at any time) and fixed planes with maximum amplitude (antinodes), which oscillate in time. Standing waves are caused in most cases from the reflection of a traveling wave at a boundary between two media A and B of different density $\rho$ and sound velocity $v$. The reflection factor $R$ is defined as the ratio of amplitudes for the reflected and the incoming wave and is given (for the pressure amplitude) by

$$R = \frac{\rho_B v_B - \rho_A v_A}{\rho_B v_B + \rho_A v_A} \tag{4}$$

This factor is dimensionless and can assume values between $-1$ and $+1$. The product $\rho v$ is called specific acoustic impedance.

Plane waves are utilized in most acoustical techniques, which determine $v$ and $\alpha$. From these values, one obtains the attenuation per wavelength $\alpha\lambda$ and the complex compressibility

$$\chi = \frac{1}{\rho} \left[ \frac{1}{v^2} - \frac{\alpha^2}{\omega^2} - i \left( \frac{2\alpha}{\omega v} \right) \right] \tag{5}$$

The molar cross section $Q$ is defined by

$$Q = \frac{2\alpha}{N \cdot C} \tag{6}$$

with Avogadro's number $N$ and concentration $C$. The term $Q$ represents the surface (normal to the direction of plane wave propagation) which transmits the same sound power that is absorbed by a single molecule.

Acoustical quantities such as pressure or intensity (which is proportional to the square of the pressure amplitude) are often expressed in a logarithmic scale. The intensity level $L$ is defined for the power ratio of two intensities $I_1$ and $I_2$

$$L = 10 \log_{10} \frac{I_1}{I_2} \ \mathrm{db} \tag{7}$$

db means "decibel," which is generally used on calibrated attenuators. The relation for a pressure (or voltage) ratio is

$$L = 20 \log_{10} \frac{p_1}{p_2} \tag{8}$$

Another level scale is based on the natural logarithm with the "neper"; one neper is equivalent to 8.69 db.

Owing to diffraction effects, a pistonlike motion of the transducer surface (i.e., an in-phase motion with equal amplitudes at all surface points) does not yield an exact plane wave field in front of the transducer. The sound field results from a superposition of a plane wave and an interference wave from the circumference of the transducer and becomes quite complex. In the near field region, also called the Fresnel region, the wave is approximately planar, but the pressure amplitude at the axis exhibits a series of maxima and minima. By means of a receiving crystal which averages over the total cross section of the sound beam, pulse measurements (with traveling waves) can be carried out in the Fresnel region with only small effects from the pressure fluctuations.

For a circular transducer of radius $r$, the far field region, also called the Fraunhofer region, begins at a distance of approximately $d \sim r^2/\lambda$ from the transducer. The sound energy spreads out at an angle determined by the ratio of sound wavelength $\lambda$ and radius $r$. In this region, the pressure, as in a spherical wave, decreases inversely proportional to the distance. For ultrasonic loss measurements in the far field region, corrections must be introduced for beam spreading.[5,11a] If the radius $r$ of the transducer is below one wavelength $\lambda$, spherical waves will be radiated. For attenuation measurements the radius $r$—as a rule of thumb—should be larger than 10 $\lambda$; this requirement again causes the far field region to be more than 10 radii away.

Until now no preference has been expressed for measuring ultrasonic absorption or dispersion. It is, of course, true that the changes of ultrasound velocity (dispersion) or attenuation coefficient (absorption) with changing frequency are both consequences of the same physical process. Taken separately, they each yield the same information concerning the basic physical process, namely the irreversible energy interchange between the acoustic wave and the medium. For the subject of this chapter, that information would be the reaction system's relaxation time, $\tau$.

Nevertheless, practical considerations dictate that a choice be made. Since the relative change in acoustic velocity over a wide frequency range is very small (typically a few parts per thousand) this technique is inherently of relatively low sensitivity. For absorption, however, the change in attenuation coefficient in the vicinity of frequencies $\omega \sim 1/\tau$ is relatively large. Consequently, most investigators have chosen to measure the absorption mode, and this technique will be featured in this chapter.

The actual determination of $\alpha$ is then very straightforward. Let $p$, in

Eq. (8) equal the pressure level in a reference system with no "chemical" absorption, for which $\alpha_{\text{chem}} = 0$ (we will return to this point in Section II,B). Insertion of Eq. (3) into Eq. (8) shows that the difference in decibels will be directly proportional to distance between two measurement points.

$$\Delta(\text{db}) = \alpha_{\text{chem}} \Delta x \qquad (9)$$

Thus, a plot of db *vs.* distance should be a straight line with slope equal to $\alpha_{\text{chem}}$. (Note that this absorption coefficient is *decadic.*)

## D. Chemical Dependence of Ultrasonic Attenuation

The ultrasonic technique is based upon the principle of perturbing a system at equilibrium and determining the response. Regarded in this way, the method is no different from temperature- or pressure-jump. It differs mainly in the form of the perturbation, which is periodic. Thus, for a single-step reaction subjected to a very sudden drop in pressure, the dynamical response to this stepwise forcing function is an exponential readjustment to the new stationary condition. If $\Delta C_i$ is the change in concentration, and $A$ is some amplitude factor depending upon the boundary conditions, the response is [assuming the molar coefficient in the balanced equation is unity; *cf.* this volume [1], Eq. (11) and [2], Eq. (11)]

$$\Delta C_i = A e^{-t/\tau} \qquad (10)$$

In Eq. (10), $\tau$ is the relaxation time characteristic of the reaction.

In contrast, if the perturbation is periodic, the dynamic response will be oscillatory instead of exponential. For the simple example we have chosen, the observed (real) response would be

$$\Delta C_i = \frac{A \cos(\omega t - \delta)}{(1 + \omega^2 \tau^2)^{1/2}} \qquad (11)$$

with $\delta = \arctan(\omega \tau)$. We are interested not so much in the response as a function of time, but rather of frequency. If the frequency is low with respect to $1/\tau$, the system responds in perfect phase to the periodic perturbation. As the frequency of the forcing function is raised, the response is inhibited and begins to lag behind the perturbation. At the same time, the amplitude of the response decreases. This description may be summed up as follows

$$\omega \tau \cong 0; \delta \cong 0; \Delta C_i \cong A \cos(\omega t)$$

$$\omega \tau = 1; \delta = 45°; \Delta C_i = \frac{A \cos(\omega t - 45°)}{\sqrt{2}}$$

$$\omega \tau \to \infty ; \delta \cong 90°; \Delta C_i \cong \frac{A \cos(\omega t - 90°)}{\omega \tau}$$

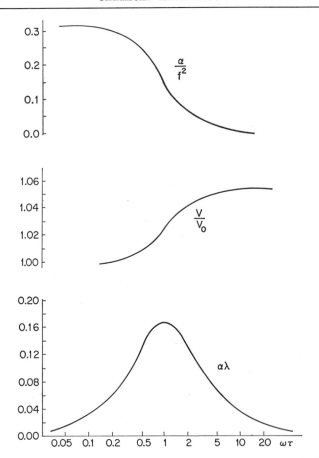

FIG. 1. Absorption per wavelength and velocity dispersion from a single relaxation process as a function of normalized frequency (relaxation strength $r = 0.1$). After Davies and Lamb.[9a]

Characteristic effects from a single relaxation process on the attenuation and the velocity dispersion are sketched in Fig. 1.

As in other relaxation techniques, the interpretation of the data requires the perturbation to be small, thus allowing all rate equations to be linearized. The level of acoustic intensities should therefore be far below 1 W cm$^{-2}$. The treatment for coupled multistep reactions leads to a spectrum of relaxation times. The analysis is more complicated than for single-step processes, although, in principle, it is identical. For details, the reader is referred to the theoretical treatments already cited (especially footnote 8). A rigorous analysis with regard to ultrasonics has also been given by Eigen and Tamm.[17]

[17] M. Eigen and K. Tamm, Z. Elektrochem. **66**, 93 (1962).

Two more points will be briefly touched upon before the equipment itself is described. These points are (1) the magnitude of the response and (2) the experimental acquisition of $\tau$.

The magnitude of the response is governed by the laws of thermodynamics (more properly, irreversible thermodynamics). It is convenient to use the product $\alpha\lambda$, the attenuation per wavelength. Defining

$$\mu \equiv \alpha\lambda$$

one obtains

$$\mu = \frac{\pi}{\mathbf{v}\chi_{A,\mathbf{s}}}\left(\frac{\partial\xi}{\partial A}\right)_{p,T}\left\{-\Delta\mathbf{v} + \Delta\mathbf{h}\frac{\alpha_{\xi,p}}{\rho C_{\xi,p}}\right\}^2\frac{C_{\xi,p}}{C_{A,p}}\frac{\omega\tau}{1 + \omega^2\tau^2} \tag{12}$$

In Eq. (12) $\mathbf{v}$, $\mathbf{h}$, and $\mathbf{s}$ are the specific (e.g., volume per mass) volume, enthalpy, and entropy, respectively; $T$ is °K; $\xi$ is the progress variable, defined as $\xi = (n_i - n_i^0)/\nu_i$, where $n_i$ is the mole number, the superscript zero denotes a time-independent reference, and $\nu_i$ is the molar coefficient in the balanced equation; $A$ is the affinity; $C_{\xi,p}$ and $C_{A,p}$ are the specific heats at constant $\xi$ and $p$, and constant $A$ and $p$, respectively. Other symbols in Eq. (12) have their usual meaning. In pure liquid solutions the enthalpy term may dominate, and frequently does, leading to a perturbation due to heating by ultrasonics. This effect has been exploited by Lamb and coworkers.[9a,b]

In aqueous media the volume term dominates. We shall therefore neglect the enthalpy term, which is a good approximation for the solutions commonly dealt with in biochemical research. As far as the frequency dependence is concerned, let us note that the absorption coefficient will be maximal when $\omega = 1/\tau$. Taking $C_{\xi,p}/C_{A,p} = 1$, and designating the frequency maximized $\mu$ as $\mu_{fm}$, we may write

$$\mu_{fm} = \frac{\pi}{2}\frac{\rho v^2}{VRT}\left(\frac{\partial\xi}{\partial\ln K_c}\right)_{p,T}(\Delta V)^2 \tag{13}$$

with $K_c$ the (concentration) equilibrium constant and $R$ the gas constant. The use of molar quantities and volumetric concentrations introduces insignificant error, so long as the reaction is associated with small volume changes. For a more extensive discussion of this point, see footnote 8. Consequently, in the form we wish to examine it, Eq. (12) becomes

$$\mu_{fm} = \frac{\pi}{2}\frac{\rho v^2 C^\circ}{RT}\Gamma_c(\Delta V)^2 \tag{14}$$

where

$$\Gamma_c = \frac{\partial\beta}{\partial\ln K_c} \tag{15}$$

and $\beta$ = degree of dissociation.

Analysis of this equation shows that the magnitude of the response is a function, primarily, of $\Delta V$ and the equilibrium quotient. Obviously, $\Delta V$ must not equal zero and it should be relatively large; biologically important systems generally have appreciable $\Delta V$ values.[18] Physically, the $C°\Gamma_c$ term shows that care must be taken to study systems in a concentration region where both reactants and products are significantly present, a requirement that will be fulfilled if $0 < \beta < 1$ and if sufficient material is present for detection of a response. The practical lower limit on $C°$ has been about $10^{-2}$ $M$ for most of this equipment. Experiments are usually carried out in the range 0.05 to 0.5 $M$. For some of the apparatus described, a lower limit of approximately $10^{-4}$ $M$ has been realized.

Once the amplitude attenuation coefficients have been measured, the relaxation times may then be determined. Two methods are in general use. In one, the functional relation is of the form

$$\mu_{\text{chem}} = \frac{B\omega\tau}{1 + \omega^2\tau^2} \tag{16}$$

and a graphical evaluation is used.[19] That is $\mu$ vs. $\omega$ is plotted on a log-log scale, and at the maximum $\tau = 1/\omega$. In this way a single relaxation curve for the absorption per wavelength can be drawn with a template, since the shape does not change; the slopes for low and high frequencies are as $\omega$ and $\omega^{-1}$, respectively. Alternatively, the expression for $\mu$ may be manipulated into a linear form. One such expression,[2] in terms of $\alpha$, is

$$\frac{1}{2v^2}\left(\frac{\omega^2}{\alpha}\right) = \frac{\omega^2\tau}{(v_\infty^2 - v_0^2)} + \frac{1}{(v_\infty^2 - v_0^2)\tau} \tag{17}$$

where $v_\infty$ and $v_0$ are, respectively, the high and low frequency limiting velocities. A plot of $\omega^2/\alpha$ vs. $\omega^2$ therefore has slope $= \tau/(v_\infty^2 - v_0^2)$ and intercept $= 1/(v_\infty^2 - v_0^2)\tau$; the quotient, slope/intercept $= \tau^2$.

Thought should be given to the form in which the data are acquired. Computer techniques are especially useful in evaluating sound absorption.[20a] For this purpose, any of the linear forms of the equation for $\mu$ would be most well suited. Advantages of this type of data acquisition and analysis will also accrue to the investigator upon encountering multistep reaction mechanisms.[20b]

[18] W. Kauzmann, A. Bodansky, and J. Rasper, *J. Am. Chem. Soc.* **84,** 1777 (1962); J. Rasper and W. Kauzmann, *ibid.*, p. 1771.

[19] In this form, $\mu$ has been corrected for dissipative, nonchemical relaxation modes ($B$ is a constant); see footnote 8, p. 959.

[20a] G. G. Hammes and H. O. Spivey, *J. Am. Chem. Soc.* **88,** 1621 (1966).

[20b] G. G. Hammes and W. Knoche, *J. Chem. Phys.* **45,** 4041 (1966).

## II. Ultrasonic Techniques

### A. Principles of Ultrasonic Methods

Ultrasonic measurements are related to three basic phenomena:[21] (1) pulse transmission; (2) continuous wave transmission; (3) resonance effects.

1. The pulse method has been introduced to ultrasonics during and after World War II, using newly developed radar techniques.[22] A short train of several sine waves travels a defined distance in the liquid; the group velocity can be determined from the time delay and the attenuation from the amplitude of the received signal. An advantage of pulse methods is the low average power of sound in the liquid with no heating effects present, also the time separation of pulses, which allows discrimination against unwanted signals from reflections in the sound path. It is also

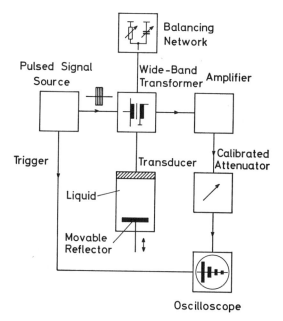

FIG. 2. Basic circuit for a single transducer pulse-echo method. Both sound velocity and attenuation can be determined in this way. The wide-band transformer with the balancing network reduces the driving pulse amplitude at the amplifier input, preventing overload problems. For the nth echo the attenuation effect is multiplied n times, thus increasing the sensitivity of the method.

[21] G. Kurtze and K. Tamm, *Acustica* **3**, 33 (1953).
[22] J. H. Andreae, R. Bass, E. L. Heasell, and J. Lamb, *Acustica* **8**, 131 (1958).

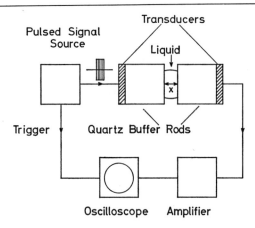

FIG. 3. Pulse technique using buffer rods. Subsequent echoes are compared in amplitude, which yields the attenuation in the liquid. The quartz buffer rods improve the discrimination between driving and received pulses.

possible to check the parallelism of both transducers (or transducer and reflector) by controlling the observed pulse pattern. Lack of parallelism is a chief source of error in this technique. Several variations of the pulse method are feasible; e.g., reflection methods use the same transducer as sender and receiver (Fig. 2). Quartz buffer rods may be employed to produce an increased time delay between transmitter and receiver pulse in order to obtain an improved separation of pulses on the screen of an oscilloscope and to avoid receiver overload by the driving pulse (Fig. 3).[23a] A differential pulse technique, which is described below, yields improved accuracy by direct determination of the excess absorption.

2. Continuous waves are employed in interferometers (invented by G. W. Pierce). In Fig. 4 the main elements of an interferometer are sketched. Standing sound waves are excited in the liquid between the source and the reflector at one particular frequency. Movements of the reflector cause periodic impedance changes for the transducer, which are transferred into the electrical driving circuit; they produce periodic changes of the driving current which can be measured by sensitive electronic circuits. The sound velocity is determined from the reflector spacings which cause extrema in the driving current and which are equivalent to half wavelengths in the liquid. The reflector must be moved maintaining absolute parallelism with respect to the driving transducer, which demands extreme care in mechanical construction and adjustment. The literature is extensive[23b,c] for this and also for measurements at elevated temperature or pressure.

[23a] G. Atkinson, S. K. Kor, and R. L. Jones, *Rev. Sci. Instr.* **35,** 1270 (1964).

[23b] V. A. Del Grosso, E. J. Smura, and P. F. Fougere, *Naval Res. Lab. Rept. No.* **4439** (1954).

[23c] W. Schaaffs, "Molekularakustik," p. 67. Springer Verlag, Berlin, 1963.

The Debye-Sears effect is often used to detect the energy density of a traveling sound wave, which must be absorbed at the end of the path to prevent formation of standing waves (Fig. 5). The light source and photomultiplier can be moved along the sound path, and the response is written by a (logarithmic) level recorder giving a straight line with a slope proportional to the attenuation coefficient. A square modulated sound wave facilitates amplification behind the photomultiplier and avoids drift problems.

3. Resonance effects are utilized in measurements of reverberation time (or rate of decay) for liquid-filled vessels and frequency half-power bandwidth for resonant cavities (e.g., spheres); the latter are similar to microwave techniques. These methods have proved useful at the lower end of the frequency range under consideration, from approximately 100 kHz down to 5 kHz, where several liters of liquid are necessary for reliable measurements.[24a,b]

Since no completeness is intended for this article, only a few selected techniques will be discussed in some detail. It must be mentioned that no commercial sound measuring devices for the scientist, which yield direct

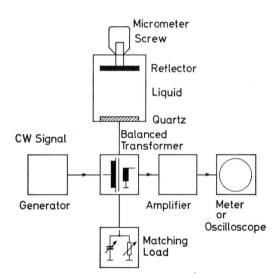

FIG. 4. Basic circuit for an interferometer. By means of the balanced transformer the impedance variations for the quartz crystal at constant frequency can be measured, while the reflector is moved maintaining a high degree of parallelism with respect to the quartz. Many other circuits for interferometers have been described in the literature including measurement of the plate current of a valve oscillator circuit; also commercial RF bridges may be used.

[24a] J. Karpovich, *J. Acoust. Soc. Am.* **26**, 819 (1954).
[24b] L. E. Lawley and R. D. C. Reed, *Acustica* **5**, 316 (1955).

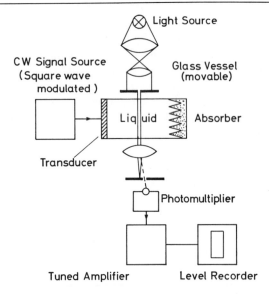

FIG. 5. Determination of sound intensity by means of the Debye-Sears effect for a propagating ultrasonic wave. This method needs transparent liquids and yields the *intensity* absorption coefficient $2\alpha$. When the sound wave is square wave modulated (chopped) with a frequency, for instance, of 1 kHz, the output from the photomultiplier can be fed into an amplifier tuned to the same frequency, thus avoiding null line drift problems. If the level recorder has a logarithmic response, a straight line is produced when the light beam moves along the sound path; the slope is proportional to the absorption coefficient.

results in sound parameters over an extended frequency range, are as yet on the market. A laboratory interested in ultrasonic work will therefore need a high-quality mechanical shop. The necessary electronic equipment can, in many cases, be acquired commercially now.

## B. Specific Techniques

1. *Pulse.* Carstensen[25a] has described a substitution method, which allows direct determination of excess absorption (i.e., measured absorption minus solvent absorption). Both sender and receiver transducer are at a fixed distance but movable with respect to the liquid container, which is divided into two compartments by a thin, sound transmitting membrane (see Fig. 6). Compartment I is filled with the liquid to be evaluated and compartment II with the solvent or another suitable "reference" liquid. A pulsed RF generator feeds the sending transducer, which radiates a sound pulse into the liquid. This pulse passes from compartment I through the

---

[25a] E. L. Carstensen, *J. Acoust. Soc. Am.* **26**, 862 (1954).

membrane into compartment II and is received by the other transducer. The generated voltage is amplified, and the pulse response is displayed on an oscilloscope screen with the x-axis triggered by the RF driving pulse. When the transducer assembly is moved a distance x in the direction of the sound propagation, the total sound path length remains constant, but the partial lengths in both compartments change in a way that the change in pulse height is directly proportional to the excess absorption for the distance x.

This method has been used by Carstensen for measurements in the frequency range from 0.3 to 10 MHz; the liquid volume needed is about 500 ml. Siegert[12b] has increased the accuracy and was able to measure absorption down to 0.01 db cm$^{-1}$.

Dispersion measurements with the same equipment are also described by Carstensen.[25b] The phase of the received signal is compared to that of the transmitter signal. When the membrane moves with respect to the transducers, the phase changes thus produced allow a direct calculation of the velocity difference between both solutions. Carstensen's method gains from the fact that the substitution takes place under fixed acoustical conditions. Disturbing effects due to beam spreading and phase or amplitude shifts in the membrane are minimized, because all parameters remain constant during one measurement. However, excellent temperature

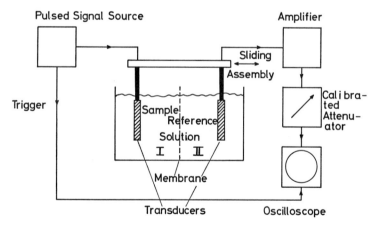

FIG. 6. Substitution method by Carstensen.[25a] This is a modified pulse technique which has been refined by Siegert[12b] to achieve high accuracy in amplitude determination. This method gains from the fact that irregularities in the sound field, beam spreading, etc., are nearly eliminated since the sound velocities of both sample and reference solution do not differ very much. The membrane may limit the upper frequency range and prohibit the use of potentially corrosive liquids.

25b E. L. Carstensen, J. Acoust. Soc. Am. 26, 858 (1954).

stability is important—true for all precision ultrasonic measurements—in both compartments. Temperature gradients, which will deflect the sound beam, must also be avoided.

Pulse propagation in a single line, filled with the liquid to be tested, has been widely used. The output pulse from the line or one of the subsequent echoes must be compared to a "standard" pulse; its height is often measured on an oscilloscope screen, while the traveling distance for the pulse is changed.[26a] In most cases the absolute value of attenuation is not as important as an accurate determination of the excess absorption. Busse[26b] has achieved this directly by means of a differential technique. The apparatus (Figs. 7A and 7B) consists of two identical propagation cells, the first filled with the test liquid and the second with the solvent. Both input transducers are fed from the same pulsed signal source through an adjustable attenuator. The pulses propagate independently in both liquids and are received from two transducers at the opposite end of the line. The outputs from both receiver crystals are subtracted in a wide-band differential transformer. The output transducers can be moved with respect to the input transducers by means of a high precision carriage. In this way both pulse-travel distances can be adjusted separately to a high degree of accuracy. It is possible to set the continuously adjustable attenuator and also the distances $x_1$ and $x_2$ exactly to the point where the output from the differential transformer is zero; i.e., both signals at the transformer primary are equal in amplitude and phase. This adjustment done, $x_1$ and $x_2$ are both increased by 1 cm (or any other suitable distance) by moving the output transducers on their carriages. The balance in the transformer is then eventually disturbed; first, because both sound velocities differ, which alters the phase between the received signals. This can be compensated for by a slight correction of $x_1$ or $x_2$ (i.e., moving one carriage), which is directly related to the difference of both sound velocities. The second reason for any disturbance is the difference in attenuation for both sound pulses; an exact null is obtained only when the variable attenuator in the receiving circuit has been readjusted. This amount of attenuation change is the excess absorption for 1-cm (or any other distance chosen) path length.

Because it is difficult to obtain continuously adjustable, calibrated attenuators for this frequency range, the differential measurement is carried out in a modified way: one uses two uncalibrated potentiometers or variable capacitors as "attenuators" in the driving circuit and determines the driving voltage at each transducer input separately by means of a calibrated piston type attenuator of high precision. This type of attenuator has a considerable insertion loss (minimum transmission loss), but one can

[26a] R. S. Brundage, Ph.D. Thesis, Brandeis University, Waltham, Massachusetts, 1969.
[26b] F. Busse, Doctoral Thesis, Technische Hochschule, Braunschweig, West Germany, 1967.

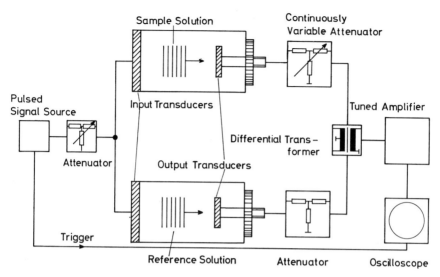

FIG. 7A. Differential pulse technique (simplified circuit). This seems to be the most sensitive method at present for the determination of attenuation increments. The amount of liquid needed is about 50 ml, which makes the equipment suitable for biochemical problems; frequency range is approximately 5–80 MHz. The differential transformer allows the evaluation of the attenuation difference between sample and reference solution. Both receiving transducers can be moved separately by high-precision carriages.

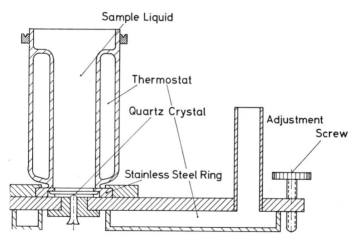

FIG. 7B. Cross section of a glass cell of Busse's equipment.[26b] The receiver crystal at the top of the vessel is not drawn; it is fixed to a stainless steel cylinder that can be moved into the vessel. The leveling is done by two adjustment screws (only one shown in the sketch) for the lower crystal. Since this apparatus consists basically of two pulse lines in parallel, this cell can be used also for single-line pulse work.

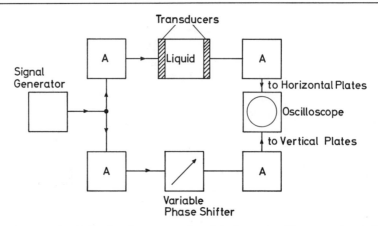

Fɪɢ. 8. Basic circuit for the phase comparison interferometer. It was constructed by Schaaffs[23b] for the precise determination of sound velocity increments (down to 1 mm/sec) at one frequency. Both transducers are at a fixed distance; changes in sound velocity are indicated by chanegs of the Lissajou ellipse on the oscilloscope screen. A, tuned amplifier.

read attenuation differences down to 0.02 db independent of frequency. The method has been used for measurements on low concentration solutions (0.001 $M$ and less) in the frequency range from 5 MHz to more than 80 MHz, depending on the magnitude of attenuation. The volume needed— about 50 ml—makes it suitable for biochemical studies.

2. *Continuous.* Interferometers have been mainly employed to determine sound velocity in liquids and gases. Schaaffs[23b] has developed a phase-comparison interferometer, which, at a constant frequency of 4 MHz, can determine incremental velocity changes down to 1 mm sec⁻¹. The block diagram of the apparatus is shown in Fig. 8. A sine wave signal generator drives (through frequency selective amplifiers) two channels: one is a sound transmission line with the liquid between the transmitter and receiver crystal, and the other is a variable phase shifter. The outputs from both channels are fed to the vertical and horizontal plates of an oscilloscope. The phase shifter is adjusted so that a straight line appears on the screen (zero phase difference between both signals). This straight line is split as soon as minute changes of the sound velocity alter the phase relation.

3. *Resonance.* Kurtze and Tamm[21] have employed spherical vessels for a resonator method. They used symmetrical radial vibration modes, which were excited through electrostatic forces with a circular electrode and picked up by small piezoelectric receivers (see Fig. 9 for a block diagram). One can either determine the half-power bandwidth $\Delta f$ for a single resonance with continuous wave excitation and find

$$\alpha\lambda = \frac{\pi \cdot \Delta f}{f} \tag{18}$$

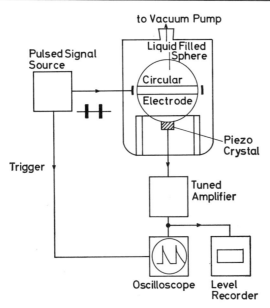

FIG. 9. Resonance decay method. The liquid in the sphere is excited to radial oscillations. The decay time of the response is measured after the driving signal is cut off. In analogy to reverberation time determination in room acoustics, the decaying signal results in a straight line on the (logarithmic) level recorder, the slope of which is proportional to the attenuation. A second possibility for measurement lies in the accurate determination of the resonance half-power bandwidth, which uses a continuous wave oscillator and a frequency counter in addition.

where $f$ is frequency. Or, one may switch off the excitation and measure the decay rate or reverberation time. The voltage from the pick-up system is displayed on an oscilloscope screen with calibrated time axis. Relaxation time $\tau$ and attenuation constant $\alpha$ follow from

$$u = u_0 \cdot e^{-(t/\tau)} = u_0 \cdot e^{-\alpha v t} \tag{19}$$

where $u$ is voltage. Since the accuracy is determined by additional energy losses (as deformation losses of the vessel wall and fluid friction at the wall, which are relatively low for symmetrical radial modes) it is advisable to employ, with several liters of liquid, large vessels made from aluminum, glass, or quartz with thin, uniform walls. They have a more favorable volume to surface ratio than small ones. Suspension of the vibrating sphere by thin steel wires in an evacuated chamber reduces the radiation losses into the surrounding air.

At frequencies higher than 50 kHz the subsequent resonances (harmonics) of a spherical vessel with dimensions necessary to obtain low loss values come rather close together and cannot be excited separately.

Skudrzyk and Meyer have therefore used filtered noise (with a small bandwidth) or short pulses to excite several resonances in a limited frequency band simultaneously.[21] The decay rate (which represents an average value for several vibrational modes) is recorded by a level recorder with logarithmic response, and the slope obtained is evaluated. These resonator methods cover a frequency range from approximately 5 kHz to 1 MHz. A second measurement—the vessel filled with the solvent—permits one to subtract wall losses and solvent absorption and to obtain the excess absorption of the test liquid.

For biochemical problems it is of special importance to have methods that need only very small liquid quantities. Andreae, Jupp, and Vincent[27] worked with a tuning fork resonator, which contains a few milliliters of the test liquid in a hollow arch. A small barium titanate transducer excites and detects the vibrations. The decay rate after cutoff is determined from an oscilloscope display. First the decay rate is measured for the liquid with an additional small vapor bubble in the arch; then the arch is completely filled and the liquid is compressed and expanded periodically, contributing to the energy losses of the vibration. The system worked around 100 Hz and could be tuned to different frequencies by attached solid masses.

A method with a cylindrical resonator has been developed by Eggers.[28] One measurement cell covers a frequency range from about 0.2 to more than 10 MHz and needs less than 40 ml of liquid. The quantity can be reduced to less than 5 ml for a smaller cell, with a frequency range from approximately 1 to 20 MHz. The resonator consists of two disk-shaped quartz transducers and the liquid enclosed by both. Figure 10 shows a cross section of the 40-ml cell. Transducer $Q1$ is driven by a sinusoidal voltage and produces standing waves in the liquid at particular frequencies $f_n$ that obey a transcendental resonance equation

$$\rho_Q \cdot v_Q \tan \pi \frac{f_n}{f_Q} = \rho_F \cdot v_F \begin{Bmatrix} -\tan \\ \cotan \end{Bmatrix} \frac{\pi}{2} \frac{f_n}{f_F} \qquad (20a,b)$$

Here $\rho_Q$ and $v_Q$ are density and sound velocity for the quartz transducers; $\rho_F$ and $v_F$ are the corresponding values for the liquid. The fundamental frequency of the transducers is $f_Q$, and that of the liquid column $f_F$. Figure 11 shows the block diagram of the equipment. At the resonance frequencies, transducer $Q2$ delivers pronounced voltage peaks, whose frequency position depends on the sound velocity and whose half-power bandwidth $\Delta f$ depends on the attenuation in the liquid.

Velocity and dispersion measurements are possible but less convenient and also sensitive to temperature drift; we shall therefore discuss only the

[27] J. H. Andreae, C. Jupp, and D. G. Vincent, J. Acoust. Soc. Am. **32**, 406 (1960).
[28] F. Eggers, Acustica **19**, 323 (1967/68).

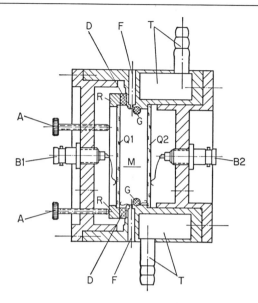

FIG. 10. Cross section of a 40-ml resonator cell, frequency range 0.2–10 MHz. Transducer quartz $Q1$ can be leveled against transducer $Q2$ by means of three adjustment screws $A$. $R$ indicates a metal frame for $Q1$ and $D$ is an O-ring which allows a slight movement of $R$ and serves as a gasket. The liquid volume $M$ is filled through channel $F$. $B1$ and $B2$ are BNC connectors, and $T$ is connected to the thermostat. An important function is played by rubber O-ring $G$, which scatters and absorbs oblique sound waves, enhancing a one-dimensional plane wave field. Recently cells of much simpler construction have been built, which need only 1 ml of liquid and operate from approximately 1 to 30 MHz.

determination of excess absorption. The signal generator—necessarily of excellent stability—is tuned manually first on top of a resonance peak and then to both half-power points (where the amplitude is 3 db down), while the frequency difference is read on an electronic counter. The output from transducer $Q2$, amplified by a frequency selective amplifier, is displayed on an oscilloscope screen or on a meter. A sweep generator proves to be very helpful for the initial mechanical adjustments of the cell, since both quartz crystals must be absolutely parallel.

The energy loss of a stationary sound field in the cell is caused by the liquid absorption (which is to be determined), and by additional losses from imperfect reflection at the transducers, wall friction losses, and acoustical beam spreading. These additional losses are eliminated by a comparison procedure. The total energy loss is first determined for the test liquid and then for the solvent or another standard liquid; both values are subtracted and the difference gives the excess absorption. The relation

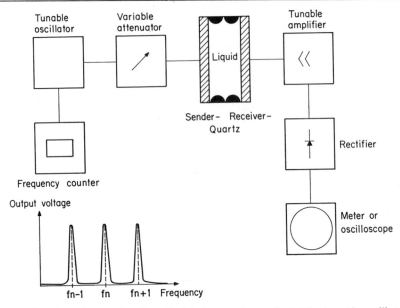

FIG. 11. Block diagram for the cylindrical resonator method. The tunable oscillator must be of extreme stability with a facility for incremental tuning. The variable attenuator has a calibrated 3 db step for determining the half-power bandwidth. A sweep generator (not shown) is most helpful for the mechanical adjustment of the cell. The tunable amplifier must have sufficient bandwidth for broader resonance peaks.

$$Q^{-1}{}_{\text{total}} = \frac{\Delta f}{f} = Q^{-1}{}_{\text{solvent}} + Q^{-1}{}_{\text{excess}} \qquad (21)$$

yields, in analogy to Eq. (18),

$$(\mu)_{\text{chem}} = \pi \frac{\Delta f_{\text{liquid}} - \Delta f_{\text{solvent}}}{f} \qquad (22)$$

with the mechanical quality $Q$ inversely proportional to the energy loss of that resonating system.

The following tabulation gives data of three model cells, built by one of the authors (F.E.) (all values are approximate).

| Cell | A | B | C |
|---|---|---|---|
| Sound beam diameter (mm) | 18 | 55 | 30 |
| Distance of transducers (mm) | 15 | 15 | 10 |
| Cell volume (ml) | 7 | 40 | 10 |
| Quartz fundamental frequency (MHz) | 6 | 2 | 5 |

The corresponding half-power bandwidth data are compiled in Fig. 12. It is obvious how the achievable accuracy deteriorates at the lower fre-

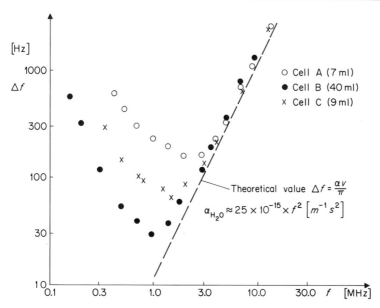

FIG. 12. Half-power bandwidths for some resonator cells filled with water at 20° obtained by Eggers.[28] At higher frequencies the water attenuation dominates; at the lower end, additional losses from the finite size of the cell limit the achievable accuracy. This diagram is intended only to give practical hints for the construction and testing of new cells.

quency side. Whether $\Delta f/10$ of these values can be safely determined depends on the stability of the acoustical system and the reproducibility and accuracy of the frequency measurement. These factors limit the minimum detectable $\alpha\lambda$ in the lower frequency range.

Resonator cells of this kind are rather uncomplicated to design and to build, and all the electronic equipment needed is commercially available. The narrow spacing of subsequent resonances (from about 45 to 200 kHz) allows close measurement points for a relaxation curve, which again helps to average out errors. The sensitivity of this method is reached through multiple reflections (up to several hundred) and "folding" of the acoustical path. Thus any disturbances from imperfect reflection and misalignment, even the smallest phase distortion of the sound wave, are multiplied many times and decrease the attainable accuracy considerably. Therefore it is extremely important not to load the transducers by the mounting and to avoid any static stress or bending. For aqueous solutions silicone rubber is helpful in fixing the transducers to their frames. When the driving voltage on the input is below 0.1 V with sufficient postamplification behind the receiver quartz, no recognizable heating effects occur.

## C. Selection of Components and Construction Details

Quartz transducers are in general made to customer specifications. Two companies, among many others, who provide excellent facilities for quartz preparation are Valpey Crystal Corporation in Holliston, Massachusetts, and Gooch and Hausego Ltd. in Ilminster (Somerset, England). Transducers with fundamental frequencies higher than 30 MHz are rather thin and difficult to handle unless they are bonded to buffer rods, which can be done with indium films or highly viscous liquids, such as silicone oil. When overtone operation becomes necessary, this should be specified on the order and the upper frequency limit be given. Quartz disks are mostly fabricated in batches; therefore it is cheaper per piece to order a few disks, depending on size. Surfaces are normally gold-plated by the manufacturer; if corrosive liquids are to contact the surface, this should be specified to the manufacturer. Piezoceramic materials in many sizes can be obtained from Brush Clevite Co. in Cleveland, Ohio or from Gulton Industries Inc. in Metuchen, New Jersey; data books for designers are distributed by these companies. Descriptions of quartz mountings and holders are given in the literature.[3,11]

In the last few years it has become easier to obtain commercially most of the necessary electronic components for ultrasonic research work. Signal generators for continuous-wave operation (output around 1 V into 50 ohms) with electronic incremental frequency tuning facility can, for instance, be bought from Hewlett-Packard, General Radio Co., Marconi, or Rhode and Schwarz. Pulsed RF sources (pulse duration around 1 μsec; output in the 100-V region) are manufactured by the Arenberg Ultrasonic Laboratory, Boston, Massachusetts and the Matec Inc., Warwick, Rhode Island. Sweep generators (Hewlett-Packard, Kay, Telonic, Texscan, and others) are very helpful for electrical tuning and mechanical adjustments of ultrasonic devices. Attenuators are available from many manufacturers; most common are either fixed or step attenuators (in 10, 1, and 0.1 db steps). Only very few continuously adjustable attenuators for the frequency range under consideration are, so far, on the market.

In many instances it is desirable to match the electrical impedance of a transducer to the preceding or following electrical network. It is known that maximum power is transferred from one network to another if both impedances are complex conjugates. Many generators and receivers exhibit 50 ohms impedance; therefore it is common practice to resonate the capacitance of a transducer with an inductance (for that particular frequency) and to transform the remaining resistance to 50 ohms. This can be done by lumped elements (e.g., coils and wide-band transformers from Anzac, North Hills Electronics, and Vary-L Co.) or, at higher frequencies, by transmission line elements as variable shorts, line stretchers, and stub stretchers (manufactured by Microlab, Weinschel, and others) which can be continuously adjusted over a limited range.

Special care should be spent on the selection of a good receiving system. Main points of interest are frequency range (also ease of tuning), bandwidth, sensitivity, linearity, and overload recovery. For improved signal to noise ratio, frequency selective amplifiers are normally employed. They incorporate resonating circuits, which are either at fixed frequency or tunable. Superheterodyne detection converts the test frequency into a fixed frequency of the intermediate frequency amplifier (often called IF-amplifier) by means of a nonlinear mixing process with an auxiliary oscillator. For continuous wave measurements the bandwidth of the receiver can be rather narrow. With pulsed signals the spectrum exhibits side bands which the amplifier must not suppress, otherwise the pulse shape will deteriorate. The sensitivity of an amplifier is mainly determined by the noise from its input stage and depends on the effective bandwidth. More important than the absolute value of sensitivity is, in many instances, the stability and linearity, especially for high precision absorption measurements. Finally, the overload recovery becomes important in pulsed systems where the driving signal feeds through into the detecting amplifier, which can be blocked by overload for some time, while the desired signal passes by. Special amplifiers have been made with extremely short time constant and rapid recovery. It may also be helpful to use a crystal switch (i.e., a network with biased diodes, which can rapidly change attenuation on a video pulse applied) in front of the receiver, which interrupts the signal path during the duration of the driving pulse.[29]

Oscilloscopes, frequency counters, voltmeters, etc. are available from many manufacturers and do not need any explanation. For an oscilloscope a frequency response up to 40 MHz is normally sufficient, because higher frequencies in ultrasonic work can easily be downconverted by superheterodyning techniques. The selection of suitable equipment will definitely gain from the assistance of an experimental physicist or an electronic engineer.

### D. Concluding Remarks

Ultrasonic methods are now adaptable to many biochemical problems and cover a wide range of relaxation times, approximately from $10^{-5}$ to

---

[29] Radio frequency mixers and IF-amplifiers can be obtained from General Radio Co., Hewlett-Packard, LEL Division Varian, RHG Electronics Lab., and many other manufacturers. For the resonator method by Eggers,[28] the level measuring set PSM-5 made by Wandel u. Goltermann in Reutlingen (Germany) is very useful; it contains a precision type oscillator (range 0.01–36 MHz) and a narrow band receiver of high sensitivity, which is automatically tuned to the oscillator frequency. For the selection of suitable electronic equipment, it is recommended to consult the Electronic Engineers Master EEM from United Technical Publications.

$10^{-9}$ seconds and shorter.[14,30] An accuracy of a few percent for ultrasonic amplitude attenuation and of 0.01% for velocity dispersion measurements may be achieved. One limitation, certain to be improved on, is the relatively large quantity of solution needed for the frequency range around 1 MHz and lower. Most mechanical parts cannot be obtained commercially; therefore a mechanical shop is necessary. On the other hand, electronic components of adequate quality are readily available.

Until now ultrasonic techniques have been indispensable for the evaluation of many very fast chemical reactions, especially because the almost universal temperature-jump method does not reach time constants much shorter than 1 $\mu$sec. It appears desirable to have several ultrasonic methods operating at one place; ultrasonic spectroscopy, which is still in the developmental stage, requires the coverage of a wide frequency range in a single instrument. Even before this goal has been attained, it is clear that numerous new biochemical problems can be successfully tackled by ultrasonic methods.

[30] F. Dunn and J. E. Breyer, *J. Acoust. Soc. Am.* **34,** 775 (1962).

# [4] Electric Field Methods

## *By* Leo C. M. De Maeyer

## I. Chemical Equilibrium in Electric Fields

### A. Introduction

The primary effect of an electric field on a material substrate is 2-fold: (1) orientation of dipolar species and deformation of polarizable molecules (dielectric polarization); (2) movement of free electric charges (ionic conduction). In both cases a coupling to chemical transformations of the species involved may exist and can sometimes be used to study chemical reaction rates and mechanisms. We will first give a short and condensed survey (limited to liquid solutions) of some theoretical notions that are required to describe adequately the chemical effects of electric fields. (We will not consider in detail electronic or atomic polarizibility, since its coupling to chemical phenomena is negligible or, at least, far beyond the limits of present experimental possibilities.)

A system containing free charges (ions) cannot be in true thermo-dynamic equilibrium in the presence of an electric field, since the latter implies a steady flow of charges. But in the absence of free charges, in a solution containing permanent dipoles or polarizable molecules, a thermo-dynamic approach to the problem of chemical coupling is suitable.

A thermodynamic derivation of the effect of a homogeneous electric field upon a chemical reaction equilibrium leads to the following expression:

$$\left(\frac{\partial \ln K_c}{\partial E}\right)_{T,P} = \frac{\Gamma_c^*}{\Gamma_c}\frac{\Delta M}{RT} \tag{1}$$

Here, $K_c$ is the equilibrium constant of the chemical reaction defined in suitable concentration variables $c$; $\Gamma_c^*/\Gamma_c$ is a correction factor near to one that accounts for the deviation of ideal behavior in terms of activity coefficients; $R$ is the gas constant, and $T$ the temperature.

The quantity $\Delta M$ is the change in the macroscopic electric moment of the system per mole equivalent transformation of reaction partners:

$$\Delta M = \left(\frac{\partial M}{\partial \xi}\right)_{T,P,E} \tag{2}$$

The variable $\xi$ is the extent of reaction, expressed by the number of mole equivalents transformed ($dN_i = \nu_i \, d\xi$ with $N_i$, mole number of component $i$; $\nu_i$ stoichiometric coefficient of component $i$ in the reaction scheme). The correction factor $\Gamma_c^*/\Gamma_c$ is given by

$$\frac{\Gamma_c^*}{\Gamma_c} = \frac{1}{1 + \Gamma_c \left(\dfrac{\partial \Sigma \nu_i \ln f_i}{\partial \xi}\right)_{T,P,E}} \simeq 1 \tag{3}$$

where $f_i$ are the activity coefficients corresponding to the concentration variables $c$ for which $K_c$ is defined. The function $\Gamma_c$ is defined by

$$\Gamma_c = \left(\frac{\partial \xi}{\partial \Sigma \nu_i \ln c_i}\right)_{T,P,E} \tag{4}$$

It can be expressed as a function of the composition of the reaction medium. The form of the expression will depend upon the stoichiometry of the reaction and the choice of the concentration variable. For reactions in dilute solutions, the concentrations are usually expressed as volume concentrations $c_i = N_i/V$. The volume change due to reaction is usually small, and its influence upon $\Gamma_c$ can mostly be neglected, as well as the influence of activity coefficients in $\Gamma_c{}^*$. If we denote by $\alpha$ the degree of transformation for a unimolecular transformation $A \rightleftarrows B$ with actual concentrations $[A] = C_0(1 - \alpha)$; $[B] = \alpha C_0$, we obtain

$$\Gamma_c = VC_0\alpha(1 - \alpha) \tag{5}$$

For a bimolecular association of the form $2A \rightleftarrows B$ with $\beta = $ degree of association and $C_0 = $ total concentration in terms of monomers the expression for $\Gamma_c$ is

$$\Gamma_c = VC_0\frac{\beta(1 - \beta)}{2(1 + \beta)} \tag{6}$$

For other reactions the equivalent expressions are easily derived.

## B. Dipole Equilibria

Equation (1) is directly applicable to the study of chemical equilibria involving molecules with permanent dipole moments.

A molecule with a permanent dipole moment $\mu$ and oriented with its dipole axis[1] along a direction making an angle $\theta$ with the direction of a uniform field $E = -\mathrm{grad}\ \phi$ has an energy

$$W = -\mu E_r \cos\theta \tag{7}$$

(In free space $E_r = E$, but in a dielectric medium $E_r$ is not equal to the external field but larger, because of the polarizing influence of the surroundings.)

In a system comprising many molecules in thermal equilibrium the number of molecules with energy $W$ is given by the Boltzmann distribution. The number of molecules having their dipole axis pointing in the direction $d\omega = \sin\theta\ d\theta\ d\varphi$ is

$$dn = C \exp(-W/kT)\ d\omega \tag{8}$$

The normalizing constant $C$ is determined considering that the total number of molecules is given by

$$N = C \int_\Omega e^{-W/kT}\ d\omega = C \int_0^{2\pi} d\varphi \int_0^\pi e^{\mu E_r \cos\theta/kT} \sin\theta\ d\theta \tag{9}$$

---

[1] It is convention to give the vectorial dipole moment $\mu$ a positive direction in the direction of its positive charge.

or

$$C = \frac{N}{4\pi \sinh\left(\dfrac{\mu E_r}{kT}\right)} \frac{\mu E_r}{kT} \simeq \frac{N}{4\pi} \text{ for } \frac{\mu E_r}{kT} \ll 1 \tag{10}$$

The electric moment per mole in the direction $\theta = 0$ resulting from this distribution is given by[2]

$$M_{\text{dip}} = \int_\Omega \mu \cos\theta \, dn = \mu N_A \left(\coth\frac{\mu E_r}{kT} - \frac{kT}{\mu E_r}\right)$$

$$\simeq N_A \frac{\mu^2 E_r}{3kT} \text{ for } \frac{\mu E_r}{kT} \ll 1 \tag{11}$$

The molecule with dipole moment $\mu$ is subjected to changes in its dipole orientation by rotation due to Brownian movement, but can also be involved in an intra- or intermolecular chemical reaction which changes its dipole moment, or the orientation of the electric moment in the molecule, or both.

For example, let us consider a molecule that undergoes an intra-molecular change leading to a modification of its dipole moment. The chemical equilibrium that exists between the two states $A_1$ and $A_2$, respectively, of the molecule is given by $K_c = k_{12}/k_{21}$

$$A_1 \underset{k_{21}}{\overset{k_{12}}{\rightleftarrows}} A_2 \tag{12}$$

These two states may be characterized by different dipole moments $\mu_1$ and $\mu_2$ and let us assume for further simplification of our example that in one state the dipole moment is vanishingly small, e.g., $0 \simeq \mu_1 \ll \mu_2$.

We denote by $\gamma_1(\theta)$ and $\gamma_2(\theta)$ the number of particles per unit volume that have their molecular axis making an angle $\theta$ with the direction of the field

$$\gamma_1 = \frac{1}{V}\frac{dn_1}{d\omega}; \qquad \gamma_2 = \frac{1}{V}\frac{dn_2}{d\omega}; \qquad d\omega = 2\pi \sin\theta \, d\theta \tag{13}$$

The molecular axis is to correspond with the direction of the dipole moment for $A_2$, whereas it must correspond to that direction in $A_1$ in which the dipole moment will appear after chemical transformation.

---

[2] The function $L(x) = \coth x - (1/x) = (x/3) + (x^3/45) + (2x^5/945)\cdots$ is known as the Langevin function. It was derived by Langevin for the orientation of magnets in a magnetic field. For small permanent dipole molecules with dipole moments of the order of a few Debye units (1 Debye unit = $3.336 \times 10^{-28}$ amp sec cm) even with electric field strengths of the order of $10^6$ volts cm$^{-1}$ the contribution of the higher order terms remains far below 1%.

For every angle $\theta$ an equilibrium constant $K_\gamma(\theta)$ can be defined which, in the presence of an external field, will be a function of $\theta$ in view of Eq. (1)

$$\frac{\partial \ln K_\gamma(\theta)}{\partial E} = \frac{\Delta M(\theta)}{RT} = \frac{1}{RT}\left(\frac{\partial M(\theta)}{\partial \xi(\theta)}\right) = \frac{N_A \mu_2 \cos \theta}{RT} \tag{14}$$

First of all we may thus note that, in the presence of an external field, the orientational equilibrium distribution of the molecular species is disturbed: the polar form now has its dipole axis preferentially in the direction of the external field, whereas the orientational distribution of the nonpolar form remains random. This state of affairs must be independent of the relative rates of rotational diffusion and chemical transformation. Even if the latter is much faster than the former, there can be no change in the orientational distribution (with a corresponding decrease in entropy due to increased ordering) of the nonpolar species. The relative concentrations of the polar and nonpolar form of the molecule lying in a given direction are a function of the external field. The change in population of polar molecules with a given orientation with respect to a field can be supplied either from rotation or from chemical transformation.

A quantitative discussion of this behavior[3] shows that in this way chemical transformation rates may eventually determine dielectric relaxation times. This will be taken up again in the next section.

We may also ask how the overall equilibrium constant $K_c$ depends upon the external field. For this purpose we can use Eq. (11), but this equation does not yet include the contributions of atomic and electronic polarizibility to the total electric moment of the system. If $\alpha_k$ and $\mu_k$ are the polarizibility and permanent dipole moments of the $k$th species in the system, the total electric moment is given by

$$M_{\text{tot}} = \sum_k N_k \left\{ \alpha_k(E_i)_k + \frac{\mu_k^2}{3kT}(E_r)_k \right\} \tag{15}$$

The internal fields $(E_i)$ and the directing fields $(E_r)$ are different for each molecule. These fields were calculated by Onsager.[4] On the basis of his results, the total electric moment of the system is given by

$$M_{\text{tot}} = \sum_k N_k \frac{\epsilon(n_k^2 + 2)}{2\epsilon + n_k^2} \left\{ \alpha_k + \frac{\mu_k^2}{3kT} \frac{n_k^2 + 2}{2\epsilon + n_k^2} \frac{2\epsilon + 1}{3} \right\} E \tag{16}$$

where $n_k$ is the refractive index of species $k$ measured on the pure compound in a high density state at a wavelength where dipole orientation has

[3] G. Schwarz, J. Phys. Chem. 71, 4021 (1967).
[4] L. Onsager, J. Am. Chem. Soc. 58, 1486 (1936); cf. C. J. F. Böttcher, "Theory of Electric Polarization." Elsevier, Amsterdam, 1952.

relaxed but electronic and atomic polarization still prevail, $\epsilon$ is the macro-scopic dielectric constant of the system.

For our purposes this equation can be simplified by taking $n_k = n_j = n$, using the Lorenz-Lorentz equation

$$\sum_k N_k \alpha_k = 3\epsilon_0 V \frac{n^2 - 1}{n^2 + 2} \tag{17}$$

it is found that

$$\frac{\partial \ln K_c}{\partial E} = \frac{1}{RT} \frac{\partial M}{\partial \xi} = \frac{1}{RT} \left\{ \frac{\epsilon^2 (n^2 + 2)^2}{2\epsilon^2 + n^4} \frac{E}{9kT} \sum_k \mu_k^2 \frac{\partial N_k}{\partial \xi} \right\}$$

$$= \frac{\epsilon^2 (n^2 + 2)^2}{2\epsilon^2 + n^4} \frac{E}{9k^2 T^2} \sum \nu_k \mu_k^2 \tag{18}$$

[This equation is easily derived from the fact that $\partial M / \partial \xi = \epsilon_0 V E (\partial \epsilon / \partial \xi)$.] The contribution of atomic and electronic polarizability must therefore remain included in Eq. (16) since the internal field is itself dependent upon the orientational polarization of the medium.)

Besides the effect of the electric field upon the chemical reaction of a dipole molecule as a function of its orientation in the field, as given in Eq. (14), there is also an effect upon the overall reaction leading to an increase in the concentration of the molecular species with the larger dipole moment. It is easily seen by integration of Eqs. (14) and (18) that the first effect is a linear function of the external field, whereas the second one leads to a quadratic expression in terms of the field.

Both effects can be used, under suitable conditions, to investigate the rate of the chemical transformations involved.

## C. Dielectric and Chemical Relaxation

In an alternating electric field the orientational distribution of dipoles must change periodically. The change in concentration of molecules lying in a specified direction can be supplied by rotational flow and by chemical transformation

$$\left( \frac{\partial \gamma_i}{\partial t} \right) = \left( \frac{\partial \gamma_i}{\partial t} \right)_{\theta, \text{chem}} + \left( \frac{\partial \gamma_i}{\partial t} \right)_{\text{rot}} \tag{19}$$

The rotational flow is driven by the electric force, tending to orient the dipole axis in the field direction, and by the Brownian movement, counter-acting the resulting gradient in angular distribution

$$\left( \frac{\partial \gamma_i}{\partial \gamma} \right)_{\text{rot}} = D_{\text{rot}\,i} \left( \text{divgrad } \gamma_i + \gamma_i \frac{\mu_i E \sin \theta}{kT} \right) \tag{20}$$

The net chemical flow is given by the difference in forward and backward rate of transformation, which must be calculated here for fixed orientation of the reaction partners. For the example of Eq. (12) we may write

$$\left(\frac{\partial \gamma_1}{\partial \gamma}\right)_{\theta,\text{chem}} = -\left(\frac{\partial \gamma_2}{\partial \gamma}\right)_{\theta,\text{chem}} = -k_{12}(\theta)\gamma_1 + k_{21}(\theta)\gamma_2 \qquad (21)$$

In the neighborhood of chemical equilibrium and for small field amplitudes ($\mu E/kT \ll 1$) the system of differential equations obtained from Eq. (19) can be linearized. The eigenvalues of the resulting system of linear differential equations are the reciprocal relaxation times with which the Boltzmann equilibrium state in the electric field is attained.

For the example quoted in Eq. (12) two relaxation times are obtained

$$\frac{1}{\tau_I} = \frac{1}{\bar{k}_{12} + \bar{k}_{21}} + 2D_{\text{rot}} = \frac{1}{\tau_{\text{chem}}} + \frac{1}{\tau_{\text{or}}} \qquad (22)$$

$$\frac{1}{\tau_{II}} = 2D_{\text{rot}} = \frac{1}{\tau_{\text{or}}} \qquad (23)$$

These relaxation times are observed in the dielectric properties of the system. The dielectric constant will be a function of the frequency of the applied field; with $j = \sqrt{-1}$ then

$$\epsilon = \epsilon^\infty + \frac{\delta\epsilon_I}{1 + j\omega\tau_I} + \frac{\delta\epsilon_{II}}{1 + j\omega\tau_{II}} \qquad (24)$$

If $\tau_{\text{chem}} \gg \tau_{\text{or}}$, only one relaxation time can be observed, since then $\tau_I \simeq \tau_{II}$. Otherwise the dielectric relaxation spectrum is characterized by two discrete relaxation times, of which the smaller one corresponds to the chemical transformation.

A variety of sufficiently fast intramolecular transitions that are accompanied by dipole moment changes (e.g., cis-trans configuration changes, rotational isomerism, tautomerism, H-bonding) could therefore be investigated by dielectric studies with small fields if the rotational motion could be slowed down sufficiently, e.g., by the use of solvents of high viscosity. In large macromolecules, because of their inherently slow rotation, this technique may be useful to study cooperative transformation processes like helix-random coil transitions or other conformation changes. The treatment may be extended to intermolecular transformations, as long as they produce dipole moment changes in a molecule whose rotational motion is sufficiently slow.

The orientational relaxation time $\tau_{\text{or}}$ can be approximately calculated from the molecular volume and the viscosity of the solvent:

$$\tau_{\text{or}} = \frac{4\pi\eta r^3}{kT} \qquad (25)$$

This relationship was derived by Debye for spherical molecules in a homogeneous nonpolar fluid and is in good agreement with experimental values if the molecular volume of the dipolar solute is more than about three times that of the nonpolar solvent molecules. Dielectric relaxation times of the order of $10^{-12}$ to $10^{-9}$ second are thus obtained for small molecules in solvents like benzene or $CCl_4$, but for large macromolecules they can be as long as microseconds or even milliseconds in relatively viscous media.

Many experimental and theoretical investigations have dealt with the dielectric relaxation behavior of dipolar species, and all the refinements and modifications that have been introduced in the theory for explaining the complex relaxation behavior often observed will not be discussed here.[5]

The coupling of the establishment of the orientational equilibrium with chemical reactions and the possibilities for the study of chemical relaxation times is discussed by G. Schwarz.[6]

In those cases where the orientational equilibrium is established practically entirely by molecular rotation ($\tau_{chem} \gg \tau_{or}$) only the nonlinear field effect, Eq. (18), can be used to study chemical relaxation with dielectric methods.

The net change in the overall concentration of dipolar species must be produced via the chemical reaction. Therefore the chemical relaxation will be observable in this case if rotational motion is fast compared to chemical transformation.

To observe the chemical relaxation, it is not necessary that high-frequency fields of high amplitudes be used. The chemical effects will be produced also by alternating fields of small amplitude provided that they are superimposed on a static field of high magnitude. The corresponding effects will still be proportional to the square of the magnitude of the static field. This will be understood by the following considerations. Integrating Eq. (18), we obtain proportionality between the relative change in concentration at equilibrium and the square of the total field applied:

$$\int_0^E \delta \ln K_c = \frac{1}{RT} \int_0^E \Delta M \, dE \sim E^2 \qquad (26)$$

If the total field consists of a static field $E_0$ and a small alternating field $\delta E \ll E_0$ we may write

$$(E_0 + \delta E)^2 \simeq E_0{}^2 + 2E_0 \delta E \qquad (27)$$

[5] See review articles by R. H. Cole, *Ann. Rev. Phys. Chem.* **11**, 149 (1960); C. P. Smyth, *ibid.* **17**, 433 (1966).

[6] G. Schwarz, *J. Phys. Chem.* **71**, 4021 (1967); *cf.* also G. Schwarz and J. Seelig, *Biopolymers* **6**, 1263 (1968).

There will thus be a periodic concentration change that is in phase with the small alternating field $\delta E$ as long as the frequency is small compared to the reciprocal chemical relaxation time. The amount of this periodic concentration change is proportional to $E_0$.

The dipolar species that are periodically produced by the small field $\delta E$ will, however, be oriented by the forces exerted by the static field $E_0$. The periodic contribution to the net electric moment of the system due to this fact will thus be proportional to $E_0{}^2$, and its relaxation time will be that of the chemical reaction.

On the other hand, there will also be a static shift in concentration, proportional to $E_0{}^2$, and the total amount of dipolar species in the system will increase. A larger number of dipoles will therefore be subjected to the orienting forces of the small periodic field $\delta E$, and the resulting periodic contribution to the net electric moment of the system will be subjected only to the relaxation of diffusional rotation.[7]

Both periodic contributions to the electric moment will be reflected by increments in the dielectric constant of the system, measured with a small alternating voltage superposed on a large static field. Both have equal magnitude, but each one is characterized by a different relaxation time.

An exact derivation leads to[8,9]

$$\epsilon_E = \epsilon_{E=0} + \frac{\Gamma_c}{\epsilon_0 VRT}\,(\Delta M)^2 \left(\frac{1}{1+j\omega\tau_{\mathrm{or}}} + \frac{1}{1+j\omega\tau_{\mathrm{chem}}}\right) \qquad (28)$$

$\epsilon_{E=0}$ is the dielectric constant as usually measured with small fields. It shows the usual dielectric dispersion due to relaxation of the mechanism of orientation. Equation (28) is therefore also written

$$\epsilon_E = \epsilon_E{}^\infty + \frac{\Gamma_c(\Delta M)^2}{\epsilon_0 VRT}\,\frac{1}{1+j\omega\tau_{\mathrm{chem}}} \qquad (29)$$

where $\epsilon_E{}^\infty$ includes a field contribution. The index $\infty$ signifies that it is the dielectric constant measured at a high superposed field at frequencies where the second term vanishes. $\Gamma_c$ and $\Delta M$ can be calculated according to Eqs. (4) and (18), respectively.

The magnitude of the second term of Eq. (29) is only of the order of $10^{-3}$ for a 0.01 $M$ solution of our reaction example, Eq. (12), even if we suppose that both forms are equally stable in a solvent with a dielectric constant of

---

[7] We have assumed that $\tau_{\mathrm{chem}} \gg \tau_{\mathrm{or}}$. Otherwise the remarks of chemical coupling to orientational distribution as discussed above for small fields will apply [Eqs. (22) and (23)].

[8] M. Eigen and L. De Maeyer, in "Techniques of Organic Chemistry," Vol. 8, Part II. Wiley (Interscience), New York, 1963.

[9] K. Bergmann, M. Eigen, and L. De Maeyer, Ber. Bunsenges. Physik. Chem. 67, 819 (1963).

about 2, and applying a static field of $10^5$ volts $cm^{-1}$, assuming that the polar form of the molecule has a dipole moment of about 4 Debye units.

In view of the smallness of $\Delta M^2$ for the dipolar species usually involved, and the limits on field strength applicable due to spark breakdown, the chemical contributions to the dielectric permittivity are extremely small. On the other hand, since $\Delta M$ involves the difference in the squares of the dipole moments involved, the chemical contributions to the dielectric constant increases practically with the fourth power of the dipole moment of a dipolar species whose moment is large compared to that of the other reaction partners.

It remains to state that all formulas used in the above derivations were given for conditions at constant pressure and temperature. For high-frequency measurements, where volume changes and heat exchange are practically excluded by mechanical inertia and slowness of thermal diffusion, the corresponding expressions at constant volume and entropy should have been used. The error produced by the use of the equations derived for constant $T$ and $P$ are minimal, however, especially for dilute solutions, since the solvent serves as a heat bath and the volume changes for reactions in solution are small. It is interesting to note, however, that theoretically at constant $S$ and $V$ high electric fields will induce chemical transformations that do not even involve dipole molecules, provided that the reaction proceeds with finite $\Delta H$ or $\Delta V$ and the solvent has a finite temperature coefficient of dielectric permittivity or a finite coefficient of electrostriction.

## D. Ionic Equilibria in Electric Fields

The dependence of the ionic dissociation equilibrium of a weak electrolyte upon an applied external electric field is known as the second Wien effect, or "dissociation-field effect." In strong fields, dissociation is enhanced. Onsager has given a theoretical derivation of this effect based upon a detailed kinetic calculation of ionic dissociation and recombination processes in an external field.[10] This treatment is limited to very small concentrations of a (neutral or charged) weak electrolyte dissociating into ions of opposite (but not necessarily equal) charge

$$K_E = K_{E=0}\left(1 + 2\beta q + \frac{(2\beta q)^2}{3} + \frac{(2\beta q)^3}{18} + \cdots\right) \qquad (30)$$

In this equation, $K_E$ is the dissociation constant at a field $E$ and $K_{E=0}$ the dissociation constant at field zero. Since this derivation applies only near infinite dilution $K_{E=0}$ is the thermodynamic equilibrium constant. The quantity $q = -z_1 z_2 e_0^2/8\pi\epsilon\epsilon_0 kT$ is known as the "Bjerrum radius"

---

[10] L. Onsager, J. Chem. Phsy. **2**, 599 (1934).

and represents the "effective range" of the ions; the energy of coulombic interaction between two ions bearing charges $z_1e_0$ and $z_2e_0$ resp. in a medium with dielectric constant $\epsilon$ becomes equal to $kT$ when they approach each other at a distance of twice the Bjerrum radius. (The ionic valencies $z_1$ and $z_2$ as written here include the sign of the charge.) The quantity $2\beta$ represents a reciprocal distance. It is given by

$$2\beta = |Ee_0(z_1u_1 - z_2u_2)/(u_1 + u_2)kT|$$

where $u_1$ and $u_2$ are the mobilities[11] of the ions. In the case of symmetric electrolytes ($z_1 = -z_2 = z$) the quantity $2\beta$ is independent of the mobilities. Two opposite charges of equal magnitude $ze_0$ must be separated by a distance $1/2\beta$ to form a dipole that acquires the energy $-kT$ if aligned in the direction of a field $E$. The ratio of two characteristic distances $q$ and $1/2\beta$ describes thus, in first approximation, the possibility of two ions to escape from an ion pair under the influence of an external field.

It is interesting to note that Onsager's derivation shows that only the dissociation rate constant is affected by the external field, whereas the recombination rate constant is not changed. The limited applicability of Eq. (30) is already borne out by the dependence on the absolute value of the electric field strength, which leads to a discontinuity at $E = 0$. The electrostatic screening effect due to interionic interaction excludes the application of this formula to low fields in systems of finite ionic concentrations.

At finite ionic concentrations $c_1$, $c_2$ and moderate fields, so that $q \ll 1/2\beta \ll 1/\kappa$, where $1/\kappa = [(z_1c_1 + z_2c_2)(N_A/1000)e_0^2/\epsilon\epsilon_0kT]^{-1/2}$ is the radius of the ionic atmosphere, Eq. (30) gives the asymptotic slope of the equilibrium displacement as a function of the field with sufficient accuracy.[12] For this region one obtains

$$\frac{d \ln K}{d|E|} = \frac{z_1u_1 - z_2u_2}{u_1 + u_2} \frac{|z_1z_2|e_0^3}{8\pi\epsilon\epsilon_0k^2T^2} \tag{31}$$

The dependence of the dissociation field effect on an even function of the field allows the use of alternating fields to perturb the equilibrium. It must, however, be pointed out that the dissociation field effect will effectively vanish in stationary alternating electric fields with frequencies approaching the reciprocal relaxation time of the ionic atmosphere, even if the absolute value of an alternating field has a constant average value. Onsager's treatment of the dissociation field effect is based upon the steady-state equations of relative ionic motion in a constant external field.

[11] $u_1$ is the mechanical mobility (reciprocal of the friction constant). The ionic (electrical) mobility in $cm^2$ $volt^{-1}$ $sec^{-1}$ is then given by $z_ie_0u_i$. The mechanical mobility is related to the diffusion coefficient by $u_i = D_i/kT$.

[12] A more exact calculation of the Wien effect at finite concentrations and low fields has meanwhile been carried out. For values of $\kappa/2\beta \leqslant 0.2$ the error in Eq. (31) becomes less than 10%. (L. Onsager, personal communication.)

For an application of the dissociation field effect to the measurement of chemical reaction rates, the following conditions must therefore be fulfilled.

1. Reaction rates can be measured only if their relaxation time is large compared to the relaxation time of the ionic atmosphere.

2. Sufficiently high fields must be applied; in aqueous media, field intensities between $10^4$ and $10^5$ volts/cm are usually required.

3. Since the conduction currents caused by such fields will be very large, short pulses must be used to avoid excessive heating of the sample.

4. High-frequency alternating field pulses may be used as long as the period of oscillation is long compared to the relaxation time $\theta$ of the ionic atmosphere. According to the theory of Debye and Falkenhagen

$$\theta = \frac{1}{\kappa^2 kT(u_1 + u_2)} \tag{32}$$

The dissociation field effect is measurable as an increase in electrical conductivity of the sample or by direct determination of changes in the concentration of the reaction components (e.g., by optical means). The dissociation field effect can be used also to induce perturbations in coupled reaction equilibria.

## E. Saturation Currents

Chemical reaction rates can sometimes limit ionic conduction currents. This is the case if ions produced by a chemical reaction are swept away from the reaction medium by a sufficiently large electric field, so that ionic recombination cannot take place. If the chemical reaction is the only supply of ionic species, the conduction current corresponding to the ionic transport is a direct measure of the rate of ion production.

The requirements for a successful application of this principle to the measurement of ionic dissociation rates are in general difficult to fulfill: the ions produced in the process must be discharged at an electrode without any time lag in order to avoid the buildup of space charges; on the other hand, the electrodes must be completely blocking, which means that they may not themselves inject charge carriers into the medium. Furthermore, relatively high field strengths are required to remove the ions as soon as they are produced, and the depletion layer in which the reaction takes place must be sufficiently thin so that the probability for recombination becomes small. That is the time required to move an ion from its origin of production to an electrode where it is discharged must be small compared to the relaxation time of the chemical process:

$$\frac{l}{uze_0E} \ll \frac{1}{k_R \bar{c}_j} \tag{33}$$

where $u$ represents the mechanical mobility ($ze_0u$ = electrical mobility in cm$^2$ volt$^{-1}$ sec$^{-1}$), $z$ the valency and $\bar{c}_j$ the equilibrium concentration of the ions produced by the dissociation process, $l$ the distance between two electrodes at which the ions are discharged, $E$ the applied field strength, and $k_R$ the recombination rate constant.[13]

Under these conditions the saturation current density between the electrodes is given by

$$i_{sat} = N_Aze_0lk_Dc_u{}^\alpha \tag{34}$$

$k_D$ is the dissociation rate constant, $c_u$ the concentration of the undissociated electrolyte, and $\alpha$ describes the molecularity of the dissociation process. In most cases $\alpha$ is equal to 1 (unimolecular dissociation process), but in some cases in media of low dielectric constant a bimolecular dissociation process may be observed.[14]

Saturation current measurements between two electrodes are restricted to media of relatively low dissociating power. In aqueous systems, the self-dissociation of the solvent is already too large to allow the direct observation of saturation currents since the thickness of the depletion layer according to Eq. (33) must be smaller than $10^{-2}$ cm with a field of $10^5$ volt cm$^{-1}$ in order to remove all ions produced by self-dissociation. The most serious obstacle, however, is the fact that metallic electrodes at which ions can be discharged will also produce charge carriers by dissociation processes at the surface. It is sometimes possible to make a distinction between bulk and surface dissociation processes by measuring the dependence of the observed currents upon the thickness of the layer between the electrodes.

If high fields are used, the dissociation field effect will cause the saturation currents to depend upon the field strength.

Saturation current measurements for studying chemical dissociation rates have been made successfully only in a few cases,[15] and a careful study of the behavior of the measured currents upon the experimental conditions is necessary to obtain reliable information.

## F. Electrochemical Methods

Different techniques have been developed in which fast chemical reactions are observed as rate-limiting processes preceding an electrochemical transformation (reduction or oxidation) at an electrode surface.

---

[13] M. Eigen and L. De Maeyer, Z. Elektrochem. **60**, 1037 (1956).

[14] H. F. Eicke, L. C. De Maeyer, Ber. Bunsenges. Physik. Chem. **70**, 92 (1966); H. F. Eicke, ibid., **71**, 384 (1967).

[15] The measurement of saturation currents in pure ice crystals enabled the determination of the self-dissociation equilibrium constant of H$_2$O in ice and led to an evaluation of proton mobility in hydrogen bond lattices, cf. M. Eigen and L. De Maeyer, Proc. Roy. Soc. (London) **A247**, 505 (1958).

A polarizable electrode (e.g., mercury drop, platinum disk, etc.) is held at a potential where the electrode reaction proceeds via the oxidation or reduction of a reducible or oxidizable substance, called the depolarizer. Under these conditions, a steady state with a constant electrical current is reached. The current is equal to the net transport of depolarizer molecules toward the electrode.

If the flow of depolarizer is limited entirely by diffusional transportation from the bulk of the medium through a concentration gradient, the limiting current is called a diffusion current. This is the basis of the usual application of the method in analytical polarography.

In some cases, however, the depolarizer can be in chemical equilibrium with a nondepolarizing substance, and may thus be produced also in the immediate vicinity of the electrode by chemical transformation. When the transformation is fast enough, the amount of depolarizer transported to the electrode by diffusion may become small compared to its rate of production by chemical transformation. The limiting current is called a kinetic current, and may be used to determine the rate of chemical production of the depolarizer.

In the two preceding cases the products of the electrode reaction have not played any role; they are merely transported into the bulk of the medium via diffusional transport, with or without being involved in further chemical equilibria that do not affect the electrode reaction.

It is possible, however, that the product of the electrode reaction is involved in an homogeneous reaction (with a reducible or oxidizable reaction partner) that leads to regeneration of the initial depolarizing substance. In such a case, a depolarizer molecule brought to the electrode surface can undergo a cycle of many conversions acting as a catalyst for the ultimate reduction or oxidation of its reaction partners. The limiting current measured under these conditions is called a catalytic current, and its magnitude depends upon the rate of the homogeneous reaction, since this rate determines the number of cycles that the depolarizer molecule will go through before again being lost by diffusion into the bulk of the medium.

Electrode reactions may thus be coupled in many ways to other chemical reactions taking place in the homogeneous phase, and polarographic and electrochemical techniques can be of great value for the investigation of fast chemical processes that can be coupled into such a reaction scheme.

Many different techniques, based upon steady-state measurements as well as upon observation of transient unstationary processes, and involving potentiostatic and galvanostatic conditions, with electrode configurations of many kinds, have been described in detail in the literature. The great variety in techniques is due to the difficulty of finding simple mathematical expressions for the coupling of diffusion and homogeneous and heterogeneous reaction phenomena at the interphase, and to the difficulty of establishing experimental conditions that sufficiently approximate the idealized description that is required for adequate mathematical formulations. The reader is referred to the original literature for further information. A survey of electrochemical techniques for the study of fast reactions has been given by Strehlow,[16] and is discussed in detail by P. Zuman in this volume [5].

## II. Experimental Techniques

### A. Measurement of Conductive and Dielectric Properties

The alternating current density $i = i_p \exp[j(\omega t + \varphi)]$ caused by an alternating electric field $E = E_p \exp(j\omega t)$ is proportional to the complex admittivity $\chi$ and the field $E$

$$i = \chi E = (\sigma + j\omega\epsilon_0\epsilon)E \tag{35}$$

where $\sigma$ is the ohmic conductivity and $\epsilon_0\epsilon$ the dielectric permittivity, comprising the different contributions to dielectric polarization. The terms $i_p$ and $E_p$ are the amplitudes (peak values) of current density and electric field intensity, respectively, and $\varphi$ is the phase angle between current and voltage.

Relaxation of one of the contributions ($\delta\epsilon$) to the dielectric constant results in a complex permittivity with a characteristic relaxation time:

$$\epsilon_0\epsilon = \epsilon_0(\epsilon' + j\epsilon'') = \epsilon_0\left(\bar{\epsilon} - \delta\epsilon\frac{j\omega\tau}{1 + j\omega\tau}\right) \tag{36}$$

where $\bar{\epsilon}$ is the low frequency dielectric constant. [If there is more than one contribution, a similar equation with a sum of relaxing contributions $\Sigma_i\delta\epsilon_i(j\omega\tau_i/1 + j\omega\tau_i)$ may be written; $\delta\epsilon_i$ and $\tau_i$ may be related to normal modes of relaxation if the relaxation times of the single contributions are not sufficiently separated.]

Equation (35) can then be expressed as

$$i = \left\{\sigma + \epsilon_0\frac{\delta\epsilon}{\tau}\frac{\omega^2\tau^2}{1 + \omega^2\tau^2} + j\omega\left(\bar{\epsilon} - \delta\epsilon\frac{\omega^2\tau^2}{1 + \omega^2\tau^2}\right)\right\}E \tag{37}$$

The imaginary part of the dielectric permittivity $\epsilon''$ contributes to the ohmic losses in the system. These losses therefore increase with the square of the frequency as long as $\omega \ll 1/\tau$.

The phase angle $\varphi$, and its complement, the loss angle[17] $\delta = (\pi/2) - \varphi$, are expressed by the ratio of the real to the imaginary part of the admittivity:

[16] H. Strehlow, in "Technique of Organic Chemistry" (A. Weissberger, ed.), Vol. 8. Part II. p. 799. Wiley (Interscience), New York, 1963.

[17] The quantity sin $\delta$ is called the power factor, and is the ratio of the energy dissipated per unit volume and per cycle to the energy density supplied per cycle to the dielectric. The quantity tan $\delta$ is sometimes called the dissipation factor. For dielectrics with vanishing ohmic conductivity, the loss angle is small and tan $\delta \cong \sin \delta$.

$$\tan \delta = \cot \varphi = \frac{\sigma + \epsilon_0 \dfrac{\delta\epsilon}{\tau} \dfrac{\omega^2\tau^2}{1 + \omega^2\tau^2}}{\omega\epsilon_0 \left( \bar{\epsilon} - \dfrac{\delta\epsilon\omega^2\tau^2}{1 + \omega^2\tau^2} \right)} \tag{38}$$

Experimental methods for the determination of the dielectric relaxation time $\tau$ depend upon accurate measurements of $\epsilon'$, $\epsilon''$ and their frequency dependence. These quantities are not directly measurable, but must be calculated from observed electrical parameters.

In the frequency region below $10^8$ cps, network and filter theory can be used, and the electrical properties of a sample contained between two conducting electrodes can be deduced from the ohmic and reactive impedance or admittance of such a two-terminal system as a network component.

Above $10^8$ cps, electromagnetic wave propagation theory is required to calculate the intrinsic electrical properties of the sample medium, usually contained between coaxial conductors or placed in a waveguide structure.

In many cases electrical measurements are carried out not with the direct purpose of studying relaxation behavior of dielectric properties, but as an analytical tool for the observation of the time course of chemical reactions that are induced in the system by other methods.

## B. Design Considerations and Experimental Procedures

At low frequencies up to about $10^5$ to $10^6$ cps the complex admittance of a two-terminal element can be obtained most easily by direct comparison with standard resistances and reactances in a bridge circuit. The common form of such a bridge circuit is the Wheatstone bridge, which consists of four impedances $Z_a$, $Z_b$, $Z_c$, $Z_u$, of which $Z_u$ is the unknown, connected in a two-port configuration. One port is driven from an alternating voltage source with internal impedance $Z_g$, and the other port is connected to an indicator, which presents to the bridge a load impedance $Z_l$ (see Fig. 1).

Since any of the four bridge impedances may be composed of an arbitrary combination of resistances, capacitances, and inductances, the number of possible bridge configurations is unlimited. Some particular configurations are especially adapted for the measurement of capacitive or inductive impedances and are referred to as Schering bridges, Maxwell bridges, etc. For some value of the four complex bridge impedances, no energy is transferred to the load. This will be the case if

$$\frac{Z_a}{Z_c} = \frac{Z_b}{Z_u} \tag{39}$$

Under these conditions the bridge is balanced, and any of the four bridge impedances is then uniquely determined in terms of the three others. For absolute measurements, the comparison impedances $Z_a$, $Z_b$, $Z_c$

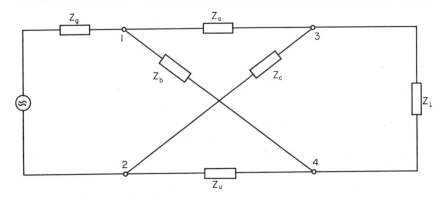

FIG. 1. Equivalent lattice network of a Wheatstone bridge: $Z_g$ represents the generator impedance, $Z_l$ is the external load impedance of the indicating instrument or amplifier. $Z_a$ and $Z_c$ form the comparison arm, $Z_c$ and $Z_u$ the measuring arm. $Z_u$ is the unknown impedance. The power delivered to the load is zero when $Z_a Z_b = Z_c Z_u$.

will be composed of accurately calibrated standard elements that can be changed by switching (decade boxes) or that are continuously variable. Only resistive and capacitive elements can be manufactured with sufficient accuracy and stability, so that the bridge configuration should mainly contain resistances and capacitances as adjustable elements. (An exception to this rule are the ratio-transformer bridges, in which accurately tapped transformers are used as adjustable elements, see below.) Wire-wound standard resistances still have a residual inductance, which is usually below a few $\mu H$ for high quality resistors. Residual inductance introduces errors in low value resistors ($< 1\Omega$), whereas for high value resistors ($> 10k\Omega$) the residual capacitance becomes important. It is therefore preferable that the compensating resistive elements of a bridge used at varying frequencies up to about $10^6$ cps remain between these values.

In a common conductivity bridge configuration the impedances $Z_a$ and $Z_c$ are purely resistive and, in the simplest case, equal. It is then easily seen from Eq. (39) that the resistive and reactive parts of the unknown impedance have to be balanced by exactly equal resistive and reactive parts in $Z_b$. This can be achieved in two ways: $Z_b$ can be composed of a resistance $R_b$ with a capacitance $C_b$ in parallel connection, or of a resistance $R_b^*$ and a capacitance $C_b^*$ in series connection (Fig. 2). If the unknown impedance $Z_u$ is that of a conducting sample between two electrodes, its equivalent circuit is a conductance $G_u = 1/R_u$ parallel to a capacitance $C_u$.

In the first case balance is obtained when $R_b = R_u$ and $C_b = C_u$. In this case the balance condition will be independent of the frequency used to drive the bridge (unless $R_u$ or $C_u$ are themselves dependent upon frequency). In the second case, however, balance is obtained only if

$$R_b^* = R_u \frac{1}{1 + \omega^2 R_u^2 C_u^2}, \quad C_b^* = \frac{1}{C_u} \frac{\omega^2 R_u^2 C_u^2}{1 + \omega^2 R_u^2 C_u^2} \tag{40}$$

It is evident that the balance condition is now frequency dependent, since for different frequencies different values of $R_b^*$ and $C_b^*$ are required.

The most useful bridge configuration for a given purpose thus depends upon the equivalent circuit representing the unknown impedance. In practically all cases of conductivity and dielectric constant measurements the equivalent circuit of the sample cell can be approximated by an ohmic resistance in parallel with a capacitance. For solutions of high conductivity (electrolytes) the electrode double layer capacitance becomes important. Since it is impractical, and often impossible, to use a comparison impedance configuration that simulates the unknown impedance in all respects, a residual frequency dependence will always exist. It is therefore very important that the oscillator driving the bridge deliver a pure sinusoidal alternating voltage, free of distortion and harmonics, since otherwise it becomes very difficult to determine the exact balancing condition for the fundamental frequency. To reduce the influence of harmonics, tuned filters, or amplifiers in the detector may be helpful. The balancing adjustments can further be simplified by phase-sensitive demodulation of the detector output. If separate phase demodulated outputs of the in-phase and quadrature components of the detector output are provided, individual and convergent adjustment of the resistive and reactive parts of the comparison impedance becomes possible. For the same purpose a Lissajous display of the bridge detector output on an oscilloscope can be used.

In many applications the purpose of the measurement is not the determination of the absolute values of conductance and susceptance of a sample, but the accurate measurement of small increments of these quantities. In such cases it is necessary to examine the magnitude of the bridge output (i.e., the voltage appearing across $Z_1$ at the detector terminals) near

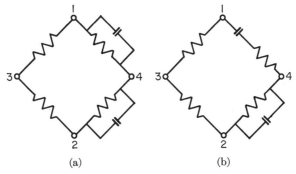

(a)                              (b)

FIG. 2. Bridge configurations with frequency-independent (a) and frequency-dependent (b) balance condition.

balance, as a function of small variations of the impedance in one of the bridge arms.

The open-circuit bridge output is given by

$$U_1 = U_g \left( \frac{Z_u}{Z_b + Z_u} - \frac{Z_c}{Z_a + Z_c} \right) \tag{41}$$

and is zero if the bridge is balanced. If the bridge is driving a load impedance $Z_1$, the right-hand term of this equation must be multiplied by the factor

$$F = \frac{Z_1}{Z_1 + \dfrac{(Z_a + Z_b)(Z_c + Z_u)}{Z_a + Z_b + Z_c + Z_u}} \tag{42}$$

$U_g$ is the voltage at the input terminals and is equal to the source voltage if $Z_g$ is small enough.

The sensitivity of the bridge can then be defined by the relative change in output voltage for a given change in the unknown impedance. This quantity is easily obtained by differentiation of Eq. (41).

$$\frac{dU_1}{U_g} = \frac{dZ_u}{Z_u} \frac{Z_b Z_u}{(Z_b + Z_u)^2} \tag{43}$$

Usually the incremental change in impedance $dZ_u$ does not equally affect the conductive and capacitive part. In many cases (e.g., electrolyte solutions) only the conductive part changes as a result of concentration changes, whereas the capacitive part of the admittance is largely due to the dielectric properties of the solvent and is much less affected by a change in ionic concentration. In several cases the bridge output near balance is recorded as a function of time to measure the time course of a chemical reaction that affects the conductivity of the sample, and that is initiated by suitable perturbations of the chemical equilibrium in the system. The bridge output is displayed directly on an oscilloscope, and the relaxation time of the chemical reaction is obtained from the envelope of the observed waveform.

It must therefore be investigated whether the bridge arrangement has a sufficient sensitivity and bandwidth for faithful recording of the change in conductance. The amplitude of the voltage $\Delta \hat{U}$ appearing at the output terminals of a previously balanced bridge after a change in conductance $\Delta G_u$ is given by

$$\Delta \hat{U} = \Delta G \hat{U}_g \frac{\sqrt{G_b^2 + \omega^2 C_b^2}}{(G_b + G_u)^2 + \omega^2 (C_b + C_u)^2} \tag{44}$$

A maximal imbalance voltage is observed if $G_b = G_u$ and $C_b = C_u$. For the recording of fast changes in conductance a carrier frequency $f = \omega/2\pi$ must be used that is much higher than the highest frequency components of the recorded transient. The sensitivity may then be quite low.

Since for maximum sensitivity equal conductances and capacitances are necessary in the unknown and comparison bridge arm, it is useful to place in the comparison arm a reference electrolytic cell with the same dimensions and filled with a solution of similar conductivity and capacity as that in the sample cell.

Another bridge configuration that is very useful for absolute conductance measurements is the ratio transformer bridge (Fig. 3). This type of bridge has recently been brought to high perfection through the use of toroidal transformers with laminated cores of very high permeability. Together with efficient winding schemes, transformers with extremely high coupling between the different windings and very low stray inductance can be made, so that the voltage appearing at the different windings is exactly in proportion of the turns ratio, independent of the loading of the different windings.

To balance the bridge the voltage on a standard capacitor and the voltage on a standard conductance in the standard arm are adjusted so that the sum of the currents through them is equal to the current flowing

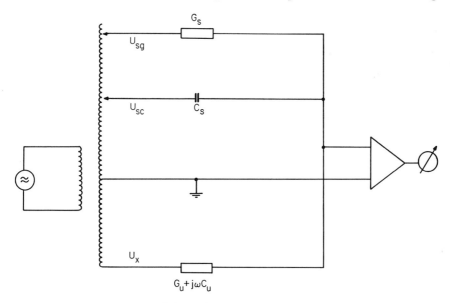

FIG. 3. Ratio-transformer bridge. $G_u + j\omega C_u = 1/Z_u$ is the unknown admittance. $G_s$ is a fixed standard ohmic conductance and $C_s$ a fixed standard capacitor. Ohmic and capacitive parts of the unknown admittance can be balanced independently by changing the voltages $U_{sg}$ and $U_{sc}$.

through the admittance in the unknown arm. The difference between both currents is detected by a detecting amplifier. At balance, therefore,

$$U_x(G_u + j\omega C_u) = U_{sg}G_s + j\omega U_{sc}C_s \qquad (45)$$

The advantage of this bridge is that switching in the standard arm is avoided and replaced by switching the transformer windings. Thereby the adverse effects of stray inductances and capacitances in the standard arm can be avoided.

For maximal sensitivity to incremental changes, it is again found that the standard conductance and susceptance should be equal to that of the unknown, and Eq. (44) applies, with $U_g = 2U_x$.

For high-frequency applications, and in general for measurements of high accuracy, shielding, and grounding of a bridge is very important. Usually one of the bridge terminals is directly grounded, that is, this point is taken as the reference point and is used as the return point for all external shielding. It is very common to ground one of the input terminals of the bridge. This is practical if the oscillator used to drive the bridge is single-ended and one of its output connections is grounded anyway. This procedure sometimes has its disadvantages, especially at higher frequencies. Since both output terminals carry alternating voltages, a difference amplifier or a transformer must be used at the bridge output. At high frequencies the rejection ratio of a differential amplifier is usually too small for accurate measurements. A transformer must have sufficient bandwidth if the modulated output (by time-varying conductance changes) is to be observed. Shielding of the bridge output terminals will introduce capacitances that are parallel to the measuring bridge arms. It is therefore preferable to ground one of the output terminals. The bridge must then be driven from a balanced source. For this purpose a simple transformer can be used, since only a single-frequency component is present at the bridge input.

The shielding capacitance of the input terminals is now parallel to the ratio arms of the bridge, but this is unimportant if these are of sufficiently low impedance.

In many cases (measurement of very high conductances or susceptances or measurements at high frequencies) the impedance of the leads connecting the measured conductance to the bridge becomes important. In such cases more elaborate configurations may be necessary. Four-terminal standards and four-terminal sample cells may be used. Two of these terminals are called the "current" terminals and the two other the "potential" terminals. The measured conductance is that between the junctions of a "current" terminal with its "potential" terminal. The location of these junctions should be permanent and independent of the position of the connecting

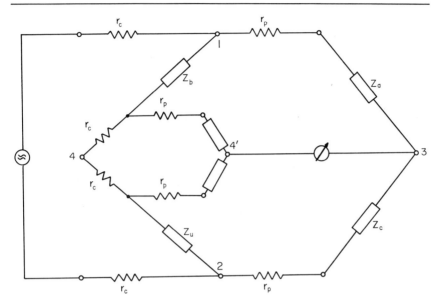

FIG. 4. High-conductance Kelvin bridge with four-terminal unknown and standard impedances $Z_u$ and $Z_b$. $r_c$ and $r_p$ are the resistances of the current and potential leads and connectors. Two-terminal standard impedances $Z_a$ and $Z_c$ are present in the comparison arm, carrying a much smaller current if these impedances are large enough. $Z_e$ and $Z_f$ are equal impedances larger than $r_p$. They prevent heavy current flow through the potential leads $r_p$.

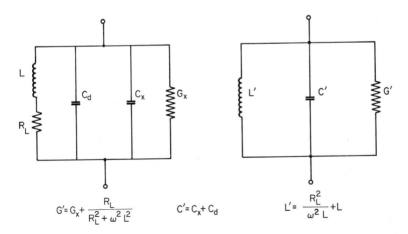

$$G' = G_x + \frac{R_L}{R_L^2 + \omega^2 L^2} \qquad C' = C_x + C_d \qquad L' = \frac{R_L^2}{\omega^2 L} + L$$

FIG. 5. Parallel resonant circuit showing separate lossy circuit elements and equivalent circuit in which all losses are represented by a single equivalent parallel conductance $G'$.

leads to avoid calibration errors. The mutual inductance between "current" and "potential" leads should be kept small. Figure 4 shows a four-terminal capacitance bridge for the measurement of high capacitances (Kelvin bridge).

Above frequencies of the order of $10^6$ to $10^7$ cps, the measurement of conductances at different frequencies becomes very tedious. Between $10^6$ and $10^8$ cps conductance and susceptance and incremental values of these quantities can be measured more easily in resonant circuits, in which a two-terminal sample cell is combined with an inductance to give parallel or series resonance. Conductance and susceptance of the sample are now derived from accurate measurements of the resonant frequency and of the width of the resonance curve.

A resonant circuit is described in terms of its resonance frequency $\omega_0$ and its $Q_0$ (quality factor). The equivalent admittance of a parallel resonant circuit (Fig. 5) is given by

$$Y_{\mathrm{eq}} = G_x + \frac{R_L}{R_L{}^2 + \omega^2 L^2} + j\omega \left( C_x + C_d - \frac{L}{R_L{}^2 + \omega^2 L^2} \right) \qquad (46)$$

$L$ is the inductance, $R_L$ represents the losses in the inductance, and $C_d$ the distributed capacitance of the inductance, $G_x$ and $C_x$ are the conductance and capacitance of the admittance to be measured. At the resonance frequency $\omega_0$, the imaginary part of $Y_{\mathrm{eq}}$ is zero, or

$$\omega_0 = \left( \frac{1}{L(C_x + C_d)} - \frac{R_L{}^2}{L^2} \right)^{1/2} \simeq \frac{1}{\sqrt{L(C_x + C_d)}} \qquad (47)$$

The error made in neglecting $R_L{}^2/L^2$ in this equation is of the order $R_L{}^2/(\omega_0{}^2 L^2)$ and is usually sufficiently small for high-quality inductances in circuits of high $Q_0$. The circuit $Q_0$ is defined by the ratio of the capacitive susceptance to the equivalent parallel conductance at resonance

$$Q_0 = \frac{\omega_0(C_d + C_x)}{G_x + \dfrac{R_L}{R_L{}^2 + \omega_0{}^2 L^2}} \simeq \frac{\omega_0(C_d + C_x)}{G_x + \dfrac{R_L}{\omega_0{}^2 L^2}} \qquad (48)$$

Measurements are carried out on a resonant circuit by coupling it very loosely to an RF-oscillator (usually with a very small capacitance, sometimes by a small mutual inductance) and observing the magnitude of the RF-voltage appearing across the resonant circuit with a high-impedance RF-voltmeter or with an oscilloscope (it must be noted that the measuring voltmeter introduces an additional admittance parallel to the resonant circuit). At resonance, the voltage across the resonant circuit goes through a maximum. If a fixed frequency oscillator is used, the resonance condition can be obtained by a continuously variable tuning capacitor which is part of the circuit and included in $C_x$. Very accurate measurements can be done with variable oscillators and digital frequency counters.

The $Q_0$ of the circuit is measured by inspection of the resonance curve (magnitude of the observed voltage across the circuit in the vicinity of the resonance frequency). The resonance curve can be inspected by variation of the adjustable capacitor at fixed frequency (reactance variation method) or by frequency variation at fixed configuration of the network. Again, if a digital counter is used for frequency measurement, the latter method can be very accurate even for very high $Q_0$. A universal resonance curve is shown in Fig. 6.

The frequency is increased beyond the resonance frequency $f_0 = \omega_0/2\pi$ (where the magnitude of the resonance voltage is $U_0$) to a frequency $f_1$, and the magnitude of the circuit voltage $U_1$ is noted. The frequency is then lowered below $f_0$ to a frequency $f_2$, where the voltage has again dropped to a value $U_2 = U_1$. The circuit $Q_0$ can then be calculated with good accuracy from

$$Q_0 = \frac{1}{2}\left(\frac{f_2 + f_1}{f_2 - f_1}\right)\left(\frac{U_1{}^2}{U_0{}^2 - U_1{}^2}\right)^{1/2} \qquad (49)$$

This equation does not involve $f_0$, which is more difficult to measure accurately since the top of the resonance curve is flat. Equation (49) is derived with the assumption that the resonance curve is symmetric, i.e., $f_2 - f_0 = f_0 - f_1$, and that the equivalent parallel conductance does not

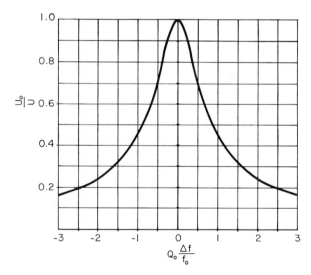

Fig. 6. Symmetrical resonance curve of a parallel resonant circuit in which the equivalent circuit elements are independent of frequency.

change significantly over this frequency range. The measurement becomes simple if $U_1 = (\sqrt{2}/2)U_0 = 0.707 \ U_0$, since then the quantity $[U_1^2/(U_0^2 - U_1^2)]^{1/2}$ becomes unity.

With one such measurement $G_x$ and $C_x$ cannot yet be determined separately. The measurement should be repeated with an additional calibrated standard capacitance connected to the circuit. For determination of the dielectric constant and conductivity of a sample, the measurements can be done with the sample cell emptied, filled with the sample, and filled with suitable reference liquids.

The resonant circuit method is very useful also for the determination of incremental changes in capacitance by measuring the shift in resonance frequency. Relative frequency changes of a few parts in $10^6$ can be measured accurately with counters or beat frequency techniques; the sensitivity for such measurements is quite good if the resonant circuit is made to be the frequency determining element of a small oscillator.

Incremental changes in conductance can be measured by observing the drop in resonant voltage

$$\frac{dU_0}{U_0} = -Q_0 \frac{dG_x}{\omega_0(C_d + C_x)} \tag{50}$$

At frequencies above a few $10^8$ cps, resonant structures with distributed capacity and inductance are used (reentrant cavities, shorted coaxial lines, etc.). The microwave measurement techniques that must be used in this frequency range will not be discussed here. Specific applications with respect to chemical kinetics have not yet been reported. Dielectric relaxation times of dipole orientation for simple molecules are usually found in this frequency region.

## III. Applications and Design Examples

### A. Introduction

Specific applications to enzyme chemistry of the high-field effects treated in this chapter are very scarce. High fields have been used often in temperature-jump relaxation studies of enzyme reactions and are treated in article [1]. The dissociation field effect has been used mainly for the study of rapid protolytic equilibria in aqueous media. The intimate mechanism of such reactions is investigated most easily using simple model substances rather than polymeric molecules with many substituents. The same is true for the study of dipole interaction and hydrogen bond formation. The elementary processes in hydrogen bond formation are being investigated using the high field-dipole reaction effect discussed before.

The effect of high fields on macromolecular conformation have not yet been studied. The importance of such studies cannot be denied, but a large amount of theoretical work will be required in this direction. Low-field studies have hitherto been undertaken mainly with respect to the determination of dielectric relaxation times. It was shown in the beginning of this chapter that the mechanism of intramolecular chemical transformations may sometimes be studied with these techniques. Here a large field of application for the investigation of macromolecular conformation changes is open, but specific applications are still in progress. For such studies, commercial equipment is readily available, and an example will not be given here. There are, finally, a number of trivial applications of electric measurements for the purpose of analysis and recording of a progressing chemical reaction. Specific examples are treated in other chapters of this volume; some of the general principles that are important for such applications have been discussed in Section II.

## B. Resonance Method for Dielectric Measurements at High Fields

The first of our examples will describe the apparatus developed for the measurement of dielectric absorption caused by the high-field dipole reaction effect of associating dipolar molecules in solution in nonpolar solvents. The specifications for such an apparatus are as follows:

A very small and frequency-dependent increase in the dielectric loss of a solution of the order of 0.001 to 0.1% is expected and to be measured in the frequency range of 1 Mc to 100 Mc at superimposed direct current (dc) fields of 100,000 volts $cm^{-1}$. A schematic circuit diagram of the experimental set up is given in Fig. 7.

A variable frequency generator covering the region of 1 Mc to 100 Mc with an output voltage of 2–3 volts is used. An adjustable fraction of this voltage is coupled via a small coupling capacitor $C_c$ to a resonant circuit that comprises an exchangeable inductance $L$, the sample cell capacitor $C_x$, and a blocking capacitor $C_b$. High voltage is applied to the sample cell via a resistor $R_b$ from an HV generator. In this generator the high voltage is obtained from cascade rectification of a high-amplitude RF voltage. The maximal high voltage output is 20 kV. The high voltage can be switched on and off by applying or removing the power to the RF oscillator in the HV generator. The HV output lead remains always connected via $R_b$ to the resonant circuit. In this way the load impedance presented by the HV generator to the resonant circuit remains essentially constant without and with a dc voltage applied to the sample cell. Shielding compartments that contain the different circuit components are shown in the diagram.

The measurement procedure is then the following: The sample cell is filled with the solution and temperature equilibrium is established with a thermostat. HV is off. A suitable inductance for a given frequency region

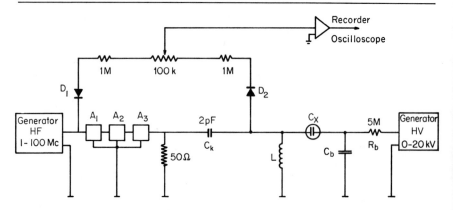

FIG. 7. Experimental arrangement for the measurement of dielectric properties in the frequency region 1–100 Mc at high superposed dc fields. The HF-generator output is attenuated by attenuators $A_1$ (10 db per step), $A_2$ (1 db per step), and $A_3$ (0.1 db per step) and coupled via a 2 pF coupling capacitor $C_k$ to the resonant circuit consisting of the inductance $L$, sample cell capacitor $C_x$ and dc-blocking capacitor $C_b$. High voltage is applied to the sample cell capacitor via a 5 M resistor. HF-voltages at the generator output and across the resonant circuit are rectified by diodes $D_1$ and $D_2$. If both voltages have equal amplitude the input to the recording amplifier is zero. The 100 k variable resistor allows fine adjustment.

is plugged into the resonance circuit. (The frequency range 1 Mc to 100 Mc is covered with 32 different inductances.) The frequency of the signal generator is varied until a maximal deflection is observed at the dc recorder. The frequency of the signal generator is then the resonance frequency of the tuned circuit and is measured with a counter. The output voltage of the signal generator is monitored and held at a constant value for all measurements in order to avoid the effect of mismatch in the peak rectifying HF-diodes $D_1$ and $D_2$. The signal coupled to the resonant circuit is then attenuated with attenuators $A_1$, $A_2$, and $A_3$ until equal HF voltages are applied to both diodes $D_1$ and $D_2$. The tuning adjustment is repeated, and exact balancing of the HF voltage is inspected, using the output oscilloscope with a high-sensitivity dc amplifier.

The high voltage is then applied to the sample cell. The output voltage will change by a small amount. HV is left on the sample cell for a short duration only (a few seconds) in order to avoid heating due to conduction currents. During about 4 seconds the following adjustment can easily be made: The frequency of the signal generator is varied with a fine-tuning adjustment until the output voltage change is minimal. The HV is then switched off again. The new resonance frequency is counted and noted. Then the frequency is again set to its previous value or adjusted to resonance (a slight drift may have occurred). The output voltage should go to its initial value which was adjusted to zero. A new adjustment of the zero is made if necessary. The HV is then switched on again, etc.; the measurement can be repeated as often as necessary.

With such measurements values of $\omega_0$, $\Delta\omega_0$, $U_0$, and $\Delta U_0$ are determined. ($U_0$ is equal to the output voltage of the signal generator since the voltage across the resonant circuit was adjusted to give zero output voltage in the comparison diode bridge.)

In the design of the sample cell and its associated elements (blocking condensor, HV terminal, shielding, thermostatting) a conduction, shown in Fig. 8, has finally been adapted that enables a very compact structure of the resonant circuit, minimizing parasitic capacitances and inductances, and allowing a maximal flexibility in the use of the cell (variable electrode distance, easy assembling and dissassembling for cleaning, etc.).

The complete cell structure is contained in a silvered brass shielding and grounded cylinder, containing the measuring cell as well as the blocking capacitor. The blocking capacitor consists of a ceramic disk of 100 mm diameter, 2.3 mm thick, and provided on both sides with an evaporated silver film. To avoid breakdown in inhomogeneous field portions at the edge, the lower electrode has a diameter of 80 mm only. The capacitance of the blocking condensor is about 1500 pF. The breakdown strength of the dielectric is 15 kV/mm, and the capacitor has been tested with 30 kV dc.

The middle electrode is hollow and can be connected to an external thermostatted oil bath. Its upper part consists of a replaceable highly polished stainless steel plate that serves as the lower sample cell electrode. The high-voltage connector contains a hollow chamber in which a 5 M$\Omega$ decoupling resistor connected to this electrode is enclosed. The upper half of the measuring cell carries the stainless steel upper electrode, which can be vertically adjusted with a precision micrometer screw. Electrodes can be exchanged so that different electrode diameters can be used to obtain optimal capacitances for each frequency region.

Insulating parts have been made from a polyacetal plastic that has very good mechanical and electrical properties.

The upper and lower parts of the cell are tightly connected by 5 screws at the periphery. Channels are provided to fill the sample chamber. Electrical connections for the inductance are brought out and allow a very easy exchange of the different inductances, which are mounted upon a small plexiglass plate.

The differential diode detector bridge compares the signal generator output with the voltage across the resonant circuit. The resonance-peaking

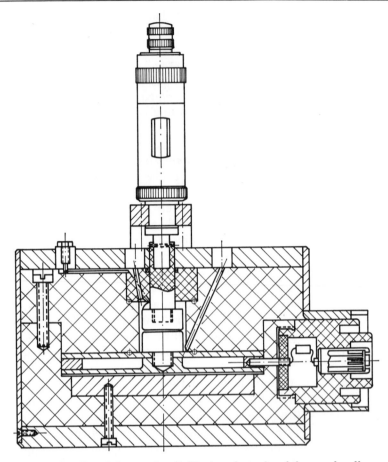

FIG. 8. Cross section of the sample cell. The two electrodes of the sample cell capacitor
are shown in the middle. The lower electrode is in contact with the upper electrode of
the blocking capacitor, which is hollow to allow cooling by an electrically insulating
thermostat fluid. This electrode is also connected via a 5 M resistor to the main high-
voltage connector at the right-hand side. The upper electrode of the sample cell capacitor
is adjustable in vertical direction with the micrometer screw on top of the cell. Channels
are provided in the inner insulating block for filling the sample cell capacitor. The lower
electrode of the blocking capacitor is connected with brass screws to the outer metal
shielding of the cell. Inductances of different size can be plugged in the connector (shown
at the upper left side of the cell) that is connected with a short silver wire to the upper
sample cell electrode.

effect of the tuned circuit is balanced by the attenuator, consisting of
three 50 Ω matched rotary switch attenuators connected in series. A total
attenuation of 61 db can be introduced in 0.1 db steps.

A fine-balancing adjustment can be done with a 100 k potentiometer after amplitude peak-detection. At balancing the voltages at the diodes thus are equal within 0.1 db. This means that they are always working (within these limits) at the same point of their characteristic. If matched diodes are selected, the balancing adjustment becomes independent from small variations in the signal generator amplitude. For all measurements this amplitude is kept constant and equal to 3 V. Vacuum tube diodes were used since they are more resistant to accidental HV transients that may occur when a spark breakdown occurs in the sample cell. The tubes are heated from a regulated dc supply. The diode measuring the resonance voltage at the resonant circuit and its associated circuitry are placed in a small shielded box that can be plugged directly into appropriate connectors at the measuring cell compartment. The reference diode and the rest of the circuitry are contained in another shielded compartment. All connections between such compartments are made with shielded coaxial cables. The whole apparatus including the cell compartment is placed in a large grounded metal container that can be completely closed.

A complete schematic diagram of the resonant network formed by the sample cell configuration described above is given in Fig. 9a, in which $L$ is the exchangeable inductance, its losses (including skin effect, etc.) are represented by a series resistance $R_L$; $C_L$ is the parasitic capacitance of the inductance, and its leads and connecting elements, and includes the

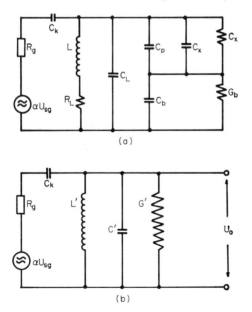

FIG. 9. Circuit diagram of the sample cell showing discrete and parasitary resistive and capacitive elements of the sample cell (a) and equivalent parallel resonant circuit (b).

parasitic capacitance of the upper sample cell electrode to the grounded shielding parts. $C_L$ also comprises any additional capacitance that may be connected directly across the inductance to extend the lower frequency range of the circuit. $C_x$ is the effective capacitance of the sample cell, $G_x$ represents the losses in the sample cell. $C_p$ is the parasitic capacitance between sample electrodes (and is independent of the dielectric medium in the sample cell). $C_b$ is the blocking capacitor; $1/G_b = R_b$ is the internal impedance of the high-voltage generator (5 megohm coupling resistor).

An equivalent network configuration for this circuit is given in Fig. 9b. Here the coupling capacitor $C_k$ and the internal impedance $R_g$ of the driving source have been included. The EMF of the driving source is $\alpha U_{sg}$, where $U_{sg}$ represents the output voltage of the signal generator, and $\alpha$ is determined by the setting of the attenuator. Since the attenuator impedance is 50 ohms, and since it is terminated in a matched load, $R_g$ is equal to 25 ohms. $G'$, $C'$, and $L'$ are given by

$$G' = \frac{R_L}{R_L^2 + \omega^2 L^2} + \frac{\begin{aligned}\{G_xG_b - \omega^2 C_b(C_p + C_x)\}(G_x + G_b) \\ + \omega^2\{G_b(C_p + C_x) + G_xC_b\}(C_b + C_p + C_x)\end{aligned}}{(G_x + G_b)^2 + \omega^2(C_b + C_p + C_x)^2} \quad (51)$$

$$C' = C_L + \frac{\begin{aligned}\{G_b(C_p + C_x) + G_xC_b\}(G_x + G_b) \\ - \{G_xG_b - \omega^2 C_b(C_p + C_x)\}(C_b + C_p + C_x)\end{aligned}}{(G_x + G_b)^2 + \omega^2(C_b + C_p + C_x)^2} \quad (52)$$

$$L' = \frac{R_L + \omega^2 L^2}{\omega^2 L} \quad (53)$$

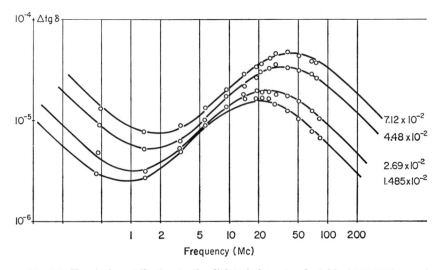

FIG 10. Chemical contribution to the dielectric loss at a dc field of 200 kV/cm and 22° at different concentrations of 1-isobutyl-6-methyluracil in benzene as a function of frequency.

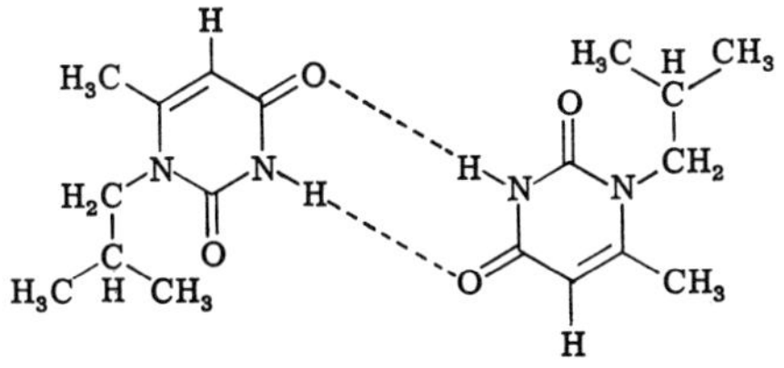

FIG. 11. Structure of the 1-isobutyl-6-methyluracil dimer.

The voltage developed across the output terminals of this circuit may be calculated using Thevenin's theorem: The voltage between any two points in a linear network is equal to the current flowing between these points if they were short-circuited, divided by the admittance measured between these points with all generators replaced by their internal impedances.

For this circuit one obtains at resonance, and with $\omega_0 C_k \ll 1/R_g$

$$U_0 = \alpha U_{sg} \frac{\omega_0 C_k}{G' + R_g \omega_0^2 C_k^2} = \alpha U_{sg} Q_0 \frac{C_k}{C' + C_k} \tag{54}$$

$$\omega_0 = \{L'(C' + C_k)\}^{-1/2} \tag{55}$$

$$Q_0 = \omega_0 (C' + C_k)/(G' + R_g \omega_0^2 C_k^2) \tag{56}$$

All expressions given above can be greatly simplified by proper choice of the circuit parameters. If

$$\omega_0^2 C_b (C_p + C_x) \gg G_x^2 + G_b^2 \tag{57}$$

$$C_b \gg C_p + C_x \tag{58}$$

$$\omega L \gg R_L \tag{59}$$

$$R_g \omega_0^2 C_k^2 \ll G' \tag{60}$$

one obtains

$$G' = \frac{R_L}{\omega^2 L^2} + G_x \tag{61}$$

$$C' = C_L + C_p + C_x \tag{62}$$

$$L' = L \tag{63}$$

Most easily by differentiation of Eq. (54) one may derive:

$$\frac{\Delta \omega_0}{\omega_0} = -\frac{1}{2} \frac{C_x}{C' + C_k} \frac{\Delta \epsilon}{\epsilon} \tag{64}$$

$$\frac{\Delta U_0}{U_0} = \frac{\Delta \omega_0}{\omega_0} - \frac{\Delta G'}{G'} = \frac{\Delta \omega_0}{\omega_0} - Q_0 \frac{C_x}{C' + C_k} \Delta \tan \delta \tag{65}$$

The field-induced change in the loss factor may therefore be written in terms of directly measurable quantities:

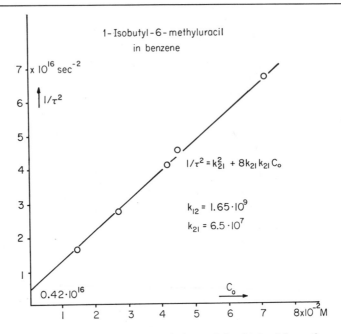

1-Isobutyl-6-methyluracil in benzene

$1/\tau^2 = k_{21}^2 + 8k_{21}k_{21}C_0$

$k_{12} = 1.65 \cdot 10^9$

$k_{21} = 6.5 \cdot 10^7$

$0.42 \cdot 10^{16}$

FIG. 12. Rate constants of dimer association. $1/\tau$ is obtained from the maxima of absorption in Fig. 10.

$$(\Delta \tan \delta)_E = \frac{\gamma}{Q_0}\left[\left(\frac{\Delta\omega_0}{\omega_0}\right)_E - \left(\frac{\Delta U_0}{U_0}\right)_E\right] \tag{66}$$

The quantity $\gamma = (C' + C_k)/C_x$ is obtained from calibration measurements involving solutions with known increments in dielectric constant. Dilute solutions of chlorobenzene in benzene may be used where $\epsilon = 2.280 + 3.404x$ ($x$ is the mole fraction of chlorobenzene). The resonance frequencies of the circuit filled with different such samples are noted, and $\gamma$ may be obtained from the slope of a plot of $\Delta\epsilon/\epsilon$ against $(-2\Delta\omega_0)\omega_0$. $Q_0$ is measured from the 0.707 points of the resonance curve as indicated in Eq. (49). These points may be measured by balancing the circuit at resonance, then removing 3 db from the attenuator and noting the upper and lower frequencies for which balance is again obtained. $(\Delta U_0/\omega_0)_E$ and $(\Delta U_0/U_0)_E$ are obtained by the measuring procedure described above.

Figure 10 shows an example of the data obtained with this method on the chemical relaxation of the association reaction of 1-isobutyl-6-methyl-uracil[18] in benzene solution. The maximum in the dielectric loss factor at a

[18] This compound was kindly supplied by Dr. R. Hopmann. The results given here are preliminary.

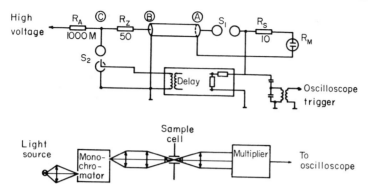

FIG. 13. Schematic diagram of a coaxial cable discharge arrangement to produce high-voltage rectangular pulses of short risetime and duration, and optical arrangement to observe concentration changes during application of the pulse to a weak electrolyte solution. The coaxial cable A–B is 300 m long. It is charged via a 1000 M resistor connected to point C. The voltage at the inner conductor is connected to the observation cell $R_M$ via a small damping and protection resistor $R_s$ when the first spark gap $S_1$ fires. This triggers the firing of spark gap $S_2$ after a variable delay. Between point B and C a 50 ohm resistor is provided to terminate the cable in a matched load when $S_2$ fires.

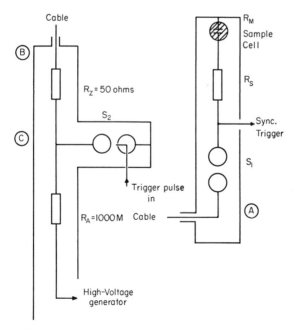

FIG. 14. Coaxial arrangement of spark gaps, resistors, and sample cell. Points A, B, and C correspond to the same points in Fig. 13.

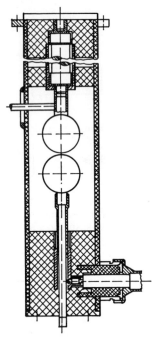

Fig. 15. Cross section of the spark gap arrangement $S_1$. Distance between the spherical spark gap electrodes is adjustable by a gliding movement of the bottom electrode support in a cylindrical sleeve. The 10 ohm resistor $R_s$ is shown at the top where it terminates in a special coaxial connector to the sample cell container.

superposed dc field of 200 kV cm$^{-1}$ is due to chemical relaxation of the hydrogen-bond association indicated in Figs. 11 and 12 for reaction (67)

$$2C_4N_2O_2H_2 \cdot CH_3 \cdot C_4H_9 \underset{k_{21}}{\overset{k_{12}}{\rightleftharpoons}} (C_4N_2O_2H_2 \cdot CH_3 \cdot C_4H_9)_2 \tag{67}$$

At lower frequencies another increase in the loss angle with decreasing frequency is observed, caused by the dissociation field effect.

## C. High Field Pulse Method for Spectrophotometric Observation of the Dissociation Field Effect

Aqueous electrolyte solutions can be subjected to electric fields of the order of $10^5$ volts cm$^{-1}$ only for a few microseconds. (At such a field strength even pure water would boil after 1 second.) In order to measure relaxation phenomena within the short duration of a field pulse, its rise time must be as short as possible, and its amplitude should remain constant until the field is rapidly removed. Rectangular pulses with such characteristics can be obtained from coaxial cable discharges. In the instrument that will be described, and that was designed by Dr. G. Ilgenfritz, this method was used; it allowed the generation of single pulses of 50,000 volts amplitude with a variable duration between 1.5 and 20 $\mu$-sec and with a rise time of the order of a few nanoseconds.

The cable arrangement and its associated elements are schematically shown in Fig. 13. The cable consists of about 330 meters of coaxial cable RG 218/U with a characteristic impedance of 50 ohms. Its total capacitance is about 33,000 pF. It is charged at one end via a high value resistor $R_A$ ($\sim$1000 M) from a HV generator that delivers up to 50 kV. Both ends are provided with spherical spark gaps contained in shielding cylinders, as shown in Figs. 14–16. As the breakdown voltage $U_0$ of spark gap $S_1$ is attained, $S_1$ fires and connects the high voltage to the sample cell via a small safety resistor $R_S$ ($\simeq$10 ohms). In most practical cases the resistance $R_M$ of the sample cell is much larger than the cable impedance. The voltage across the sample cell is given by

$$U_M = U_0 \frac{R_M}{Z + R_M + R_S} \simeq U_0$$

This voltage remains constant, until the voltage step $\Delta U_0 = U_0 - U_M$, after twice traveling the cable (delay time 2 $t_0 \approx 3$ $\mu$sec) again reaches

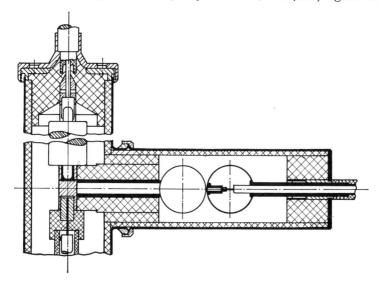

Fig. 16. Cross section of spark gap $S_2$. The right-hand electrode is adjustable in horizontal direction and is provided with a trigger electrode. At the left top is the cable connection, followed by the 50 ohm terminating resistor. At the left bottom is the connection to the 1000 M charging resistor.

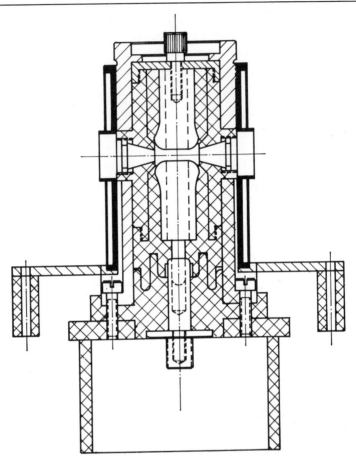

Fig. 17. Cross section of the coaxial sample cell and container. The external container is a double-wall cylinder connected to a thermostat. The sample cell is removable and clamped in place with a metal collar shown on top. The electrode distance in the sample cell is 5 mm. Quartz windows are provided for optical observation.

point $(A)$. The cable is thus discharged through the sample cell in staircase form.

If now, at an arbitrary moment, spark gap $S_2$ is fired, the potential at point $(C)$ drops to zero. At $(B)$ the voltage drops to $U_B/2$ since between $(B)$ and $(C)$ a resistance $R_2$ matched to the cable impedance is connected. [$U_B$ is the voltage at $(B)$ when $S_2$ is fired.] This voltage drop runs through the cable as a negative step, and is reflected with equal (negative) phase at $(A)$. Therefore the voltage at $(A)$ is brought to zero with a delay time $t_0$ after the firing of $S_2$.

The firing of $S_2$ can be synchronized with the self-firing of $S_1$ in such a way that the length of the rectangular high-voltage pulse corresponds to the necessary duration of the measurement. A synchronizing signal is derived from $S_1$, amplified and delayed if necessary and used to trigger $S_2$. Without delay the duration of the rectangular pulse at $(A)$ is equal to the single delay time of the cable, which is about 1.5 $\mu$sec. The design of cable endings and spark gaps is shown in Fig. 16.

The sample cell is shown in Fig. 17. Electrodes are platinum disks that have been soldered upon and shaped around a connecting brass rod. The cell is provided with quartz windows to allow spectrophotometric detection of concentration changes during the field pulse. Since the apparatus is designed for observation of very rapid changes, the transient spectrophotometer must have a bandwidth in excess of 10 Mc. This necessitates

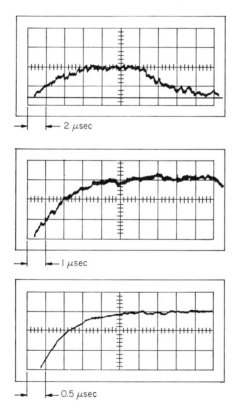

Fig. 18. High-field equilibrium displacement and relaxation of the protolytic color indicator bromcresol purple at different concentrations ($c_0 = 7.5 \times 10^{-6}$ $M^{-1}$, $3.8 \times 10^{-5}$ $M^{-1}$, $2.35 \times 10^{-4}$ $M^{-1}$) in unbuffered aqueous solution. A rate constant for protonation in the reaction equilibrium $I^{2-} + H^+ \rightleftharpoons IH^-$ of $8 \times 10^{10}$ $M^{-1}$ sec$^{-1}$ is obtained from these measurements.

high-intensity light sources with high surface brightness, large aperture of the monochromator, photomultipliers with large photocathode and few dynode stages, a small anode load impedance, and sufficient postamplification. A detailed description will not be given here since the design considerations for spectrophotometric measurements are discussed in other chapters of this book (e.g., Chapter [1]). An example of the time resolution that can be obtained with this arrangement is shown in Fig. 18.

# Section II
# Electrochemistry

# [5] Polarographic Methods

## By PETR ZUMAN

## I. Introduction

As a method used to study the kinetics of enzymatic reactions, polarography is one whose advantages are not widely appreciated. It is the aim of the present contribution to demonstrate some of the possibilities that polarography offers and also to point out certain limitations that restrict its use in kinetic studies of enzymatic systems. In some cases it is possible only to indicate potentialities, in others it is possible to demonstrate the usefulness of the method by some examples. In order not to assume any special knowledge, the principles of polarography will be briefly discussed first.

## II. Principles of Polarography

Polarography is an electrochemical method based on electrolysis with a dropping mercury electrode (or with some other types of mercury electrodes, with a periodically renewed surface). To a cell containing such a dropping mercury electrode immersed in the solution to be electrolyzed and connected to another electrode (so called reference electrode the potential of which should not change during electrolysis), a regularly increased direct current (dc) voltage is applied. The current resulting from the electrolysis at the surface of the mercury drop is recorded as a function of the applied voltage. The current flow takes place in solutions containing an electroactive substance. As electroactive, we denote substances that in the potential range studied either undergo reduction or oxidation, or form compounds with mercury, or catalyze some other oxidation or reduction processes. In the presence of an electroactive compound the recorded current-voltage

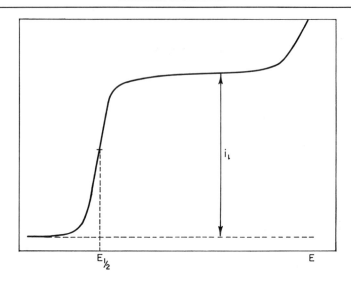

Fig. 1. Polarographic curve. Dependence of current $(i)$ on the applied voltage $(E)$. Half-wave potential $(E_{1/2})$ and limiting current (wave height, $i_l$) marked.

curve (called the polarographic curve) shows a stepwise increase of the current with increasing applied voltage (Fig. 1). This stepwise increase of current is called a polarographic wave.

Detailed descriptions of the techniques used in realization of the dropping and reference electrodes, description of vessels and polarographic apparatus used and of the procedures used to record the polarographic curves as well as the interpretation and evaluation of polarographic curves are well known.[1-6a] It is therefore unnecessary to enumerate many details, and only some of the principal procedures will be mentioned.

A dropping mercury electrode is usually used as the indication electrode, with a drop-time of about 3 seconds and with an outflow velocity of

[1] I. M. Kolthoff and J. J. Lingane, "Polarography," 2nd ed. Wiley (Interscience), New York, 1952.

[2] O. H. Müller, "The Polarographic Method of Analysis," 2nd ed. Chem. Education Publ. Co., Easton, Pennsylvania, 1951.

[3] P. Zuman and I. M. Kolthoff (eds.), "Progress in Polarography," Vols. I and II. Wiley (Interscience), New York, 1962.

[4] P. Zuman, "Organic Polarographic Analysis." Macmillan (Pergamon), New York, 1964.

[5] L. Meites, "Polarographic Techniques," 2nd ed. Wiley, New York, 1966.

[6] J. Heyrovský and J. Kůta, "Principles of Polarography." Academic Press, New York, 1966.

[6a] J. Heyrovský and P. Zuman, "Practical Polarography." Academic Press, New York, 1968.

mercury of about 2 mg/sec. Such an electrode can be realized by a thick-walled (barometer) capillary with an inner diameter of 0.05 to 0.09 mm. When such a capillary is connected to a mercury reservoir placed so as to ensure that the mercury level is 40–80 cm above the orifice of the capillary, the above electrode characteristics can be achieved.

The solution to be investigated by means of polarographic electrolysis is placed into a polarographic vessel, and the capillary is immersed in the solution. Polarographic vessels used in the study of kinetic problems differ from those used for analytical purposes. Various types of constructions are possible, and it is usually not of great importance which one is chosen. The vessel used should permit function with a separated reference electrode; it should also allow fast mixing of the solutions, temperature control, and exclusion of atmospheric oxygen.

In studies of reactions at the electrode surface (Section III), the separation of the reference electrode is best carried out by liquid junction; whereas for the reactions taking place in the bulk of the solution (Section IV), separation by a sintered glass frit is best. Temperature control is assured by a water jacket or immersion of the vessel into a temperature-controlled bath. To exclude the effects of oxygen the polarographic electrolysis is carried out in an air-tight system and the oxygen dissolved in the solutions used is removed by purging with an inert gas (usually nitrogen).

The increase of the voltage applied to the dropping mercury electrode can be plotted on the $y$-axis and voltage on the $x$-axis. The recording of the changes in current with the applied potential can be carried out manually, photographically, or by pen-recording instruments. Because of convenience, the latter type of recording predominates. The apparatus can be easily built from available units, but to be sure that the apparatus will fulfill all the criteria necessary for a proper record of current-voltage curves, the use of manufactured instruments called polarographs can be recommended. For study in aqueous solutions and other media of low ohmic resistance, the two-electrode system is usually sufficient; for the studies in nonaqueous media, a potentiostatic (three-electrode) polarograph is necessary. It is preferable, in particular for the studies of fast reactions taking place at the electrode surface, to use polarographs that record mean values of current (i.e., records with frequency characteristics corresponding to a critically damped galvanometer with a period of swing of about 10 seconds) rather than the so-called "peak current," as with but few commercially available instruments is the value of this current strictly defined. Oscillations on polarographic curves indicate a properly working capillary electrode. They can sometimes indicate the type of the process involved, and their mean value is easily determined.

Polarographic curves offer various types of information that are of interest for detailed electrochemical studies. In the solution of problems of reaction kinetics, two quantities measured from polarographic curves are of particular importance: The limiting current ($i_{lim}$) corresponding to the increase in current resulting from the electrolytic process, and the half-wave potential ($E_{1/2}$) which equals the potential on the polarographic curve, where the current reaches half its limiting value.

The limiting current (or wave height) is usually measured at a selected potential as the difference between the limiting current studied and the current observed in the pure supporting electrolyte (before the substance to be electrolyzed has been added). In samples of biological origin it is sometimes difficult to obtain a reference sample in which the last traces of the electroactive substance would be absent—and at the same time all other components present. In these cases it is difficult to subtract the current due to supporting electrolyte in the measurement of the wave height. Moreover, the surface-active substances present in biological material can affect the current of the supporting electrolyte. In such cases, empirical methods of wave-height measurements based principally on the prolongation of linear parts of the curve before and after the current rise have proved useful.

Polarographic curves show oscillations resulting from the growth of the surface of the mercury drop during the lifetime of each drop. Because the growth is periodic and regular, the shape of the recorded oscillations is also regular. It has been shown best to measure the mean value of the oscillations when determining the wave height.

The limiting current is governed by the rate of transport of the electroactive species to the surface of the electrode. The most common type of governing process is diffusion. In addition to the rate of diffusion, the limiting current can be affected by various types of adsorption phenomena, or by the rate of a chemical reaction which takes place in the vicinity of the surface of the electrode. The diffusion controlled limiting current (usually called "diffusion current") is the most common type of limiting current. In most cases it is linearly proportional to the concentration of the electroactive species in the electrolyzed solution. By following the time change of the wave height of a diffusion-controlled process it is possible to obtain directly the time dependence of the concentration of the studied electroactive species. On this principle, polarography can be used for the study of kinetics of slower homogeneous reactions taking place in the bulk of the solution. On the other hand, limiting currents corresponding to electrode processes that involve chemical reactions taking place in the vicinity of the surface of the mercury electrode depend on the velocity of these reactions. When the equilibrium constants of these reactions can be determined by an independent measurement, it is possible from measurement of the height of such waves to calculate values of the rate constants of fast reactions, corresponding to the reestablishment of the equilibrium which has been perturbed in the vicinity of the surface of the dropping mercury electrode by electrolysis.

The second quantity important in all applications of polarography is the half-wave potential, which under certain conditions can be a measure of the free energy required for the fulfillment of the electrolytic process (activation energy or standard energy—for irreversible or reversible processes). The value of the half-wave potential depends on the nature of the electroactive species, but also on the composition of the solution in which the electrolysis is carried out. The composition of the solution electrolyzed being kept constant, and consisting of the electroactive substance and a proper supporting electrolyte, often buffered, it is possible to compare the half-wave potentials of various substances. When the mechanism of the electrode process is similar for all compounds compared, the half-wave potential can be considered as a measure of the reactivity of the compound toward the electrode. Hence the half-wave potentials are physical constants that characterize quantitatively the electrolyzed compound, or the composition of the electrolyzed solution. In the application of polarography to reaction kinetics the half-wave potentials are of importance both for slow and fast reactions. For slower reactions differences in half-wave potentials makes a simultaneous determination of several components of the reaction mixture possible. In some advantageous cases it is possible in one single measurement to determine reactant, intermediates, and products simultaneously. Moreover, in some cases, the values of half-wave potentials, and their change with composition of the electrolyzed solution, in particular with pH, enable the identification of products or intermediates. For fast reactions a difference of half-wave potentials of both components which are in the rapidly established equilibrium is a condition necessary for the determination of rate constants of these reactions.

It seems to be useful to indicate the analogy with spectrophotometry. The limiting current at a given potential corresponds to absorptivity measured at a given wavelength, as both depend on concentration; thus, if the concentration changes with time, they both change with time. The wavelength of an absorption maximum corresponds to the half-wave potential, as both depend on the nature of the substance studied and on the composition of the media in which the measurement is carried out. When during a reaction, a new compound is formed showing an absorption band at a wavelength different from those of the reactants, it happens quite often that on polarographic curves a new wave appears at a potential different from those of the reactants.

For the application of polarography to the study of both faster and slower reactions, at least one component taking part in the reaction—one of reactants, intermediates, or reaction products—must be polarographically active, i.e., give a polarographic wave in the available potential range. Information on the polarographic behavior can sometimes be found in the literature.[1,4-6a] For an organic compound that has not been studied before, it is not easy to predict whether it will be polarographically active or not, because the presence of a polarographic wave in the potential range studied depends not only on the presence of the bond which is cleaved or formed during the polarographic electrolysis, but also on the molecular frame on which this bond is situated and on the type and number of atomic groupings in the vicinity of the attacked bond. Nevertheless, as a simple guiding rule, polarographic reductions (these are more frequently studied than oxidations) occur with strongly polarized bonds, in particular multiple bonds (such as in $C{=}O$, $C{=}N$, $C{=}S$, $C{=}C$, $C{\equiv}N$, $C{\equiv}C$, $N{=}N$, $N{\equiv}N^+$, $NO$, $NO_2$, etc.) especially when they are conjugated, and with some single bonds (such as $C{-}I$, $C{-}Br$, $O{-}O$, $S{-}S$). Some less easily polarizable bonds (such as $C{-}F$, $C{-}Cl$, $C{-}N$, $C{-}S$, $C{-}O$, $C{-}C$) must be activated by the presence of some electron-donating groups in the vicinity.

If the polarographic behavior of the given substance has not been described in the literature, it is not sufficient to record the wave in some few haphazardly chosen supporting electrolytes, but a systematic polarographic study must precede the application in kinetic studies. It is necessary to have information in sufficient detail available before the kinetic application, e.g., how the changes in the composition of the polarographed solution affects the polarographic curves. In particular it is important to find out how changes of acidity (usually pH), buffer type and concentration, solvent, neutral salts, temperature, and illumination affect the wave-heights, half-wave potential, and shape of the polarographic wave.

As mentioned above, polarography can be applied to the study of two types of systems, viz., of the fast reactions taking place in the vicinity of the surface of the electrode and of slower reactions taking place in the bulk of the solution. As the techniques used for elucidation of the kinetic laws and evaluation of rate constants involved are substantially different, a separate discussion is preferred.

### III. Fast Reactions

### A. Principle

For substances of which the oxidized or reduced form is made or cleaved by a chemical reaction, this chemical reaction will be competitive with the transport of the substance to the surface of the electrode. Such chemical reactions as result in the formation or change of the electroactive species can be antecedent, parallel, or consecutive to the electrode process proper or can be interposed between two consecutive electrode processes. General information on the treatment of such systems can be found in monographs[1,5-8] and reviews.[9-11]

To demonstrate how chemical reactions affect polarographic curves, a system involving a reaction preceding the electrode process proper was chosen, because this type of system is most commonly observed. The electroactive species C is formed by this type of reaction as in the reversible reaction (1)

$$A + B \underset{k_{-1}}{\overset{k_1}{\rightleftharpoons}} C \tag{1}$$

The electroinactive form A is transported from the bulk of the solution toward the surface of the electrode. The equilibrium between the electroinactive form A and the electroactive form C is established by a final rate under participation of an electroinactive reagent B, which is often a component of the reaction media. To simplify the treatment the experimental conditions are adjusted in such a way that compound B is present in excess and/or its concentration can be considered constant. The equilibrium (1) is then simplified to (2)

$$A \underset{k_{-1}}{\overset{k'_1}{\rightleftharpoons}} C \tag{2}$$

The formal rate constant $k'_1$ is defined as $k'_1 = k_1[B]$.

In order to apply the polarographic method to the study of systems of the type (2) the condition must be fulfilled that the potential at which the reduction[12] of the substance C takes place ($E_C$) is sufficiently different from the potential at which the substance A is reduced ($E_A$). In this contribution the convention is adopted that substance C undergoes reduction at more

[7] P. Delahay, "New Instrumental Methods in Electrochemistry." Wiley (Interscience), New York, 1954.

[8] S. G. Mairanovskii, "Catalytic and Kinetic Waves in Polarography." Nauka, Moscow, 1966. (In Russian) English translation. Plenum Press, New York, 1968.

[9] R. Brdička, Collection Czech. Chem. Commun. 19 (Suppl. II), S41 (1954).

[10] R. Brdička, V. Hanuš, and J. Koutecký, in "Progress in Polarography" (P. Zuman and I. M. Kolthoff, eds.), Vol. I, p. 145. Wiley (Interscience), New York, 1962.

[11] H. W. Nürnberg and M. von Stackelberg, J. Electroanal. Chem. 2, 350 (1961).

[12] The discussion is restricted here to reduction processes only. A corresponding discussion of an oxidation process would differ only in the relative sequence of potentials. What for reduction is more positive, would be for an oxidation process more negative, and vice versa.

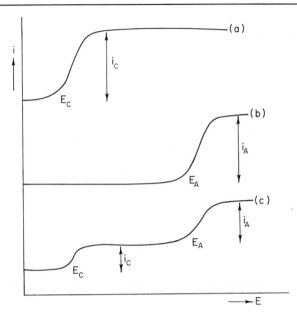

FIG. 2. Polarographic curves corresponding to an equilibrium between forms A and C: curve $a$, form C predominating; $b$, form A predominating; $c$, concentrations of A and C comparable. $E_C$, $E_A$, potentials at which waves of C and A appear (e.g., half-wave potentials); $i_C$ and $i_A$, limiting currents. [From P. Zuman, *Advan. Phys. Org. Chem.* **5**, 31 (1967).]

positive potentials than the substance A. This sequence is in fact involved in the assumption that, in the given potential range, species C is electroactive whereas the form A is electroinactive. This supposition does not exclude the possibility that species A can be electroactive at more negative potentials.

The shape of polarographic curves is first described for the case when the equilibrium between A and C is established slowly when compared with the drop-time, which is usually about 3 seconds. Equilibria established during some 10 to 30 seconds are in this context described as slowly established. The heights of waves at potentials $E_C$ ($i_C$) and $E_A$ ($i_A$) are in this case limited only by diffusion. In such cases three possibilities exist:

1. At a great excess of component B the equilibrium is shifted toward formation of the substance C. In the solution practically only compound C is present, showing a wave $i_C$ at the potential $E_C$ (Fig. 2, curve $a$).

2. At a small concentration of the component B and for a proper (suitable) value of the equilibrium constant $K = k_1/k_{-1}$ the equilibrium is oppositely shifted toward A. On the polarographic curve only wave $i_A$ at the potential $E_A$ is observed (Fig. 2, curve $b$).

3. Finally at such a concentration of the substance B that the value of the product $k_1$ [B] is comparable with the value of $k_{-1}$, concentrations of species A and C are comparable. On the polarographic curve two waves are observed: the height of the wave $i_C$ at the potential $E_C$ is proportional to the equilibrium concentration of the substance C. The height of the wave $i_A$ at the potential $E_A$ is similarly proportional to the equilibrium concentration of the compound A (Fig. 2, curve c).

Next the same system will be discussed under conditions when the establishment of the equilibrium between substances A and C does not take place slowly (in the above meaning of "slow" relative to the drop-time). Hence at potentials at which the reduction of the substance C takes place, the equilibrium between A and C is perturbed because the substance C is cleaved by electrolysis at the surface of the electrode. The scheme (3) is followed

$$A \underset{k_{-1}}{\overset{k'_1}{\rightleftharpoons}} C \overset{\text{el}}{\rightarrow} \text{products} \tag{3}$$

The equilibrium perturbed by the electrolytic process (el) tries to be reestablished. The compound C is replenished (produced) in the vicinity of the electrode by a chemical reaction with the rate constant $k'_1$ by which the compound C is formed from the species A. In this way the amount of substance C that can undergo reduction at the surface of the electrode is increased and the wave $i_C$ becomes higher as compared with the conditions for a slowly established equilibrium (Fig. 3).

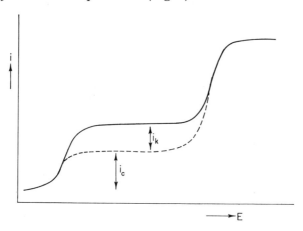

FIG. 3. Effect of fast chemical reactions on polarographic waves: $i_C$, diffusion-controlled limiting current of substance C; $i_k$, kinetic current governed by the rate of chemical reaction by which substance C is formed in the vicinity of the electrode. [From P. Zuman, *Advan. Phys. Org. Chem.* **5**, 32 (1967).]

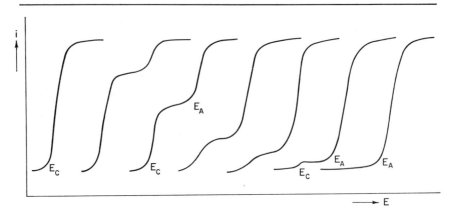

FIG. 4. Change of polarographic waves of the Form C and A with decreasing concentration of substance B. [From P. Zuman, *Advan. Phys. Org. Chem.* **5**, 33 (1967).]

The magnitude of this increase depends, for given experimental conditions, on the value of the formal rate constant $k'_1$, i.e., on the magnitude of the rate constant $k_1$ and on the concentration of the component B present in excess.

If the wave corresponding to the equilibrium concentration of C is so small that it cannot be distinguished on polarographic curves in the particular range of concentration of the substance B, then with decreasing concentration of the component B polarographic curves change as shown in Fig. 4.

1. At higher concentrations of B the rate of formation of the compound C is so great that all the substance A present in the vicinity of the electrode can be transformed during the life of a single drop into the form C. Under such conditions, even when the reducible species C is formed by a chemical reaction, the wave height is governed solely by the rate of diffusion of the species A. On the polarographic curve only the wave $i_C$ is observed.

2. At small concentrations of the component B the rate of formation of the substance C is low and the chemical reaction has not a sufficient yield to produce measurable amounts of the compound C. Substance A, which predominates in the bulk of the solution, diffuses toward the electrode surface and is reduced at potentials $E_A$ corresponding to wave $i_A$ (if $E_A$ is not too negative so that this wave is overlapped by the current of the supporting electrolyte). Only the more negative wave $i_A$ limited by the rate of diffusion is observed on the polarographic curve at low concentrations of B (Fig. 4).

3. Finally in a proper range of concentration of the component B the rate of formation of the compound C from A is such that only a certain portion of the particles of the substance A present in the solution is transformed into the electroactive form C. On polarographic curves under

such conditions, two waves are observed ($i_C$ and $i_A$, if the wave for $i_A$ is not obscured by the current of the supporting electrolyte). The ratio of the heights of these waves $i_A$ and $i_C$ changes with the concentration of the component B.

If the ratio $i_C/(i_C + i_A)$ is plotted against log[B] the plot resembles that for a dissociation curve, obtained potentiometrically (Fig. 5).

The polarographically measured dependence of the type given in Fig. 5 is next compared with the corresponding dependence for the cases in which the equilibrium between species A and C is established slowly or for the same system as studied polarographically but investigated by other methods (such as spectrophotometry) that do not perturb the equilibrium. It is observed that the polarographically found "dissociation curve" is shifted toward smaller values of log[B] when compared with the dissociation curve, determined for equilibrium conditions. Even when the shapes of the polarographic and equilibrium dissociation curves are approximately the same, they are shifted along the log[B] axis. The value of $-\log[B]$ (denoted as pB) at which the heights of waves $i_C$ and $i_A$ are equal, is denoted as polarographic $pK'$. The value of $pK'$ for systems, in which the equilibrium between the species A and C is established in a period compa-

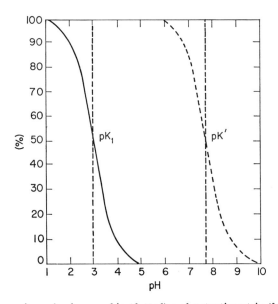

FIG. 5. Comparison of polarographic (dotted) and potentiometric (full) dissociation curves. Polarographic dissociation curve corresponds to a dependence of ratio of wave-heights $i_k/i_d$ (expressed as percentage of the total limiting current) on pH. The shift of polarographic $pK'$-value when compared with potentiometric $pK$ results from recombination reaction (A + B $\rightleftharpoons$ C). [From P. Zuman, *Advan. Phys. Org. Chem.* **5**, 34 (1967).]

rable with that of a drop-life, is greater than the value of the equilibrium p$K$. From the difference between the values of the polarographic p$K'$ and that of the equilibrium p$K$ it is possible (using expressions that depend on the system involved and that are given in Section III,C) to compute the value of the rate constant $k'_1$ or $k_1$, respectively, and—using the value of the equilibrium constant $K$—also of the rate constant $k_{-1}$. Since [B] can be of the order of $10^{-3}$, $10^{-5}$, or even $10^{-8}$ $M$, it is possible to determine in this way second-order rate constants $k_1$ in the range from $10^5$ to $10^{11}$ l mole$^{-1}$ sec$^{-1}$.

The fundamental condition for this type of application of polarography for the determination of rate constants of fast reactions is that at least the compound C is polarographically active, i.e., gives a polarographic wave, in the potential range available.

## B. Distinguishing of Kinetic Currents

Before applying any of the procedures described in Section III,C that enable us to decide to which reaction scheme the studied system corresponds and to see the proper expression for the computation of the rate constant, it is possible to give some few general experimental hints. The characteristics of the capillary (i.e., the drop-time and the outflow velocity) should be measured for the capillary used in the potential range in which the studied wave appears. Because numerous currents limited by the rate of a chemical reaction show a large temperature coefficient, it is useful to control the temperature of the electrolyzed solution by immersing the cell into a water bath. Cells with a reference electrode separated by a liquid junction have proved most useful. The concentration range is usually 2 to 5 $\times$ $10^{-4}$ $M$. Only when the equilibrium is shifted too far toward the electroinactive form (e.g., formaldehyde, glucose), are concentrations of two or even three orders of magnitude higher needed. The supporting electrolyte should ensure sufficient conductivity of the solution and control the conditions of electrolysis. As most organic electrode processes are accompanied by a proton transfer, it is often useful to keep the solutions well buffered. The more dilute buffer component should be present at a concentration 20 times higher than the electrolyzed substance. When the effect of pH change is investigated, it is useful to keep the concentration of one buffer component constant and to vary only the other one. It is further necessary to prove whether or not the polarographic curves recorded correspond to a process involving a chemical reaction taking place in the vicinity of the electrode. The vast majority of polarographic currents of organic compounds are diffusion-controlled currents, kinetic currents being less frequent. It is hence of importance to understand how kinetic currents can be identified.

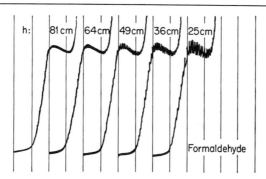

FIG. 6. Effect of the height of mercury head ($h$) on the limiting current of formaldehyde. Borate buffer pH 9.2, $3.7 \times 10^{-3}\,M$ formaldehyde. Curves starting at $-1.2$ V, 200 mV/absc., saturated calomel electrode, full-scale sensitivity 5 $\mu$A.

Methods of identification based on classical polarographic curves are discussed first. Kinetic currents often depend strongly on the pH of the supporting electrolyte, because the rate of the chemical reactions involved are pH dependent. The dependence frequently possesses the shape of a dissociation curve as depicted on Fig. 5 or of a curve with a maximum (see Section III,C).

One of the important proofs for the kinetic character of the limiting current is the dependence on mercury pressure. When the polarographic curve is recorded at varying heights of the mercury reservoir, the height of the kinetic current remains unchanged (Fig. 6) whereas a diffusion-controlled current is proportional to the square root of the height of the mercury reservoir. It has proved best to plot the wave height measured at various mercury pressures against the square root of the difference between mercury level in the reservoir and the capillary tip. The plot for a diffusion-controlled current must be both linear and go through the origin of the scale (Fig. 7A). Kinetic current shows a linear plot parallel to the mercury pressure (Fig. 7B) axis. Plots that are linear and show an intercept on the current axis that is significantly greater than the error of measurement (Fig. 7C) may indicate that chemical reaction affects the limiting current in addition to diffusion.

It is of extreme importance to decide under which conditions the dependence of mercury pressure is to be studied. Only when studied under conditions where the measured current $i_C$ corresponds to not more than 15% of the total limiting current ($i_A + i_C$), is the kinetic current virtually independent of mercury pressure. The higher the current, the less characteristic the dependence on mercury pressure, until when $i_C$ reaches the total wave height ($i_A + i_C$) and wave $i_A$ is negligible, the current depends on the square root of reservoir height as for diffusion-controlled currents. Actually,

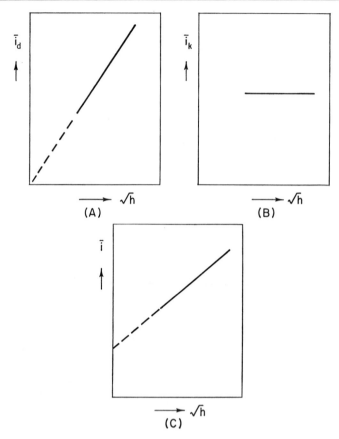

FIG. 7. Dependence of the diffusion ($\bar{i}_d$), kinetic ($\bar{i}_k$), and composed ($\bar{i}$) current on the square root of mercury head ($h$). (A) Diffusion current. (B) Kinetic current. (C) Chemical reaction affects the limiting current in addition to diffusion.

under these conditions the transformation of A into C is fast, and the wave height is limited by the rate of the diffusion of species A. Hence, the effect of mercury pressure is unequivocal, but only when the wave $i_C$ is small.

In some cases the kinetic currents even at their maximum height are small (sometimes 10% or even less) when compared with waves of the equimolar solutions that are diffusion controlled. This type of behavior was observed in particular in systems in which the waves of the compound A ($i_A$) were obscured by the current of the supporting electrolyte. An example of this behavior is the small kinetic wave of glucose, which even at a 10-fold concentration of this aldose is only some 20% of the diffusion-controlled wave of fructose (Fig. 8).

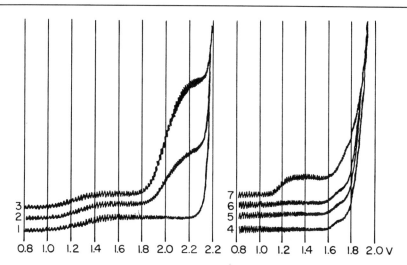

Fig. 8. Comparison of waves of fructose and glucose. Supporting electrolytes: 0.1 $M$ LiOH (curves $1-3$) or 0.02 $M$ LiCl (curves $4-7$). ($1$) Pure 0.1 $M$ LiOH; ($2$), 0.005 $M$ fructose; ($3$), 0.01 $M$ fructose; ($4$), pure 0.02 $M$ LiCl; ($5$) 0.005 $M$ glucose; ($6$) 0.01 $M$ glucose; ($7$) 2 $M$ glucose. All curves in the presence of air, starting at $-0.8$ V, 200 mV/absc., Hg-pool electrode; $h = 55$ cm, full-scale sensitivity 75 $\mu$A.

But not all kinetic currents are considerably smaller than diffusion currents. Therefore whereas small waves may indicate kinetic currents, currents of the normal height (when compared with diffusion currents) do not allow us to exclude the role of chemical reactions. On the other hand, some catalytic currents are much greater than diffusion currents. Similarly the effect of temperature on the limiting current is only of limited diagnostic value. Some kinetic currents increase with increasing temperature considerably more ($3-20\%$ deg$^{-1}$) than diffusion-controlled currents ($1.5-2.0\%$ deg$^{-1}$). The sensitivity toward temperature rise corresponds to the great temperature coefficient of the rate of the chemical reaction accompanying the electrode process. Therefore, in cases in which a current strongly dependent on temperature is observed, it can be assumed that the studied current is kinetic. But, when a temperature coefficient similar to those found for diffusion currents is obtained, the participation of the chemical reaction in the electrode process cannot be excluded, because some of the kinetic currents (e.g., in systems in which acid-base equilibria are involved) show temperature coefficients comparable in magnitude with those for diffusion currents. An example of the temperature-sensitive kinetic wave is the wave of glucose (Fig. 9).

When using a mechanical device the drop-time ($t_1$) is artificially controlled (and at the same time the outflow velocity of mercury is kept

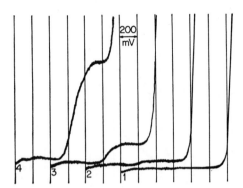

FIG. 9. Effect of temperature on the limiting current of glucose. Supporting electrolytes: 0.02 $M$ LiCl, 0.2 $M$ glucose. Curves recorded at temperatures: (1) 20°C; (2) 40°C; (3) 60°C; (4) 80°C. Curves were recorded in the presence of air, starting at $-1.0$ V, 200 mV/absc., Hg-pool electrode; $h = 75$ cm, full-scale sensitivity 20 $\mu$A.

constant), the mean limiting current ($\bar{\imath}$) is found to be proportional to $t_1^{2/3}$ for kinetic currents, but to $t_1^{1/6}$ for diffusion currents. The slope of a $\log \bar{\imath} - \log t_1$ plot is the best diagnostic test.

Another simple way to study the effect of drop-time on limiting current is to record the waves and measure the currents and the drop-time with the same capillary, first in the normal, vertical position ($\bar{\imath}'$; $t'_1$), and then in a horizontal position ($\bar{\imath}''$; $t''_1$). If the tip of the capillary remains in the same position relative to the level of the mercury in the reservoir, the outflow velocity remains practically unchanged. The change in the position of the capillary therefore only results in a change of the drop-time. For kinetic currents the relation $\bar{\imath}'_k/\bar{\imath}''_k = (t'_1/t''_1)^{2/3}$ is obeyed whereas for diffusion currents $\bar{\imath}'_d/\bar{\imath}''_d = (t'_1/t''_1)^{1/6}$ holds true.

In addition to the classical $\bar{\imath}$-$E$ polarographic curves, other techniques can also be used for distinguishing between kinetic and other currents. Perhaps the most important are the records of the changes of the instantaneous current with time on a single mercury drop ($i$-$t$ curves). Preferably these curves are recorded on the first drop (after the particular potential was applied), using an oscilloscope or a string galvanometer to follow the fast changes of the current. The currents limited by a rate of a chemical

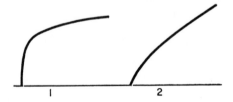

FIG. 10. Typical $i$-$t$ curves: (1) diffusion currents; (2) kinetic currents.

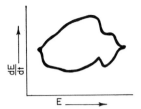

FIG. 11. Curve $dE/dt = f(E)$ of formaldehyde recorded in oscillographic polarography; 1 $M$ LiOH, 0.3 $M$ formaldehyde, $Q = 0.75$.

reaction follow a $\frac{2}{3}$ parabola, whereas diffusion controlled currents show a $\frac{1}{6}$ parabola (Fig. 10). Again, a graphical plot of log $i$ against log $t$ is the best test for the verification of the parabolic shape of the $i$-$t$ curve and for the determination of the exponent. To distinguish kinetic currents from slow electrode processes it is preferable to record the $i$-$t$ curves in the potential region corresponding to the limiting current.

In the single-sweep $i$-$E$ curves, kinetic currents are shown by the shape of the curves, which are steplike, whereas for diffusion currents they show peaks. Similarly the incissions on the $dE/dt = f(E)$ curves recorded in oscillographic polarography are L-shaped for kinetic currents whereas for diffusion currents they are V-shaped. This is shown for the case of formaldehyde (Fig. 11).

## C. Reaction Types

To calculate the value of the rate constant, it is necessary first to elucidate the scheme of the chemical reaction participating in the overall electrode process corresponding to the given type of the kinetic current and to determine the sequence of chemical and electrochemical steps. For this purpose it is necessary to record the waves of substance C at various concentrations of the reagent B at a given temperature (because of the high temperature coefficient of some kinetic currents, a rigorous control of temperature is important). To compute the rate constants it is moreover necessary to determine by an independent experiment the value of the equilibrium constant $K = k_1/k_{-1}$.

Several reaction types can be distinguished according to the sequence of chemical and electrochemical steps. Even though the division between the individual groups is not rigid, the following main types can be distinguished: reactions that take place antecedent, parallel, or consecutive to an electrode process and those that are interposed between two electrode processes.

### 1. *Antecedent Reactions*

Electrode processes that are accompanied by a chemical reaction antecedent to the electrode process proper belong to the most frequently studied examples of kinetic currents. The system involved has been discussed qualitatively in Section A above and can be described by Eq. (1')

$$A + B \underset{k_{-1}}{\overset{k_1}{\rightleftharpoons}} C \overset{el}{\rightarrow} P \tag{1'}$$

When the concentration of the component B is kept constant during the process either by the use of a buffer or by the use of an excess of substance B, the scheme (1') simplifies into (3)

$$A \underset{k_{-1}}{\overset{k'_1}{\rightleftharpoons}} C \overset{el}{\rightarrow} P \tag{3}$$

where A is the electroinactive form, C electroactive form, $k'_1 = k_1[B]$. The mass transport toward the electrode surface can be described according to Koutecký[13] by differential equations (4) and (5)

$$\frac{\partial[C]}{\partial t} = D\frac{\partial^2[C]}{\partial x^2} + \frac{2x}{3t}\frac{\partial[C]}{\partial x} + k'_1\left([A] - \frac{K_d[C]}{[B]_{eq}}\right) \tag{4}$$

$$\frac{\partial[A]}{\partial t} = D\frac{\partial^2[A]}{\partial x^2} + \frac{2x}{3t}\frac{\partial[A]}{\partial x} - k'_1\left([A] - \frac{K_d[C]}{[B]_{eq}}\right) \tag{5}$$

The first term on the right-hand side corresponds to diffusion (Fick's law), the second expresses the correction for the growth of the dropping electrode during the life of each drop, and the third term takes into account changes in concentration caused by the chemical reaction with constant $k'_1$.

The boundary conditions are:

$$t = 0, x > 0 \quad : \quad [A] = [A]_b \qquad [C] = [C]_b$$

$$t > 0, x = 0 \quad : \quad [C] = 0, \qquad \frac{\partial[A]}{\partial x} = 0$$

$$k'_1 \gg 1$$

where the index b indicates the concentration in the bulk of the solution.

For a very fast chemical reaction, a steady state can be established in the immediate vicinity of the electrode surface between the competitive processes which are the chemical reaction and diffusion. Differential equations (4) and (5) simplify to (4a) and (5a)

$$D\frac{d^2[C]}{dx^2} + k'_1\left([A] - \frac{K_d[C]}{[B]_{eq}}\right) = 0 \tag{4a}$$

$$D\frac{d^2[A]}{dx^2} - k'_1\left([A] - \frac{K_d[C]}{[B]_{eq}}\right) = 0 \tag{5a}$$

[13] J. Koutecký, *Collection Czech. Chem. Commun.* **18**, 597 (1953).

Under conditions when the diffusion-controlled current of compound C is very small as compared with the observed kinetic current, diffusion of substance A can be neglected, and when $[C] \ll [A]$, Eq. (4a) can be solved. The result can be expressed in the form (6), when we further introduce $k'_1 = k_1[B]$

$$[C] = \frac{[A][B]}{K_d}\left[1 - \exp\left(-x\Big/\sqrt{\frac{D}{K_dk_1}}\right)\right] \tag{6}$$

Combining Eq. (6) with the equation for current (7):

$$i = nF_qD\left(\frac{\partial[C]}{\partial x}\right)_{x=0} \tag{7}$$

we obtain

$$i = nF_qD[A]\sqrt{D/K_dk_1} \tag{8}$$

Solution of Eqs. (4) and (5) for a fast chemical reaction $(k'_1 \gg 1)$ (i.e., under conditions when a steady state is established between the chemical reaction and the diffusion of substance C) is possible, when the method of dimensionless parameters is used.[13,14] For the ratio of the mean limiting kinetic current $\bar{\imath}_k$ to the mean limiting diffusion current $\bar{\imath}_d$, expressions (9) and (10) were derived

$$\frac{\bar{\imath}_k}{\bar{\imath}_d} = \frac{0.886[B]\sqrt{k_1t_1/K_d}}{1 + 0.886[B]\sqrt{k_1t_1/K_d}} \tag{9}$$

$$\frac{\bar{\imath}_k}{\bar{\imath}_d - \bar{\imath}_k} = 0.886[B]\sqrt{k_1t_1/K_d} \tag{10}$$

To prove the validity of Eqs. (9) and (10) or oppositely to demonstrate that the system involved corresponds to the scheme (1') it is possible either to plot the values $\bar{\imath}_k/(\bar{\imath}_d - \bar{\imath}_k)$ measured at various concentrations against concentration [B] or to plot the value of the ratio $\bar{\imath}_k/\bar{\imath}_d$ recorded at various concentrations of B against log[B]. Experimental points are then compared with the theoretical dissociation curve calculated by means of Eq. (9) (Fig. 12).

To calculate the rate constant $k_1$, a simplified procedure can be used: the concentration of reagent B is determined at which the kinetic current is just half that of the total limiting diffusion current,[15] i.e., when $\bar{\imath}_k = \bar{\imath}_d/2$. The

[14] J. Weber and J. Koutecký, *Collection Czech. Chem. Commun.* **20**, 980 (1955).

[15] The knowledge of the magnitude of the total limiting diffusion current $(\bar{\imath}_d)$ is essential for the calculation of the rate constant $k_1$. If the value of $\bar{\imath}_d$ is not accessible experimentally, e.g., because the current $\bar{\imath}_k$ does not reach the value of $\bar{\imath}_d$ in the available range of log[B] and since the potential of reduction of the species A lies outside the available potential range, it is necessary to estimate the approximate value of $\bar{\imath}_d$ from

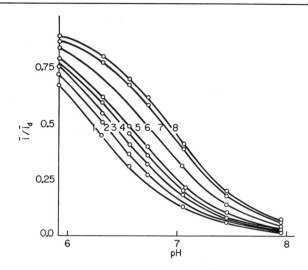

FIG. 12. Dependence of the ratio $\bar{i}_k/\bar{i}_d$ on pH. Reduction of phenylglyoxylic acid in phosphate buffers. Curves *1–8* obtained for various drop-times. Experimental points, theoretical curves. According to V. Hanuš (Thesis, 1950).

value of $-\log[B]$, at which the condition $\bar{i}_k = \bar{i}_d/2$ is fulfilled, is denoted as p$K'$. The value of the constant $k_1$ can then be calculated using Eq. (11)

$$\log k_1 = 2\mathrm{p}K' - \mathrm{p}K_d - 2\log 0.886 - \log t_1 \tag{11}$$

Equilibria corresponding to scheme (1′) can be either acid-base reactions or another type of equilibrium.

*a. Acid-Base Equilibria.* Acid-base reactions preceding the electrode process proper, for which the equilibria are established in a period comparable with the drop-time, can involve a monobasic, dibasic, or polybasic acid. Either only the proton (or the hydroxonium ion) or generally all acid components of the solution participate in the establishment of the acid-base equilibrium.

The simplest case of the reaction of a monobasic acid (HA)[16] with protons only will be considered first

---

the wave height of a model substance. This model substance is chosen so that the number of electrons consumed in the reduction of the model substance and the compound C is the same. Moreover, the model substance and compound C should be molecules of approximately the same molecular mass and of similar shape so that the values of the diffusion coefficient are similar.

[16] Symbol HA stands for the conjugate acid, symbol $A^-$ for the conjugate base; they are not aimed to indicate the charge of the species. Hence HA can be a neutral molecule, a cation or an anion, and similarly also the base $A^-$.

$$HA \underset{k_r}{\overset{k_d}{\rightleftharpoons}} A^- + H^+ \tag{12a}$$

$$HA + n_1e \overset{E_1}{\rightarrow} P_1 \tag{12b}$$

$$A^- + n_2e \overset{E_2}{\rightarrow} P_2 \tag{12c}$$

In order to apply the present treatment to a given acid-base equilibrium, it is necessary that at least one of the potentials $E_1$ (or $E_2$) be in the available potential range, i.e., that the acid or conjugate base give a polarographic wave. Moreover, a condition must be fulfilled that the half-wave potentials of the acid and the base are sufficiently different (i.e., that $E_1 \neq E_2$).

If these conditions are fulfilled, two waves appear on polarographic curves (Fig. 4). The wave which corresponds to the reduction of the acid form HA ($i_{HA}$) appears at potential $E_1$, that corresponding to the reduction of the conjugate base $A^-$ ($i_A$) can be observed at potential $E_2$ which is usually more negative than $E_1$. Sometimes the potential of the wave $i_A$ is so negative that the wave coalesces with the final rise of current of the supporting electrolyte. In this case only the more positive wave $i_{HA}$ is observed.

The more positive wave $i_{HA}$ corresponds to a transfer of $n_1$ electrons, the more negative wave $i_A$ to a transfer of $n_2$ electrons. If $n_1 = n_2$, the height of the sum of both waves remains constant, whereas the ratio of the wave heights $i_{HA}/i_A$ changes with pH as corresponds to Eq. (9), provided that solutions are either well buffered or, at both extremes of the acidity scale, contain an excess of acid or base to keep the concentration of $H^+$ or $OH^-$ constant even at the surface of the dropping electrode during the electrolysis. In these cases $[B] = [H^+]$ in Eq. (9). An example of polarographic behavior of this type is the reduction of phenylglyoxylic acid (Fig. 13).

To prove that polarographic curves recorded at various pH values correspond to a system similar to (12), experimentally measured values of $i_{HA}/(i_{HA} + i_A)$ are plotted against pH and compared with the theoretical curve, computed using Eq. (9) (Fig. 12). Rate constants can be calculated using Eq. (11).

So far no enzymatic reactions of the type (12) have been studied, but some of the systems studied participate in enzymatic processes. Among this group belong α-ketoacids,[17,18] pyridine carboxylic acids,[19,20] and some

[17] R. Brdička and K. Wiesner, Collection Czech. Chem. Commun. **12**, 138 (1947).
[18] R. Brdička, Collection Czech. Chem. Commun. **12**, 212 (1947).
[19] J. Volke and V. Volková, Collection Czech. Chem. Commun. **20**, 1332 (1955).
[20] J. Volke, Collection Czech. Chem. Commun. **22**, 1777 (1957).

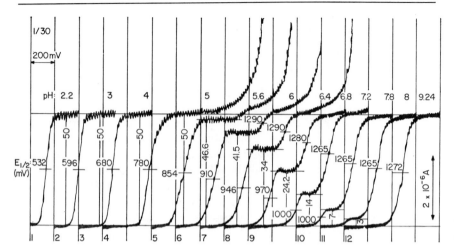

FIG. 13. Effect of pH on polarographic curves of phenylglyoxylic acid. McIlvaine buffers, pH given on the polarogram, $4.5 \times 10^{-4} M$ phenylglyoxylic acid. Curves starting at: (1)–(4), $-0.4$ V; (5)–(9), $-0.6$ V; (10)–(12) $-0.8$ V; 200 mV/absc., half-wave potentials given against the NCE, $h = 50$ cm; full-scale sensitivity 7 μA.

carbonyl compounds.[21,22] These systems are nevertheless complicated by the possible solvation of the carbonyl and carboxylic groups.

Moreover, the state of the compounds in the electrolyzed solution can be affected by tautomeric changes, and the concentration of the species participating in the reaction in the vicinity of the dropping electrode can be affected by adsorption of the acid or conjugate base form and/or of the buffer compounds. It is due to the latter factors that proton-transfer reactions can occur as surface reactions, in addition to the volume reaction which takes place in the so-called reaction volume in the vicinity of the electrode. All the above-mentioned factors can complicate the theoretical treatment, in particular the computation of the rate constants $k_1$. The presence of adsorption, desorption, attraction, and repulsion phenomena can result in a decrease of the limiting current with increasingly negative potentials.

In some other systems, e.g., iodophenols or iodoanilines, some of the complicating factors, like solvation changes and the existence of several tautomers, can be excluded, but the effects of adsorption are difficult to eliminate. To distinguish whether the reaction is only a volume reaction or takes place also as a surface reaction, the shape of the wave and the effects of neutral salts are followed. If the limiting current of a kinetic wave is not

[21] P. Zuman, J. Tenygl, and M. Březina, *Collection Czech. Chem. Commun.* **19**, 46 (1954).
[22] S. G. Majranovskii and V. N. Pavlov, *Zh. Fiz. Khim.* **38**, 1804 (1964).

decreasing considerably with increasingly negative potentials,[23] over a potential range of several tenths of a volt, it can be assumed that a volume reaction is predominating. The limiting currents and the shape of the plot of their dependence on pH for a process taking place predominantly as a volume reaction are usually only slightly dependent on the concentration and kind of an added neutral salt. The solvolysis of the tropylium ion[24] is one of the acid-base reactions for which it has been proved, using the above diagnostic proofs, that the chemical reaction accompanying the electrode process proper takes place as a volume reaction.

When the half-wave potentials $E_1$ and $E_2$ differ so little that waves $i_{HA}$ and $i_A$ merge, it proved useful to plot $\log(i_d - i)/i$ against potential $E$ for curves recorded at various pH values. These plots (called logarithmic analysis) show two linear sections of different slope (Fig. 14), the ratio of which changes with pH. The ratio of the two linear sections is plotted

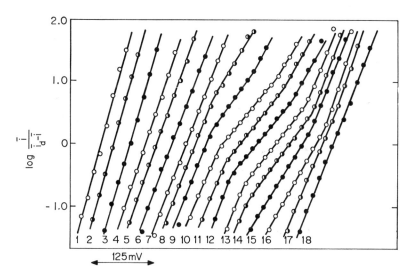

FIG. 14. Effect of pH on logarithmic analysis of polarographic waves of $p$-diacetyl-benzene. Dependence of $\log i/(i_d - i)$ on potential for different pH values. $2 \times 10^{-4}\,M$ $p$-diacetylbenzene, 30% ethanol. Curve (1), 5 N sulfuric acid; (2) 1 N sulfuric acid; curves (3)–(18), Britton-Robinson buffers of pH: (3) 1.95, (4) 2.8, (5) 3.8, (6) 4.8, (7) 5.5, (8) 6.1, (9) 6.5, (10) 7.0, (11) 7.2, (12) 7.4, (13) 7.7, (14) 7.9, (15) 8.65, (16) 9.15, (17) 9.8, (18) 10.35.

[23] To rule out the effect of the change in the drop-time with applied voltage, it is preferable to check the independence of the limiting current of potential when using a mechanically controlled drop-time.

[24] P. Zuman, J. Chodkowski, and F. Šantavý, *Collection Czech. Chem. Commun.* **26,** 380 (1961).

against pH and can be evaluated in a similar way to the $i_{HA}/(i_{HA} + i_A)$ — pH plot using Eq. (9).

In some instances the heights of wave $i_{HA}$ depend not only on the pH value, but also on the composition of the buffer used. This indicates that the conjugate base $A^-$ reacts not only with the hydroxonium ion as the sole proton donor, but also with other types of proton donors. Scheme (13) is then operating

$$HA + B_1 \underset{k_r{}^1}{\overset{k_d{}^1}{\rightleftharpoons}} B_1H + A^- \tag{13a}$$

$$HA + B_2 \underset{k_r{}^2}{\overset{k_d{}^2}{\rightleftharpoons}} B_2H + A^- \tag{13b}$$

$$HA + B_n \underset{k_r{}^n}{\overset{k_d{}^n}{\rightleftharpoons}} B_nH + A^- \tag{13c}$$

$$HA + n_1e \overset{E_1}{\rightarrow} P_1 \tag{13d}$$

$$A^- + n_2e \overset{E_2}{\rightarrow} P_2 \tag{13e}$$

In the most commonly used buffers the change in pH results from the simultaneous change in the concentration of the acid or base or both of the buffer components. Sometimes complex (so called universal) buffers are used that contain several acids or bases. In such buffers the wave $i_{HA}$ shows the tendency to decrease when the pH value is increased. Nevertheless, the shape of such a dependence is not easily evaluated. Simple buffers prepared by keeping the concentration of one buffer component constant and changing only the concentration of the other component are therefore preferred. When the wave height $i_{HA}$ depends on the concentration of all the acid present, the buffer is prepared by keeping the concentration of the base component constant and the pH is changed by changing the concentration of the acid component. Only under these conditions does the dependence of $i_{HA}/(i_{HA} + i_A)$ on pH possess the form of a dissociation curve corresponding to Eq. (9). The inflexion point of this curve, $pK'$, corresponds to the pH for which $i_{HA} = \frac{1}{2}(i_{HA} + i_A)$ but is shifted to higher pH values with increasing concentration of the base buffer component. Extrapolation to zero buffer concentration allows determination of $k_{H^+}$ unaffected by $k_{B_n}$. The value of $pK'$ depends also on the buffer type.[25]

When, on the other hand, the buffer is prepared by keeping the acid component concentration constant and changing the concentration of the base component, the shape of the pH dependence of the current $i_{HA}$ is more

[25] P. Zuman and B. Turcsanyi, *Collection Czech. Chem. Commun.* **33**, 3090 (1968).

complicated than described by Eq. (9), as it is in accordance with the theory.[25]

The simplest proof that the height of the wave $i_{HA}$ depends on the concentration of all acids, and not only on hydrogen ion concentration, is the increase of this wave $i_{HA}$ with increasing buffer concentration at a given pH. At a given ratio of the acid and base buffer component, the concentration (capacity) of the buffer is changed. An increase in the value of $i_{HA}$ indicates participation of the acid buffer component in reaction (13e). From the slope of the dependence of $i_{HA}$ on $[B_n]$ it is possible to determine the rate constant $k_d$. Independence of the wave height $i_{HA}$ of buffer concentration indicates a reaction in which specifically only H⁺ ions participate.

Another proof for the participation of other proton donors is based on comparison of wave $i_{HA}$ recorded in buffers of the same pH, prepared from various acids or bases. If, for example, in acetic acid-sodium acetate buffer pH 4.7 and in pyridine–pyridinium chloride buffer pH 4.7 heights of wave $i_{HA}$ are identical, only hydrogen ions participate in reaction (13a). Different heights indicate the participation of various acids according to (13c).

For dibasic acids three waves $i_{H_2A}$, $i_{HA}$, and $i_A$ are sometimes observed. In strongly acid media, the wave $i_{H_2A}$ predominates. With increasing pH the height of the wave $i_{H_2A}$ successively decreases and wave $i_{HA}$ increases. At still higher pH values, $i_{HA}$ decreases and $i_A$ increases. Usually the sum of the heights $(i_{H_2A} + i_{HA} + i_A)$ remains practically constant.

For the wave $i_{HA}$ the dependence of $i_{HA}/(i_{H_2A} + i_{HA} + i_A)$ on pH has the usual shape of a dissociation curve corresponding to Eq. (9), the slope of which corresponds to a decrease in the value of $i_{HA}/(i_{H_2A} + i_{HA} + i_A)$ from 0.9 to 0.1 over about 2 pH units. The decrease of the wave $i_{H_2A}$ with increasing pH is steeper, and the value of $i_{H_2A}/(i_{H_2A} + i_{HA} + i_A)$ decreases from 0.9 to 0.1 over about 1 pH unit.

Such behavior is shown only by those dibasic acids which are reduced according to scheme (14) and which moreover fulfill the condition that $pK_2 > (pK_1 + 2)$, where $K_1$ and $K_2$ are the first and second thermodynamic dissociation constants. The second condition is $pK'_1 > pK_2$, where $pK'_1$ corresponds to the pH value at which the limiting current $i_{H_2A}$ reaches half of the value of the diffusion current, corresponding to $(i_{H_2A} + i_{HA} + i_A)$.

$$H_2A \underset{k_r^1}{\overset{k_d^1}{\rightleftharpoons}} HA^- + H^+ \qquad K_1 = \frac{k_d^1}{k_r^1} \qquad (14a)$$

$$HA^- \underset{k_r^2}{\overset{k_d^2}{\rightleftharpoons}} A^{2-} + H^+ \qquad K_2 = \frac{k_d^2}{k_r^2} \qquad (14b)$$

$$H_2A + n_1 e \overset{E_1}{\rightarrow} P_1 \qquad i_{H_2A} \qquad (14c)$$

$$\text{HA}^- + n_2 e \xrightarrow{E_2} \text{P}_2 \qquad\qquad \bar{i}_{\text{HA}} \qquad\qquad (14\text{d})$$

$$\text{A}^{2-} + n_3 e \xrightarrow{E_3} \text{P}_3 \qquad\qquad \bar{i}_{\text{A}} \qquad\qquad (14\text{e})$$

The system can be studied quantitatively, if potentials $E_1$, $E_2$, and $E_3$ differ sufficiently to allow measurement of waves $i_{\text{H}_2\text{A}}$, $i_{\text{HA}}$, and $i_{\text{A}}$. If furthermore the number of electrons are the same, $n_1 = n_2 = n_3$, it can be shown[26,27] that the wave height can be described by Eq. (15)

$$\frac{\bar{i}_k}{\bar{i}_d - \bar{i}_k} = 0.886 \sqrt{\frac{k_r^{~1} t_1}{K_1}} \frac{[\text{H}^+]^2}{\left([\text{H}^+] + K_2 \dfrac{D_{\text{H}_2\text{A}}}{D_{\text{HA}}}\right)^{1/2} ([\text{H}^+] + K_2)^{1/2}} \qquad (15)$$

Assuming equal diffusion coefficients $D_{\text{H}_2\text{A}} = D_{\text{HA}}$, the value of the first recombination constant $k_r^{~1}$ can be calculated using Eq. (16)

$$\log k_r^{~1} = \text{p}K'_1 - \text{p}K_1 + \tfrac{1}{2}\log(K'_1 + K_2) - \tfrac{1}{2}\log 0.886 - \log t_1 \quad (16)$$

Among biochemically important compounds that are dibasic acids, this treatment was applied to maleic and fumaric acids,[26] to phthalic acid,[28] and to pyridoxine derivatives[29] which can exist in the completely protonated pyridinium form, as neutral molecule (or zwitterion) and in the anionic form, with unprotonated pyridine ring and dissociated phenolic group.

*b. Other Reactions.* Among other types of organic reactions antecedent to the electrode process, proper equilibria between hydrated and non-hydrated as well as cyclic and acyclic forms were studied in some detail. The rate of these reactions is in most cases pH-dependent and therefore the kinetic currents of this type change with pH. The pH-dependence of the rate of the chemical reaction governing the electrode process, by which the electroactive form AH is produced, from the electroinactive BH, can principally be caused by three types of systems: (a) the current governing the reaction is preceded by a rapidly established acid-base equilibrium (17); (b) the electroactive form is generated by an acid-base reaction the rate of which is governing the current (18); (c) the rate of the chemical reaction is subject to a general acid-base catalysis (19):

$$\text{BH} \underset{\text{fast}}{\overset{\longrightarrow}{\longleftarrow}} \text{B} + \text{H}^+ \qquad\qquad (17\text{a})$$

$$\text{BH} \xrightarrow[\text{slow}]{k} \text{AH} \xrightarrow{E} \text{P} \qquad\qquad (17\text{b})$$

[26] V. Hanuš and R. Brdička, *Chem. Listy* **44**, 291 (1950); *Khimiya* (*Prague*) **2**, 28 (1951).
[27] J. Koutecký, *Collection Czech. Chem. Commun.* **19**, 1093 (1954).
[28] A. Ryvolová and V. Hanuš, *Collection Czech. Chem. Commun.* **21**, 853 (1956).
[29] O. Manoušek and P. Zuman, *Collection Czech. Chem. Commun.* **29**, 1432 (1964).

$$BH \underset{slow}{\overset{\longrightarrow}{\longleftarrow}} B + H^+ \tag{18a}$$

$$BH \xrightarrow[fast]{} AH \overset{E}{\rightarrow} P \tag{18b}$$

$$BH \xrightarrow[general\ catalysis]{k_i} AH \overset{E}{\rightarrow} P \tag{19}$$

For reactions belonging to the type (17) the current depends on the position of the equilibrium (17a). The height of the polarographic current depends on pH in the shape of a dissociation curve; the pH value at the inflection point of this curve is identical with the $pK$ value of equilibrium (17a). For reactions corresponding to the type (18), the wave height increases linearly with the concentration of hydrogen ions.

For general catalyzed reactions of type (19) an increase of current toward both the acid and the alkaline region is observed and depends on buffer composition. The observed pH dependence can show a U-shape, is bell-shaped, goes through a maximum or can possess the form of a dissociation curve. The types (17) and (18) can also occur simultaneously or be accompanied by the acid-base reactions of the type mentioned in Section $a$ above.

The most thoroughly studied examples of type (19) are hydration-dehydration equilibria. These reactions can be characterized by a general equation (20)

$$A \underset{k_{-1}}{\overset{k_1}{\rightleftharpoons}} C \overset{el}{\rightarrow} P \tag{20}$$

This scheme is analogous to (3) with the exception that reagent B in Eq. (1') is the solvent. Hence its concentration is considered as unity. In the relation $k' = k_1[B]$ the concentration of the solvent is included in the value of $k_1$ and hence $k' = k_1$. Whereas in Eq. (3) the rate constant involved is $k'_1$, in Eq. (20) it is $k_1$. Therefore Eqs. (9) and (10) are simplified into Eqs. (21) and (22):

$$\frac{\bar{\imath}_k}{(\bar{\imath}_d)_A} = \frac{0.886 \sqrt{k_1 t_1 / K_d}}{1 + 0.886 \sqrt{k_1 t_1 / K_d}} \tag{21}$$

$$\frac{\bar{\imath}_k}{(\bar{\imath}_d)_A - \bar{\imath}_k} = 0.886 \sqrt{k_1 t_1 / K_d} \tag{22}$$

In Eqs. (21) and (22) $(i_d)_A$ stands for the hypothetical diffusion current of the form A. For the general acid-base catalyzed process, the value of $k_1$ is defined as for other catalyzed reactions by Eq. (23):

$$k_1 = k_s + k_{H_3O^+} [H_3O^+] + k_{OH^-}[OH^-] + \sum_i k_{B_i}[B_i] + \sum_j k_{HA_j}[HA_j] \quad (23)$$

where $k_s$ corresponds to the reaction with the solvent, $k_{H_3O^+}$ with hydrogen ions and $k_{OH^-}$ with hydroxyl ions, $k_{B_i}$ with base $B_i$, $k_{HA_j}$ with acid $HA_j$. For the case when $k_{H_3O^+}[H_3O^+] \gg (k_s + k_{OH^-}[OH^-] + \Sigma k_{B_i}[B_i] + \Sigma k_{HA_j}[HA_j])$ or when $k_{OH^-}[OH^-] \gg (k_s + k_{H_3O^+} [H_3O^+] + \Sigma k_{B_i}[B_i] + \Sigma k_{HA_j}[HA_j])$, Eq. (21) changes into (9) for $[B] = [H_3O^+]$ or $[OH^-]$. In such case, the shape of the pH dependence of $\bar{\imath}_k/(\bar{\imath}_d)_A$ resembles that of a dissociation curve reaching a limiting value when $\bar{\imath}_k = (\bar{\imath}_d)_A$. This is fulfilled when 1 can be neglected against $0.886 \sqrt{k_1 t_1/K_d}$.

The example studied in most detail is the polarographic behavior of aliphatic aldehydes, in particular of formaldehyde. A considerable proportion of these substances exists in solution as the electroinactive hydrated form $RCH(OH)_2$, from which the electroactive aldehydic form $RCHO$ is generated by a reaction (24) with a constant $k_1$, the rate of which governs the heights of the polarographic wave. The plot of the pH dependence of the limiting current shows a maximum (Fig. 15). This indicates a general base catalysis corresponding to scheme (19) which can be expressed for this particular case as (24)

$$RCH(OH)_2 \underset{k_{-1}}{\overset{k_1}{\rightleftharpoons}} RCHO \overset{el}{\rightarrow} P \quad (24)$$

For the calculation of the rate constant $k_1$ the dehydration equilibrium constant $K_d$ defined as $K_d = k_1/k_{-1}$ must be determined using an independent method, e.g., single-sweep techniques.[30] The curve (Fig. 15) calculated using the determined specific rate constants for the reaction catalyzed by hydroxyl, phosphate, and borate ions is in a good agreement with the experimental points. Moreover, values of the dehydration rate constants $k_1$ determined by other methods[31] are in a relatively good agreement with values of rate constants obtained by polarography.

In some instances the acid-base catalyzed dehydration is combined with acid-base equilibria, either preceding or consecutive to the dehydration process. In such a way the plots of the pH dependence of waves of pyridine carboxaldehydes[29,32,33] and of glyoxalic acid[34] were interpreted. Different dehydration rates were obtained for the protonated and the free base forms of pyridine derivatives and similarly for the neutral glyoxalic acid molecule

[30] P. Valenta, *Collection Czech. Chem. Commun.* **25,** 855 (1960).
[31] R. P. Bell and P. G. Evans, *Proc. Roy. Soc. (London)* **A291,** 297 (1966).
[32] J. Volke, *Z. Physik. Chem. (Leipzig), (Sonderheft)* 268 (1958).
[33] J. Tirouflet and E. Laviron, *Ric. Sci.* **29** (Suppl. 4), 189 (1959).
[34] J. Kůta, *Collection Czech. Chem. Commun.* **24,** 2532 (1959).

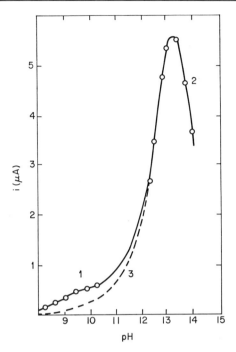

Fig. 15. pH dependence of limiting currents of formaldehyde. $4 \times 10^{-3}$ $M$ formaldehyde in borate buffers (curve *1*) and sodium hydroxide solutions (curve *2*). Dashed curve *3* corresponds to current due to the reaction with hydroxyl ions in borate buffers. Lines correspond to theoretical values, points indicate experimental data. [From P. Zuman, *Advan. Phys. Org. Chem.* **5**, 43 (1967).]

and its anion. For numerous aldehydic substances the shape of the pH-dependence indicates an effect of the hydration-dehydration equilibria, but the quantitative treatment has not been used so far.

Another more complex system involving dehydration is found[35] for aliphatic $\alpha$-dicarbonyl compounds, such as diacetyl (R = CH₃) or methylglyoxal (R = H), for which two consecutive dehydrations take place (25)

$$RC(OH)_2C(OH)_2CH_3 \underset{H_2O}{\overset{k_1}{\rightleftarrows}} RCOC(OH)_2CH_3 \tag{25a}$$

$$RCOC(OH)_2CH_3 \underset{H_2O}{\overset{k_2}{\rightleftarrows}} RCOCOCH_3 \tag{25b}$$

$$RCOC(OH)_2CH_3 \overset{el}{\underset{i_2,E_2}{\longrightarrow}} P_1 \tag{25c}$$

[35] M. Fedoronko, private communication (1966).

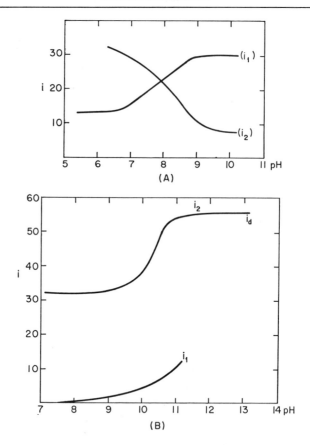

Fɪɢ. 16. Dependence of limiting currents ($i$) of (A) diacetyl and (B) methylglyoxal on pH. $i_1$ is the more positive wave of the fully dehydrated form; $i_2$ is the more negative wave of the monohydrated form. According to M. Fedoronko.

$$\text{RCOCOCH}_3 \xrightarrow[i_1, E_1]{\text{el}} \text{P}_2 \tag{25d}$$

The diketone is reduced at potentials ($E_1$) more positive than the species with one hydrated carbonyl group ($E_2$). The differing course of the pH dependence of waves $i_1$ and $i_2$ (Fig. 16) according to the nature of the group R can be interpreted as the effect of group R on the position of equilibria (25a).

The second type of chemical equilibria preceding the electrode process proper are those between cyclic and noncyclic forms. These systems can be studied when the half-wave potentials of the cyclic and open-chain forms differ. The rate of the establishment of the equilibria between cyclic and open-chain forms can again be acid-base catalyzed. Equilibria between open-chain and acetal form of pyridoxal were considered,[29,32] but this system is complicated by the possibility of hydration of the aldehydic group. Greatest interest has been paid to monosaccharides, even when their behavior is rather complex. The suggested treatment is based on the assumption that only the open-chain form of the sugar bearing an aldehydic group is electroactive, while all the cyclic forms are electroinactive.[36] For glucose a theoretical treatment has been developed[37,38] based on scheme (26)

$$\alpha\text{-pyranose} \underset{k_{-1}}{\overset{k_1}{\rightleftharpoons}} \diagup\hspace{-0.3em}C{=}O \underset{k_{-2}}{\overset{k_2}{\rightleftharpoons}} \beta\text{-pyranose} \tag{26}$$

According to this scheme two cyclic pyranose forms are in equilibrium with the noncyclic form bearing a free carbonyl group. From the pH dependence of the wave of the carbonyl form, as well as from the time-change in the height of the reduction wave after dissolving the sugar, it was possible to compute the rate constant of the mutarotation $(k)$ according to Eq. (27)

$$k = \frac{k_1 k_{-2} + k_{-1} k_2}{k_{-1} + k_{-2}} \tag{27}$$

When the value for the ratio of equilibrium concentrations of $\alpha$- and $\beta$-pyranose is used, it is possible to calculate the values of all four rate constants $k_1$, $k_2$, $k_{-1}$, and $k_{-2}$. The proportion of the free aldehyde form was found to be 0.003% of the analytical concentration of glucose.

## 2. Parallel Reactions

In parallel reactions the product of an electrode process (C) reacts rapidly with a component of the solution (B) to regenerate the original electroactive form (A). On polarographic curves an increase of the current at potentials $(E)$ corresponding to electroreduction of species A is observed. This system can be described by scheme (28)

$$A + ne \overset{E}{\rightleftharpoons} C \tag{28a}$$

$$C + B \underset{k_{-1}}{\overset{k_1}{\rightleftharpoons}} A \tag{28b}$$

When the component B is present in great excess Eq. (28b) becomes (29)

[36] K. Wiesner, *Collection Czech. Chem. Commun.* **12**, 64 (1947).

[37] J. M. Los, L. B. Simpson, and K. Wiesner, *J. Am. Chem. Soc.* **78**, 1564 (1956).

[38] J. Paldus and J. Koutecký, *Collection Czech. Chem. Commun.* **23**, 376 (1958).

$$A + ne \overset{E}{\rightleftharpoons} C \qquad (29a)$$

$$C \overset{k'_1}{\underset{k_{-1}}{\rightleftharpoons}} A \qquad (29b)$$

where $k'_1 = k_1[B]$. The current governed by the rate of reaction (29b) is usually called a regeneration or a catalytic current. Two conventions exist for the description of the catalyst: Either the fact is taken into consideration that at potential E species A is reduced. Substance B is then called a catalyst, increasing catalytically the number of particles E undergoing the reduction. Or a shift in the potential at which the reduction of B occurs is taken into consideration (even when at the potential E a chemical and not a direct electrochemical reduction occurs). Then the couple A/C is considered to be the catalyst for the reduction or oxidation of the substance B, which is—within the given potential range—polarographically inactive.

Koutecký[39] assumed equal diffusion coefficients of forms A and C, neglected the chemical reaction in the opposite direction with rate constant $k_{-1}$ (i.e., when $k_{-1} \ll k'_1$), and considered Eqs. (30) and (31).

$$\frac{\partial[A]}{\partial t} = D \frac{\partial^2[A]}{\partial x^2} + \frac{2x}{3t} \frac{\partial[A]}{\partial x} + k'_1[C] \qquad (30)$$

$$\frac{\partial[C]}{\partial t} = D \frac{\partial^2[C]}{\partial x^2} + \frac{2x}{3t} \frac{\partial[C]}{\partial x} - k'_1[C] \qquad (31)$$

Again, as for (4) and (5), the first term on the right-hand side corresponds to diffusion, the second expresses the correction for the growth of the dropping electrode, and the third term expresses the perturbation caused by the chemical reaction with first-order rate constant $k'_1 = k_1[B]$. The boundary conditions are:

$$t = 0, x > 0 \quad : \quad [A] = [A]_b; \qquad [C] = 0$$

$$t > 0, x = 0 \quad : \quad \frac{\partial[A]}{\partial x} + \frac{\partial[C]}{\partial x} = 0; \qquad [A] = 0$$

where the index b indicates the concentration in the bulk of the solution. Koutecký[39] solved these equations using the method of dimensionless parameters and tabulated the ratio of the mean limiting current ($\bar{\imath}_{lim}$) to the mean diffusion current ($\bar{\imath}_d$) (Table I)

$$\bar{\imath}_{lim}/\bar{\imath}_d = F(k'_1 t_1) \qquad (32)$$

For $k'_1 t_1 > 10$ the function can be expressed in an asymptotic form (33).

$$\bar{\imath}_{lim}/\bar{\imath}_d = 0.81 \sqrt{k'_1 t_1} \qquad (33)$$

[39] J. Koutecký, Collection Czech. Chem. Commun. 18, 311 (1953).

TABLE I
DEPENDENCE OF THE RATIO $i_{\lim}/i_d$ ON THE PRODUCT
$k'_1 t_1$ FOR REGENERATION PROCESS, EQ. (29)

| $F(k'_1 t_1)$ | $i_{\lim}/i_d$ | $F(k'_1 t_1)$ | $i_{\lim}/i_d$ | $F(k'_1 t_1)$ | $i_{\lim}/i_d$ |
|---|---|---|---|---|---|
| 0 | 1.0 | 1.2 | 1.297 | 4.0 | 1.84 |
| 0.05 | 1.013 | 1.4 | 1.342 | 5.0 | 2.01 |
| 0.1 | 1.027 | 1.6 | 1.386 | 6.0 | 2.17 |
| 0.2 | 1.054 | 1.8 | 1.427 | 7.0 | 2.31 |
| 0.4 | 1.104 | 2.0 | 1.47 | 8.0 | 2.45 |
| 0.6 | 1.154 | 2.5 | 1.56 | 9.0 | 2.57 |
| 0.8 | 1.204 | 3.0 | 1.66 | 10.0 | 2.69 |
| 1.0 | 1.250 | 3.5 | 1.75 | — | — |

Comparison of the function $F(k'_1 t_1)$ (Fig. 17, curve $1$) with the asymptotic solution according to Eq. (33) (Fig. 17, curve $2$) indicates that both treatments give virtually the same result provided that the limiting catalytic current is at least three times greater than the limiting diffusion current of substance A, in the absence of substance B.

To prove that the system studied corresponds to scheme (28) measured values of $i_{\lim}/i_d$ determined at various [B] are plotted either against the function $F(k'_1 t_1)$ (Table I) or against $\sqrt{k'_1 t_1}$, having in mind that $k'_1 =$

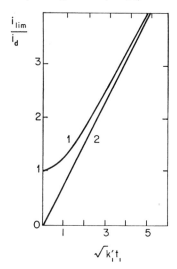

FIG. 17. Dependence of $i_{\lim}/i_d$ on $F(k'_1 t_1)$. Curve $1$ corresponds to the exact solution of a system $A + ne \overset{E}{\rightleftharpoons} C; \ C \overset{k'_1}{\underset{k_{-1}}{\rightleftharpoons}} A$; curve $2$ corresponds to the asymptotic solution for $k'_1 t_1 > 10$, when $F(k'_1 t_1) = 0.81 \, (k_1 t_1)^{1/2}$.

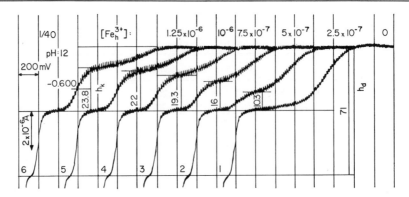

Fig. 18. Catalytic waves observed when hemin was added to oxygen-containing solutions. Britton-Robinson buffer pH 12, hemin ($Fe_h^{3+}$) added in concentrations shown on the polarogram. Curves started at 0.0 V, 200 mV/absc., Hg-pool electrode; full-scale sensitivity 8.5 $\mu$A; wave heights in mm, 11 mm corresponding to 1 $\mu$A.

$k_1[B]$. The plots can be at the same time used for the determination of the best value of the rate constant $k_1$.

Most of the reactions of this type studied belong to inorganic systems, such as Fe(III)/Fe(II) in the presence of hydrogen peroxide. In a similar way the reaction of hemin[40] and hydrogen peroxide was studied. In this case a preparation of hemin ($\alpha$-chloroprotohemin or $\alpha$-hematin) was added to buffered aqueous or alkaline solutions in which air was dissolved under atmospheric pressure. In aqueous solutions containing dissolved oxygen the molecule of oxygen is reduced in two 2-electron steps. In the first step at about $-0.1$ V oxygen is reduced to hydrogen peroxide. In the second, drawn-out wave at about $-0.9$ V, hydrogen peroxide is reduced further to water. Hence at potentials between about $-0.2$ V and $-0.7$ V hydrogen peroxide is formed electrochemically at the surface of the electrode.

After addition of hemin a wave was observed between the first and second oxygen wave; this wave increased with hemin concentration, and occurred at potentials similar to those of hemin reduction (Fig. 18). At potentials more negative than that of the first oxygen wave, all the oxygen at the surface of the electrode is transformed electrochemically into hydrogen peroxide. In the presence of hemin, ($Fe^{III}h$) this hydrogen peroxide is consumed in the reaction with the reduced form of hemin ($Fe^{II}h$), according to scheme (34)

$$O_2 + 2e + 2H^+ \xrightarrow{E_1} H_2O_2 \tag{34a}$$

$$Fe^{III}h + H^+ \rightleftharpoons Fe^{III}hH^+ \tag{34b}$$

[40] R. Brdička and K. Wiesner, *Collection Czech. Chem. Commun.* **12**, 39 (1947).

$$\text{Fe}^{III}h\text{H}^+ + e \underset{\longleftarrow}{\overset{E_2}{\rightharpoonup}} \text{Fe}^{II}h \tag{34c}$$

$$\text{Fe}^{II}h + \text{H}_2\text{O}_2 \xrightarrow{k_1} \text{Fe}^{III}h + \text{OH}^- + \cdot\text{OH} \tag{34d}$$

$$\text{Fe}^{II}h + \cdot\text{OH} \xrightarrow{\text{fast}} \text{Fe}^{III}h + \text{OH}^- \tag{34e}$$

$$\text{H}_2\text{O}_2 + 2e + \text{H}^+ \xrightarrow{E_3} \text{H}_2\text{O} + \text{OH}^- \tag{34f}$$

On Fig. 18 the first wave of oxygen occurs at $E_1$, the catalytic one in the vicinity of $E_2$, and the hydrogen peroxide wave at $E_3$. The participation of the proton transfer (34b) can be deduced from the shift of half-wave potential of hemin with pH (cf. p. 144). It cannot be excluded that the decrease of the hemin wave at pH > 9 can be affected by equilibrium (34b). Because the hydroxyl radical •OH can undergo termination reaction by other reactions than the reaction (34e), considered in the original paper,[40] this reaction will be neglected. If considered, the factor 2 would be involved in further calculation, as one molecule of hydrogen peroxide would produce 2 molecules of the oxidized form of hemin (34g)

$$2\text{Fe}^{II}h + \text{H}_2\text{O}_2 \xrightarrow{k_1} 2\text{Fe}^{III}h + 2\text{OH}^- \tag{34g}$$

The experimental material available[40] does not allow one to check the application of Eqs. (32) and (33). Namely, the reaction was carried out in many instances under conditions when a greater part of the hydrogen peroxide present was consumed during the process and hence the condition of its unchanged concentration, [B] = const in Eq. (28b), was not fulfilled and the wave height is not governed solely by the reaction rate. When we restrict ourselves to conditions when $i_k < (i_{\text{H}_2\text{O}_2}/2)$, it can be assumed in the first approximation that the above condition is fulfilled and reaction (34d) can be expressed in the approximate form (34h)

$$\text{Fe}^{II}h \xrightarrow{k'_1} \text{Fe}^{III}h + \text{products} \tag{34h}$$

where $k'_1 = k_1[\text{H}_2\text{O}_2]$. Further difficulty in verification of the validity of Eqs. (32) and (33) for the system hemin–hydrogen peroxide is caused by the fact that the experimental data available[40] do not give the dependence of the kinetic current ($\bar{i}_k$) on the hydrogen peroxide concentration (i.e., the waves were not recorded at various oxygen concentrations).

An approximate calculation of the rate constant under the assumption of scheme (29b)—corresponding to (34a), (34b, fast), (34c), and (34h)—will be discussed next. Because the reduction wave of hemin ($\bar{i}_d)_{\text{He}}$ is about $10^3$ times smaller than the observed $\bar{i}_k$, conditions for application of Eq. (33) rather than (32) are fulfilled. Introducing $(\bar{i}_d)_{\text{He}} = \kappa C_{\text{He}}$, Eq. (33) can be expressed as (35)

$$\bar{\imath}_k/c = \kappa 0.81 \sqrt{k'_1 t_1} \qquad (35)$$

At pH $< 11.6$ for hemin concentrations below $7.5 \times 10^{-7} M$ the value $\bar{\imath}_k/c$ was extrapolated (for current in amperes and concentration in moles/ml) to zero hemin concentration and a value $\bar{\imath}_k/c_{He} = 6.5 \times 10^3$ was found. For $t_1 = 3.4$ sec, $n = 1$, $m = 0.0028$ g/sec and an estimated value $D_{He} = 6.6 \times 10^{-7}$ cm$^2$ sec$^{-1}$ the value of $\kappa = 1.2$ was obtained, leading to $k'_1 = 1.3 \times 10^7$ sec$^{-1}$. Because $k'_1 = k_1[H_2O_2] = 2.7 \times 10^{-4} k_1$, the approximate value $k_1 = 5 \times 10^{10}$ l mole$^{-1}$ sec$^{-1}$ was obtained.

The actual scheme can nevertheless be more complicated than is indicated in scheme (34). The limiting value reached by $\bar{\imath}_k$ with increasing hemin concentration (Fig. 19) decreases at pH above 11.5 in the form of a dissociation curve (Fig. 20). This was attributed[40] to the dissociation of hydrogen peroxide (36) under the assumption that only the undissociated peroxide molecules react with hemin.

$$H_2O_2 \rightleftharpoons HO_2^- + H^+ \qquad (36)$$

It was shown that the decrease of the limiting value of current $\bar{\imath}_k$ on pH closely follows the dissociation curve of hydrogen peroxide, calculated for[41]

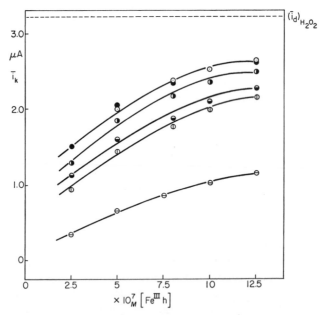

Fig. 19. Dependence of the limiting current of the catalytic wave ($\bar{\imath}_k$) on hemin (Fe$^{III}h$) concentration at various pH values Britton-Robinson buffers pH: ⊙ 8.7; ● 9.9; ◕ 11.2; ◖ 11.6; ⦶ 12.0; and 0.1 ⊖ $N$ NaOH.

[41] R. A. Joyner, Z. Anorg. Allgem. Chem. **77**, 103 (1912).

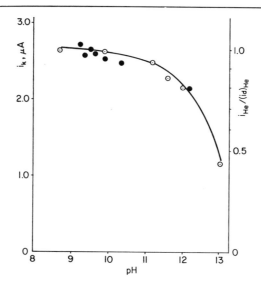

Fig. 20. Dependence of limiting currents on pH. ⊙ Catalytic wave for $1.25 \times 10^{-4}\ M$ hemin where its height reaches practically its limiting value; ● the height of the hemin reduction wave expressed as ratio $i_{He}/(i_d)_{He}$ when the current at pH $< 9.6$ was put equal to the diffusion current $(i_d)_{He}$.

$K = 1.2 \times 10^{-12}$. Because, on the other hand, reduction waves of hemin decrease in a similar way, the role of equilibrium (34b) cannot be excluded.

Half-wave potentials of the catalytic wave $i_k$ change with pH in a similar way to the half-wave potentials of hemin (Fig. 21) (provided that $i_k \ll i_{H_2O_2}$) nevertheless they are shifted toward more positive values. The shift is the greater in accordance with the theory,[40] the higher is the concentration of hemin. For sufficiently high hemin concentration, a shift of 0.058 V for a 10-fold increase in concentration was predicted. The increase from $2.5 \times 10^{-6}$ to $2.5 \times 10^{-5}\ M$ caused a shift of 0.060 V; that from $2.5 \times 10^{-5}\ M$ to $2.5 \times 10^{-4}\ M$, a shift of 0.072 V (but the wave was already shifted into the region of first oxygen wave). It cannot be excluded that these and some of the above complications are caused by the fact that polymerization of hemin cannot be completely neglected.

In an analogous way, catalytic waves were observed[40] in the presence of some other complexes of iron, such as salicylaldehyde–ethylenediimine ferric complex and the analogous hydrazine complex. The rate of the reaction of the reduced form of the former complex with hydrogen peroxide is comparable with the rate of hemin.

The effects of hemoglobin and cytochrome $c$ on the reduction waves of oxygen were essentially the same as that of hemin.

Another complex system was observed when to a solution of a defined

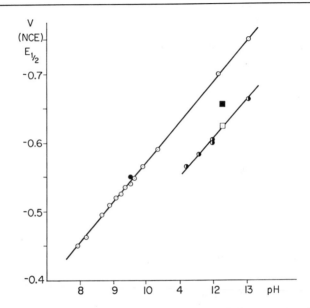

FIG. 21. Dependence of half-wave potentials on pH. ⊙ $1 \times 10^{-3}$ $M$ hemin, reduction wave, Britton-Robinson buffers; ● $1 \times 10^{-3}$ $M$ hemin, reduction wave, ammoniacal buffers; ◑ $1 \times 10^{-6}$ $M$ hemin, Britton-Robinson buffers, catalytic wave; ▫ $2.5 \times 10^{-6}$ $M$ hemin 0.02 $N$ NaOH, catalytic wave; ■ $2.5 \times 10^{-7}$ $M$ hemin, 0.02 $N$ NaOH, catalytic wave.

pH, containing oxygen dissolved at atmospheric pressure, a preparation of catalase from beef liver was added. On polarographic curves an increase of the first reduction wave of oxygen was observed[42] with increasing catalase concentration (Fig. 22). The chemical reaction responsible for this increase in current is assumed to be the interaction of catalase with the hydrogen peroxide produced at the surface of the electrode during the oxygen reduction. The system involved can be described by scheme (37)

$$O_2 + 2e + 2H^+ \rightleftharpoons H_2O_2 \tag{37a}$$

$$H_2O_2 + \text{catalase} \xrightarrow{k_1} H_2O + \tfrac{1}{2}O_2 \tag{37b}$$

This system is described by the set of differential equations, (30) and (31), in which A stands for oxygen and C for hydrogen peroxide. The initial and the boundary conditions are the same as for Eqs. (30) and (31). Because in the reaction (37b) from two molecules of the reduction product (hydrogen peroxide) only one molecule of the oxidized form (oxygen) is regenerated,

[42] R. Brdička, K. Wiesner, and K. Schaferna, *Naturwissenschaften* **31**, 391 (1943).

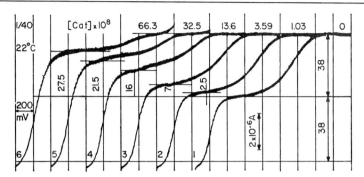

FIG. 22. Effect of catalase on polarographic waves of oxygen reduction phosphate buffer pH 7.1, saturated with air, catalase concentration given on the polarogram. Curves starting at 0.0 V, 200 mV/absc., SCE; $m = 3.28$ mg sec$^{-1}$, $t_1 = 2.92$ sec; full scale sensitivity 9 $\mu$A.

the last term in Eq. (30) becomes $\frac{1}{2}k'_1[C] = \frac{1}{2}k_1[\text{catalase}][H_2O_2]$. For the ratio of the total height of the first reduction wave of oxygen in the presence of catalase ($\bar{\imath}_{\lim}$) to the diffusion current of oxygen in the absence of catalase ($\bar{\imath}_d$) expression (38) was derived:[43]

$$\frac{\bar{\imath}_{\lim}}{\bar{\imath}_d} = 1 - \sum_{i=1}^{\infty} W_i(k'_1t_1)^i \tag{38}$$

where the values of the coefficients $W_i$ are given in Table II.

TABLE II
VALUES OF COEFFICIENTS $W_i$

| $i$ | $W_i$ |
| --- | --- |
| 0 | 1 |
| 1 | $-0.13476418$ |
| 2 | $0.022407078$ |
| 3 | $-0.003405055$ |
| 4 | $0.0004593166$ |
| 5 | $-0.0000551396$ |
| 6 | $0.0000059398$ |
| 7 | $-0.0000005792$ |
| 8 | $0.0000000516$ |
| 9 | $-0.00000000423$ |
| 10 | $0.000000000319$ |

[43] J. Koutecký, R. Brdička, and V. Hanuš, *Collection Czech. Chem. Commun.* **18,** 611 (1953).

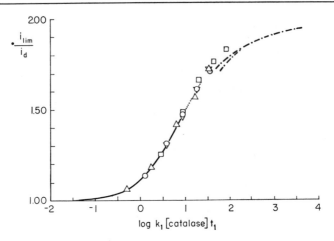

F<small>IG</small>. 23. Variation of the catalytic current ($i_{lim}/i_d$) with catalase concentration, rate constant ($k_1$), and drop-time ($t_1$). ———, Curve calculated using the rigorous treatment and Eq. (38); · · · · · , extrapolation of the rigorous treatment; – – –, extrapolation according to Eq. (39); – · – · – · , extrapolation according to Eq. (40). Experimental points from various sets of measurements.

For cases when the value of the product ($k'_1 t_1$) is large, the change in the ratio $\bar{i}_{lim}/\bar{i}_d$ with catalase concentration can be approximately expressed in the form (39)

$$\frac{\bar{i}_{lim}}{\bar{i}_d} = \frac{1.74 \sqrt{k'_1 t_1}}{1 + 0.87 \sqrt{k'_1 t_1}} \tag{39}$$

The approximations made in derivation of Eq. (39) are of such a type that the experimentally found values of $\bar{i}_{lim}/\bar{i}_d$ should be greater than those obtained using Eq. (39). The expression indicates the trend of the values $\bar{i}_{lim}/\bar{i}_d$, but is not suitable for the calculation of the rate constant $k'_1$ because a small error in the measured value of $\bar{i}_{lim}/\bar{i}_d$ causes a large error in $k'_1$. The comparison of the dependence of the value of $\bar{i}_{lim}/\bar{i}_d$ on $\log(k'_1 t_1)$ computed using Eqs. (38) and (39) with the experimental data is shown in Fig. 23.

For values of $k'_1 t_1$ (Table III) between 0 and 10 an approximate relation (40) can be used with an error smaller than 1%

$$\frac{\bar{i}_{lim}}{\bar{i}_d} = 1 + \frac{k_1[\text{catalase}]t_1}{7.42 + 1.25 k_1[\text{catalase}]t_1} \tag{40}$$

which gives for the rate constant the value (41)

$$k_1 = \frac{7.42}{t_1[\text{catalase}]} \frac{\bar{i}_{lim}}{\bar{i}_d - 1.25\bar{i}_{lim}} \tag{41}$$

TABLE III
DEPENDENCE OF THE RATIO $i_{\lim}/i_d$ ON THE PRODUCT
$k'_1 t_1 = k_1[\text{CATALASE}]t_1$, FOR THE PROCESS, EQ. (37)

| $k'_1 t_1$ | $i_{\lim}/i_d$ | $k'_1 t_1$ | $i_{\lim}/i_d$ | $k'_1 t_1$ | $i_{\lim}/i_d$ |
|---|---|---|---|---|---|
| 0 | 1.000 | 1.5 | 1.161 | 6.0 | 1.40 |
| 0.25 | 1.033 | 2.0 | 1.201 | 8.0 | $1.45_5$ |
| 0.50 | 1.062 | 3.0 | 1.267 | 10.0 | 1.50 |
| 0.75 | 1.090 | 4.0 | 1.317 | — | — |
| 1.00 | 1.116 | 5.0 | 1.363 | — | — |

The rate constant can be determined either using Eq. (41) within the limits of applicability of this expression or from the graph in Fig. 23. In the latter case, for the ratio $\bar{i}_{\lim}/\bar{i}_d$ determined from the experimental data the value of $k'_1 t_1 = k_1[\text{catalase}]t_1$ is read off from the curve in the graph.

Average values of rate constants $k_1 = 1.7 \times 10^7 \, \text{l mole}^{-1} \, \text{sec}^{-1}$ obtained in this way in phosphate buffers pH 7.1 are in good agreement with the value $k = 1.73 \times 10^7 \, \text{l mole}^{-1} \, \text{sec}^{-1}$ obtained in phosphate buffers in the presence of catalase in concentrations of the order of $10^{-10} \, M$, when unreacted hydrogen peroxide was determined manganometrically, after the reaction had been quenched by addition of sulfuric acid.

### 3. Reactions Interposed between Two Electrode Processes

In some electrode processes the product of an electrode process (C) reacts rapidly with a component of the solution (B) to produce the form D which undergoes further electron transfer according to the scheme (42)

$$A + ne \overset{E_1}{\rightleftharpoons} C \tag{42a}$$

$$C + B \underset{k_{-1}}{\overset{k_1}{\rightleftharpoons}} D \tag{42b}$$

$$D + me \overset{E_2}{\rightleftharpoons} P \tag{42c}$$

On polarographic curves two waves are observed at potentials $E_1$ and $E_2$. If the potentials $E_1$ and $E_2$ differ little, the two waves can coalesce. The height of the second wave at potentials $E_2$ depends on the rate of establishment of equilibria (42b).

When the substance B is present in excess and $k'_1 = k_1[\text{B}]$ it is possible[44] to formulate the differential equations (43) introducing $K_1 = k_{-1}/k'_1$.

---

[44] I. Tachi and M. Senda, *Proc. 2nd Intern. Congr. Polarography, Cambridge, 1959,* Vol. 2, p. 454. Macmillan (Pergamon), New York, 1960.

$$\frac{\partial[A]}{\partial t} = D\frac{\partial^2[A]}{\partial x^2} + \frac{2x}{3t}\frac{\partial[A]}{\partial x} \tag{43a}$$

$$\frac{\partial[C]}{\partial t} = D\frac{\partial^2[C]}{\partial x^2} + \frac{2x}{3t}\frac{\partial[C]}{\partial x} - k'_1([C] - K_1[D]) \tag{43b}$$

$$\frac{\partial[D]}{\partial t} = D\frac{\partial^2[D]}{\partial x^2} + \frac{2x}{3t}\frac{\partial[D]}{\partial x} + k'_1([C] - K_1[D]) \tag{43c}$$

The last two right-hand terms in Eqs. (43b) and (43c) express the effect of the chemical reaction (42b).

Initial and boundary conditions are:

$$t = 0, x > 0 \quad : \quad [A] = [A]_b; \ [C] = [D] = [P] = 0$$

$$t > 0, x = 0 \quad : \quad D\frac{\partial[A]}{\partial x} = -D\frac{\partial[C]}{\partial x} = f_1(t)$$

where index b indicates the concentration in the bulk of the solution. Further when $K_1 \gg 1$ and $(k'_1 + k_{-1})t_1 \gg 1$, it is possible[44] to derive for the height of the second wave $(\bar{i}_2/\bar{i}_{d_2})$ expression (44)

$$\frac{\bar{i}_2}{\bar{i}_{d_2}} = 1 + \frac{\sqrt{\dfrac{(k'_1 + k_{-1})t_1}{K_1}}}{1.12 + \sqrt{\dfrac{(k'_1 + k_{-1})t_1}{K_1}}} \tag{44}$$

When the concentration of B is changed, the value of $k'_1$ is changed and the ratio $\bar{i}_2/\bar{i}_{d_2}$ changes in the shape of a dissociation curve.

The reactions governing the height of the second wave can be either an acid-base reaction or some other type of reaction.

a. *Acid-Base Equilibria.* An example of acid-base equilibria interposed between two electrochemical processes is the reduction of phthalimide,[45] following the scheme (45):

$$\tag{45a}$$

$$\tag{45b}$$

[45] A. Ryvolová, *Collection Czech. Chem. Commun.* **25**, 420 (1960).

$$(45c)$$

$$(45d)$$

$$(45e)$$

The rate of the protonation of the radical (45c) with constant $k_1$ is in acid media sufficiently fast to transform all the phthalimide radical into the protonated form. Because the potential corresponding to the transfer of the second electron $E_2$ differs by less than 0.1 V from the half-wave potential $E_1$ corresponding to the first electron uptake, only one two-electron wave is observed in acid media (Fig. 24). This corresponds to a sequence (45a), (45b), (45c), (45d). With increasing pH the height of this two-electron reduction wave decreases, due to the decrease in the rate of protonation of the radical (with constant $k_1$). Simultaneously, a new wave is observed at more negative potentials $E_3$. The more positive wave gradually decreases until it reaches a value corresponding to a one-electron step at potential $E_1$, whereas the more negative wave at potential $E_3$ gradually increases until it also reaches a height corresponding to a one-electron transfer. At these higher pH values the rate of the establishment of equilibrium (45c) is negligibly slow and the process follows the sequence (45a), (45b), (45e).

The decrease of the more positive wave $\bar{\imath}_1$ at potential $E_1$ (and $E_2$, respectively) from a value corresponding to a two-electron process ($\bar{\imath}_d$) to a value corresponding to a one-electron process is given by Eq. (46):

$$\bar{\imath}_1/\bar{\imath}_d = 1 + \frac{0.886(k_1t_1/K_1)^{\frac{1}{2}}[H^+]}{1 + 0.886(k_1t_1/K_1)^{\frac{1}{2}}[H^+]} \qquad (46)$$

The principal validity of scheme (45) has been proved[45] for $5 \times 10^{-4}$ M

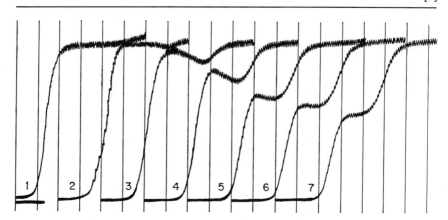

Fig. 24. Dependence of polarographic waves of $1 \times 10^{-3} M$ phthalimide on pH. Britton-Robinson buffer of pH: (1) 3.95; (2) 4.7; (3) 5.65; (4) 6.35; (5) 6.87; (6) 7.1; (7) 7.45. Curves starting at $-0.7$ V, SCE, 200 mV/absc.; $h = 60$ cm; full-scale sensitivity, 24 $\mu$A. [From P. Zuman, Advan. Phys. Org. Chem. **5**, 46 (1967).]

phthalimide, where the experimental points were in a good agreement with the shape of the pH dependence of $\bar{i}_1/\bar{i}_d$ ensuing from Eq. (46). The actual mechanism seems to be even more complicated, because the plot of the dependence of $\bar{i}_1/\bar{i}_d$ on pH is shifted with phthalimide concentration, even when the shape of the plot remains practically unchanged. Consequently, the limiting current of the first wave between pH 6 and 7 is a nonlinear function of phthalimide concentration. One of the possible explanations is dimerization, which could take place either parallelly or consecutively to the proton transfer (45c).

The computation of constants $k_1$ and $k_{-1}$ in reaction (45c) using Eq. (46) is prevented by the fact that the dissociation constant of the protonated radical in reaction (45c) has not been determined.

A scheme, similar to (45), is also followed for the reduction of carbonyl compounds in alkaline media. In these cases also a two-electron wave is observed in the medium pH range, this gradually decreases with increasing pH value until it reaches a height corresponding to a one-electron process. Simultaneously a wave increases at more negative potentials, finally reaching the height of a one-electron step. A difference exists between the first reduction process in which no protonation takes place (47a) and the following steps which are influenced by the protonation (47b)

$$>\!CO + e \overset{E_1}{\rightleftharpoons} >\!CO^{(-)} \tag{47a}$$

$$>\!CO^{(-)} + H^+ \overset{k_1}{\underset{k_{-1}}{\rightleftharpoons}} >\!\dot{C}OH \tag{47b}$$

$$>\overset{\cdot}{C}OH + e \overset{E_2}{\rightarrow} P_1 \qquad (47c)$$

$$>CO^{(-)} + e \overset{E_3}{\rightarrow} P_2 \qquad (47d)$$

At lower pH values the more positive wave at potential $E_1$ (or $E_2$, respectively) corresponds to a sequence (47a), (47b), (47c), whereas at higher pH values, when both waves reached the 1-electron limit, the more positive wave at $E_1$ corresponds to (47a), whereas the more negative wave at $E_3$ corresponds to (47d).

The quantitative treatment of this system is prevented by the possibility of dimerization of the radical formed in reaction (47b), by the possibility of its reaction with mercury or solvent, by the possibility that reaction (47b) takes place at a surface, and not as a volume reaction, by the difficulties in experimental measurement of the dissociation constant $K_a = k_{-1}/k_1$ and finally by the fact that the radical anion $>CO^{(-)}$ combines not only with protons but also with cations of alkali metals.

A similar scheme may operate also for the reduction of aromatic hydrocarbons in solutions containing a limited amount of proton donors.

*b. Other Reactions.* Among interposed reactions that correspond to the reaction (42b) in scheme (42) quantitative treatment has been applied only for the reduction of *p*-nitrophenol.[44,46] The reduction of this compound proceeds principally in one step that in the medium pH range corresponds to a 2-electron reduction. Both to higher and to lower pH values the height of this wave gradually increases until at sufficiently high and low pH values it reaches limiting current corresponding to a 4-electron process. The quantitative treatment is made possible by the reversibility of reaction (48a)

$$(48a)$$

$$(48b)$$

[46] G. S. Alberts and I. Shain, *Anal. Chem.* **35**, 1859 (1963).

$$\text{(structure)} + 2e + 2H^+ \underset{\phantom{E_2}}{\overset{E_2}{\rightleftharpoons}} \text{(structure)} \qquad (48c)$$

In this case $E_2$ is even more positive than $E_1$ so that at potentials corresponding to the limiting current (i.e., more negative than $E_1$), all the quinone imine formed is immediately reduced according to (48c). Therefore only a single wave with changing height is observed on $i$-$E$ curves.

Principal validity of the scheme (48) was verified by the good agreement between experimental points and the $i/i_d$-pH dependence calculated according to Eq. (44). The confirmation of the scheme (48) was carried out using chronopotentiometric and potentiostatic methods.[46] Similar schemes have been considered for the reductions of $o$- and $p$-nitrophenols and $o$- and $p$-nitroanilines, but the theoretical treatment is more involved due to the irreversibility of the first, 4-electron step in the reduction of the nitro groups.

Similarly, a chemical reaction is interposed between the first and second electrode process proper in the reductions of $\alpha,\beta$-unsaturated ketones. In the first electroreduction step enolate is formed which is electroinactive in the studied potential range. To be further reduced in the second step the enolate must be first transformed into the electroactive keto form. The rate of this enol-keto transformation, is acid-base catalyzed and limits the wave heights of the more negative wave which corresponds to the electroreduction of the corresponding saturated ketone. Therefore the current-voltage curves in solutions of $\alpha,\beta$-saturated ketones show a reduction wave following the first, 2-electron step. The height of this wave is smallest in the medium pH range and increases both in acid and alkaline region, due to the acid and base catalysis of the chemical transformation. The rate of the transformation depends not only on pH, but also on the structure of the $\alpha,\beta$-unsaturated ketone involved. Experimental evidence seems to indicate that extension of the conjugated systems as in $C{=}C{-}CO{-}C{=}C$ results in slowing down the rate of the transformation.

The reaction interposed between two electrode processes can also be a photochemical reaction. An example of this type is the reduction of ketones in illuminated solutions,[47] following the scheme (49)

$$R_2COH^+ \rightleftharpoons R_2CO + H^+ \qquad (49a)$$

$$R_2COH^+ + e \underset{\phantom{E_1}}{\overset{E_1}{\rightleftharpoons}} R_2\dot{C}OH \qquad (49b)$$

[47] H. Berg, *Abhandl. Deut. Akad. Wiss. Berlin Kl. Chem. Geol. Biol.* **1**, 128 (1964).

$$R_2CO + h\nu + (CH_3)_2CHOH \xrightarrow{k} R_2COH + P_1 \qquad (49c)$$

$$R_2\dot{C}OH + e \xrightarrow{E_2} P_2 \qquad (49d)$$

Because in this case the illumination induces a reaction that is parallel with the first step of the electrode process, it results in an increase of the second wave at the potential $E_2$ and because the ketone $R_2CO$ is consumed in reaction (49c), also in an increase of the first wave at potential $E_1$. The qualitative treatment of this system is complicated by the fact that in addition to the reaction (49c) taking place in the vicinity of the electrode surface, this type of photochemical reaction also takes place in the bulk of the solution. Furthermore, the ketyl $R_2\dot{C}OH$ can undergo—both in the solution and in the vicinity of the electrode—further photochemical cleavage, dimerization, and reaction with the solvent.

### 4. Consecutive Reactions

Polarographically it is possible to study only those reactions that take place consecutively to such electrode processes proper as are reversible. Whereas in all the cases mentioned above the wave heights were measured, in these cases the wave heights and wave shape remains unchanged when compared with waves of reversible electrode processes which are not affected by consecutive chemical reactions. Information about the reactions subsequent to the electrode proper can be obtained from the measurement of half-wave potentials, their comparison with the values of equilibrium oxidation-reduction potentials (measured e.g., by potentiometry, spectrophotometry or by other methods) and their dependence on the composition of the solution and on capillary characteristics, such as drop-time. First- and second-order consecutive reactions have been studied so far.

*a. First-Order Reactions.* The first-order consecutive reactions follow the scheme (50):

$$A + ne \rightleftharpoons C \qquad (50a)$$

$$C + B \underset{k_{-1}}{\overset{k_1}{\rightleftharpoons}} D \qquad (50b)$$

On polarographic curves one single wave is observed, but at potentials more positive than the equilibrium oxidation-reduction potential. When substance B is present in excess, or when its concentration does not change during the process, the scheme (50) degenerates to (51):

$$A + ne \rightleftharpoons C \qquad (51a)$$

$$C \underset{k_1}{\overset{k'_1}{\rightleftharpoons}} D \qquad (51b)$$

For this system, under the assumption that $k'_1 \gg k_{-1}$ [i.e., when equilibrium (51b) is displaced considerably in favor of the electroinactive form D] and that $k'_1 \gg 1$, it was possible to derive for the half-wave potential equation (52)

$$E_{\frac{1}{2}} = E^0 - \frac{RT}{2F} \ln 0.886 \sqrt{\frac{D_A}{D_C} k'_1 t_1} \qquad (52)$$

This indicates the dependence of the half-wave potential on drop-time for a 2-electron process.

A system, in which a scheme corresponding to (51) is supposed to operate, is the system ascorbic acid–dehydroascorbic acid. For this system it was assumed that the anodic wave corresponds to the oxidation of ascorbic acid leading to an unstable electroactive intermediate (C) that is transformed by a chemical reaction of the type (51b) into the electro-inactive form D. It was assumed that the electroinactive form is the hydrated form of dehydroascorbic acid.

It has been proved experimentally[48] that the half-wave potential of the anodic wave is practically independent of ascorbic acid concentration. This indicates that the deactivation reaction (51b) is first order in ascorbic acid. The wave form of the anodic wave is the same as for a reversible process, in agreement with the system (51). Moreover the half-wave potential has been shown[48] to depend on $t_1^{\frac{1}{2}}$ as predicted by Eq. (52). Nevertheless, the calculation of rate constant $k'_1$ cannot be carried out using Eq. (52), because the value of $E^0$ is unknown. Potentiometrically determined potentials undoubtedly differ from the standard potential $E^0$ due to reaction (51b) which has even more time to affect the measurement than in polarography. Moreover, the sluggish establishment of potentials of platinum electrodes in ascorbic acid solutions makes the use of the poten-tiometrically determined values even more doubtful.

Complications in the application of scheme (51) to this system were aroused when the polarographic behavior of dehydroascorbic acid was described.[49] The theory[50] predicted the height of the cathodic kinetic wave of dehydroascorbic acid (governed by the rate of dehydration with constant $k_{-1}$) and indicated that provided that the limiting current of the reduction of dehydroascorbic acid is small, the half-wave potential of the cathodic reduction wave should be equal to the half-wave potential of the anodic oxidation wave of ascorbic acid. Experimentally it has been found that the height of the reduction waves of dehydroascorbic acid is considerably smaller (about one thousand times) than predicted by the theory and the

[48] D. M. H. Kern, *J. Am. Chem. Soc.* **76,** 1011 (1954).
[49] S. Ono, M. Takagi, and T. Wasa, *J. Am. Chem. Soc.* **75,** 4369 (1953).
[50] J. Koutecký, *Collection Czech. Chem. Commun.* **20,** 116 (1955).

half-wave potentials of the cathodic waves of dehydroascorbic acid are more negative than those of the anodic waves of ascrobic acid by about 0.5 V. These discrepancies indicate that the reaction scheme for ascorbic acid oxidation may be more complex than scheme (51).

b. *Second-Order Reactions.* The second-order consecutive reactions follow the scheme (53):

$$A + ne \rightleftharpoons C \tag{53a}$$

$$2C \xrightarrow{k_2} P \tag{53b}$$

It has been shown[51] that in this case the wave height is unaffected and remains diffusion controlled, but the wave shape differs from that for a simple reversible system, in the absence of the subsequent reactant, like (53a). Whereas for a simple reversible reaction Eq. (54) is required, for the system (53) Eq. (55) operates:

$$E = E^0 - \frac{RT}{nF} \ln \frac{\bar{\imath}}{\bar{\imath}_d - \bar{\imath}} \tag{54}$$

$$E = E^0 - \frac{RT}{nF} \ln \frac{\bar{\imath}^{\frac{2}{3}}}{\bar{\imath}_d - \bar{\imath}} \bar{\imath}_d^{\frac{2}{3}} + \frac{RT}{3nF} \ln \frac{C_A k_2 t_1}{1.51} \tag{55}$$

In contrast to systems involving first-order reactions, described by (52), the half-wave potential depends not only on drop time, but also on depolarizer concentration.

An example of system (53)—for which the wave form was analyzed and the dependence of half-wave potentials on drop-time and concentration of the electroactive substance was studied[52]—is the reduction of compounds with a benzoyl grouping $C_6H_5COR$, following, in acid media, scheme (56)

$$\begin{array}{c} C_6H_5CR \underset{k_{-1}}{\overset{k_1}{\rightleftharpoons}} C_6H_5COR + H^+ \\ \parallel \\ OH^+ \end{array} \tag{56a}$$

$$\begin{array}{c} C_6H_5CR + e \rightleftharpoons C_6H_5\dot{C}R \\ \parallel \qquad\qquad\quad | \\ OH^+ \qquad\qquad\quad OH \end{array} \tag{56b}$$

$$\begin{array}{c} \qquad\qquad\qquad OH \\ \qquad\quad k_2 \qquad\quad | \\ 2C_6H_5\dot{C}R \rightarrow C_6H_5CR \\ \quad | \qquad\qquad | \\ \quad OH \qquad C_6H_5CR \\ \qquad\qquad\qquad | \\ \qquad\qquad\qquad OH \end{array} \tag{56c}$$

[51] V. Hanuš, *Chem. Zvesti* **8**, 702 (1954).

[52] S. G. Majranovskii, *Izv. Akad. Nauk. SSSR, Otd. Khim. Nauk*, p. 2140 (1961).

Determination of the standard potential $E^0$ and the effect of adsorption, in particular of the radicals formed in reaction (56b), complicate calculation of the rate constant $k_2$.

## D. Conclusions

For fast reactions with second-order rate constants in the range from $10^4$ to $10^8$ l mole$^{-1}$ sec$^{-1}$, measurement of polarographic waves provides a simple and fast method of determination of rate constants for some of the reaction types discussed above. For such systems the reaction takes place in the vicinity of the dropping electrode in a reaction volume the diameter (width) of which is considerably greater than that of the electrical double layer. For such systems an agreement between the values of rate constants obtained from polarographic data and from other types of measurement was found, e.g., for the reaction of tropylium ions[24] or for the dehydration of aliphatic aldehydes.[31]

For faster reactions, the values of rate constants obtained from polarographic measurements are in worse agreement with data obtained either by other methods or predicted by theory. In particular, the values of rate constants calculated using some of the expressions given in Section III,C are greater than the maximum value of a diffusion-controlled rate constant, taking place under conditions when every collision is effective, which is of the order $10^{11}$ l mole$^{-1}$ sec$^{-1}$. It is assumed that in these cases the specific properties of the electrode, in particular its electrical field and adsorption phenomena, can affect the data obtained. It is supposed that either the laws that govern the transport (diffusion) of the particles in solution are affected by the electrical field in the vicinity of the electrode; or the values of rate or equilibrium constants, or both, are drastically changed in this field; or the electroactive species, the electroinactive form, the reagent present in excess or in a constant concentration (substance B in schemes in Section III,C), the intermediates or the products or all of them are adsorbed on or desorbed from the electrode surface. The increase (or decrease) in concentration of these reactants at the surface of the electrode when compared with their concentration in the bulk of the solution results in an increase (or decrease) of the corresponding rate taking place at the surface of the electrode as a heterogeneous reaction, when compared with the rate of the same reaction taking place as a volume reaction, in the homogeneous solution.

It is therefore essential for application of polarography in the study of very fast reactions to prove that the reaction does take place as a homogeneous, volume reaction. The choice of experimental evidence was described in Section III,C. Furthermore it is advisable wherever possible, to check the value of the rate constant and the reaction scheme by two or more independent methods.

So far polarography has been applied in the study of enzymatic processes taking place in the vicinity of the electrode only in a few instances. On the other hand, many of the substances given as examples in Section III,C participate in various enzymatic reactions and therefore it was considered to be of interest to indicate certain chemical properties of these compounds. Moreover, the techniques used in the elucidation of schemes and determination of rate constants using polarographic techniques in the studies of organic reactions have been described bearing in mind that some of them could in future be applied to the study of enzymatic reactions.

## IV. Slower Reactions

### A. Principles

As mentioned in Section II, the essential condition for application of polarography in the study of slower reactions is that at least one of the reactants, intermediates or products is electroactive, i.e., gives a polarographic wave in the available potential range. In these types of applications the change of the wave height with time is then measured.

The quenching techniques used in flow methods have not yet been adapted to polarographic measurements. When no measurable waves can be obtained under the conditions under which the reaction was carried out, it is necessary to take a sample off the reaction mixture and adjust the solution composition for polarographic examination. As all this can hardly be carried out sufficiently rapidly for the study of fast reactions, it is usually necessary to carry out the polarographic measurement directly in the reaction mixture. In such cases it is necessary that the electroactive component (reactant, intermediate, or product) give a wave in the reaction mixture under conditions under which the reaction takes place. Moreover, the wave used for the study of kinetics must be dependent on the concentration of the electroactive species and no interfering substances should be present.

The effect of the concentration of the electroactive species on the wave height is studied first, under conditions as similar as possible to those existing in the reaction mixture. The limiting current, if possible, should be linearly proportional to the concentration of the electroactive species. It is also necessary to examine the character of the limiting current in order to understand the effect of changes of reaction mixture on wave heights. Diffusion-controlled currents are most frequently used for the study of homogeneous reactions taking place in the bulk of the solution. Principally other types of polarographic currents such as kinetic or catalytic currents (the latter only over a limited concentration range) can be used for kinetic studies of that type as well. The techniques used in distinguishing the type of the process governing the limiting current have already been discussed in Section III,B; in summary, diffusion currents are linearly proportional to concentration over a wide concentration range, are linearly proportional to the square root of the height of the mercury column, and are usually independent of buffer concentration. Their height increases with increasing temperature with a coefficient of 1.5 to 2.0% deg$^{-1}$ and is usually little pH dependent. The changes in the instantaneous current ($i$-$t$) correspond to a parabola of $\frac{1}{6}$-order.

In some instances, another component of the reaction mixture—in addition to the component for which the time changes of concentration are being studied—is electroactive and gives a polarographic wave. It can be either one of the components the concentrations of which change with time during the reaction (i.e., another reactant, intermediate, or reaction product); or one of the components added the concentration of which remains unchanged; or an impurity. The impurity can usually be removed from the reaction mixture. The component the concentration of which remains unchanged in the course of the reaction is usually a buffer, a neutral salt, or a solvent and is invariably in large excess over the electroactive component followed. The current caused by this substance is therefore usually considerably greater than the investigated wave. If the wave of the medium component is by some 0.2 or 0.3 V more negative than the examined wave, the presence of this component does not interfere and the measurement of the studied wave can be carried out with the usual accuracy of about ±3%. When the wave of the medium component is more positive, it interferes. A change in the composition of the reaction mixture (e.g., change in buffer composition keeping the pH constant) or in the polarographic instrumentation is necessary. For the case when alteration in buffer composition is used, the following two examples can be quoted: Phthalate buffers, can be used for some kinetic studies but are contraindicated in some polarographic studies, when the reduction waves at more negative potentials should be measured, because of the reduction current due to the electrolysis of phthalic acid. Acetate and phosphate buffers can be used instead. Similarly, barbital buffers are not recommended when the kinetics are followed by measurement of anodic waves, as barbital gives anodic waves corresponding to mercury salt formation. Borate buffers are preferable. The instrumentation changes involve the use of differential or subtraction techniques, possibly in connection with single-sweep methods.

When the concentration of more than one of the components of the reaction mixture (i.e., reactants, intermediates, or products) that is

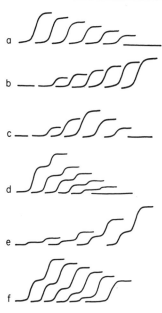

Fig. 25. Time-dependence of polarographic waves. Schematically plotted polarographic curves recorded after selected time intervals, when: (a) the reactant is electroactive; (b) the product is electroactive; (c) the intermediate is electroactive; (d) two electroactive reactants, or one compound with two electroactive groups, are cleaved; (e) two electroactive products, or one compound with two electroactive groups, are formed; (f) one of the electroactive compounds or groups remains unchanged during reaction, or there is compensation of two waves at the potential of the more negative wave—one decreasing, one increasing, both corresponding to the same number of electrons transferred.

electroactive changes with time, it is usually necessary to obtain well separated waves for kinetic applications. To obtain sufficiently separated waves, the half-wave potentials of the electroactive components must differ by some 0.1 to 0.2 V. The necessary difference depends on the number of electrons transferred in the electrode process and on the shape of the wave, and for some drawn-out waves the difference must be even greater. As usually these components are present in the reaction mixture in a comparable concentration, and because the wave heights of many waves in equimolar solutions are similar, the wave heights of various components of the reaction mixture are comparable and the relative position of the waves (i.e., which is more positive and which more negative) is not essential for their separation.

The wave of a reactant principally decreases with time during reaction (Fig. 25a), the wave of a product increases (Fig. 25b), and the wave of an intermediate first increases and then decreases (Fig. 25c). Many intermediates, being chemically reactive compounds, are also electrochemically active and show a separate wave on polarographic curves. The detection of such intermediates and investigation of their time changes, is one of the advantages polarography offers for the study of the kinetics of such systems. In some cases two or more waves are observed on the $i$-$E$ curve recorded in the reaction mixture that decrease (or increase) with time in the same proportion (Fig. 25d,e). When the ratio of such waves does not change with time, they may correspond to one single substance that is electroreduced or electrooxidized in several successive steps. With a decrease of this reactant (or an increase of this product) all its waves decrease (or increase) in the same proportion. The second possibility exists that the two waves correspond to two reactants (or to two products) the concentrations of which remain in the same ratio during the course of the reaction. In the former case, of a single electroreduced or electrooxidized substance, the occurrence of several steps may result from the presence of several electroactive forms (e.g., protonated and unprotonated, hydrated and nonhydrated, cyclic and acyclic), from adsorption phenomena, or from a system of consecutive electrode processes.

When the height of a wave remains unchanged during the course of reaction (Fig. 25f), it can be caused either by the presence of an electroactive substance that does not participate in the reaction, or by compensation of two wave heights. The latter can happen when one of the waves of a reactant coalesces with one of the waves of products. When the number of electrons transferred in the reduction of the reactant and product are the same, when their diffusion coefficients do not substantially differ, and when the waves are both diffusion controlled, the total height of the observed wave does not change. Such waves cannot be used for kinetic studies.

When, on the other hand, the waves, the half-wave potentials of which differ so little that the waves merge, differ in the number of electrons transferred in the electrode process, in the value of diffusion coefficients, or in the character of the current governing process, the changes of the heights of the observed single wave can still be used for the study of kinetics.

Such less usual types of application of polarography are best demonstrated by examples. The difference of the number of electrons consumed is exploited in the hydrolysis of $p$-nitrophenyl acetate or $p$-nitroacetanilide.[53] These compounds are reduced in alkaline media in a 4-electron process, whereas the hydrolysis products, $p$-nitrophenol and $p$-nitraniline are

[53] L. Holleck and G. Melkonian, Z. *Elecktrochem.* **58**, 867 (1954).

reduced in 6-electron steps. The increase of the 4-electron wave into a 6-electron wave can be applied for the evaluation of the rate of hydrolysis. The different diffusion coefficients make possible the study of hydrolysis of esters of 3,5-dinitrobenzoic acid with hydroxy steroids. The waves of the nitrogroup reduction in the steroid ester are significantly smaller than those of 3,5-dinitrobenzoic acid.[54] The measurement of the increase in wave height made the evaluation of the hydrolysis possible.

The differences in character of the current governing process can cause a difference in wave heights, e.g., when one of the electroactive components of the reaction mixture gives a diffusion-controlled wave and the other electroactive component shows a kinetic wave. The difference between the high diffusion controlled current of pyridoxal 5-phosphate (at pH 2–5) and the low kinetic wave of its hydrolysis product, pyridoxal, can be used in the study of the hydrolysis of the former.[55]

## B. Techniques

The classical polarographic apparatus can be used for reactions with half-times between 15 seconds and 5 minutes (slower reactions will not be explicitly discussed here), which still allow rather fast reactions to be followed provided that a high dilution of one of the reactants is used. Faster reactions, for which special equipment is to be used, are discussed separately.

For classical polarographic methods, water-jacketed polarographic cells (Fig. 26) are used. Interruption of the flow of water from the thermostat during the recording of the polarographic current or grounding of the outer surface of the cell proved useful to prevent the effects of static electricity.

The reaction can be started either by changing a physical factor (such as temperature, illumination, or application of a voltage on an electrode generating a reagent), or, more usually, by the addition of one component to the reaction mixture. The volume of the added component is either small as compared with the total volume of the reaction mixture or comparable. In the first case, the reaction mixture prior to the addition of this component is deoxygenated and the temperature adjusted. The introduction of a gas for deoxygenation can result in the change in temperature in the cell, in particular with the more volatile solvents. The temperature should be always measured in the reaction mixture, not in the bath. The concentration of the stock solution of the added component is relatively higher, to make it possible to keep the volume small. The addition is made using a pipette or a syringe. The thorough mixing of the reaction

[54] H. Berg and H. Venner, *Ric. Sci.* **29** (Suppl. 4), 181 (1959).
[55] P. Zuman and O. Manoušek, *Collection Czech. Chem. Commun.* **26,** 2314 (1961).

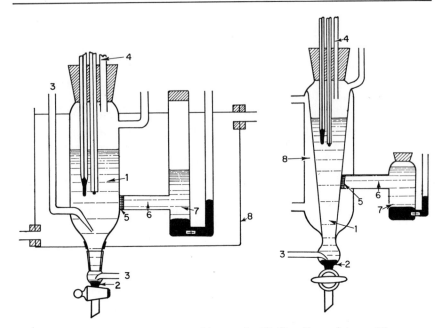

Fig. 26. Thermostated polarographic vessel. (1) Reaction mixture, (2) mercury collected from drop electrode; (3) nitrogen inlets into solution; (4) nitrogen inlet above surface of solution; (5) sintered glass plate; (6) saturated potassium chloride solution; (7) reference electrode; (8) thermostated water bath of water jacket. [From P. Zuman, *Advan. Phys. Org. Chem.* **5**, 2 (1967).]

mixture is achieved by a lively stream of gas. In the construction of the cell care should be taken to prevent the occurrence of spaces in which the mixing would be slow. When comparable volumes are mixed, both volumes are first deoxygenated and adjusted to a proper temperature. After the connection between these two solutions is secured, they are brought together using either the weight of the solution of the added component, or the pressure of a gas or of a sealing liquid such as mercury.

The time when half of the volume of the solution of the component is added is taken as the beginning of the reaction. The time measurement starts and is often synchronized with the beginning of the recording of the polarographic current.

To measure the changes in the heights of polarographic waves for faster reactions, the applied voltage is kept constant at a chosen value. This applied voltage usually corresponds to the potential range in which the wave of the particular reactant or product reaches its limiting value. The change in the limiting current with time at this potential is recorded continuously (Fig. 27). Markings can be recorded after selected time intervals. In most polarographs, however, the shift of the recording paper is so regular that it suffices to determine the rate of this shift and to make occasional checks. It is recommended that, after a particular run is terminated, a whole polarographic $i$-$E$ curve be recorded. In this way it is possible to verify that the applied voltage chosen really corresponds to the limiting current of the wave one wants to follow. It can be also detected whether not an unexpected reaction has taken place which would be manifested by the occurrence of new waves on the polarographic curve recorded.

For these measurements almost any type of a pen or photographically recording polarograph can be used. To evaluate the recorded current–time curves any of the usual numerical or graphical methods for calculation of the rate constants can be used. It is possible also to use analog computers both for verification that a proper equation for reaction rate was chosen (i.e., that the proposed reaction scheme is correct) and for determination of the best value for the rate constants.

Because usually in kinetic studies a large number of experimental data are handled, it is essential to simplify mathematical operations. Many reaction rate equations are modified so that the measured current can be directly inserted as a variable, without first calculating the concentration. Graphical methods are introduced wherever it is possible. They are particularly advantageous in cases where a function of the measured current can be plotted against time and the rate constant is determined from the slope of the plot.

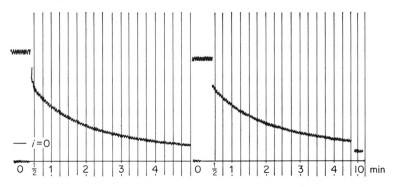

FIG. 27. Continuous recording of the limiting current. Oxidation of threo-1,2-diphenylethyleneglycol by periodate; decrease in the periodic acid concentration with time. 1 $M$ Acetate buffer of pH 4.3, $5 \times 10^{-5} M$ KIO$_4$, $5 \times 10^{-5} M$ diol. Current recorded at $-0.4$ V (mercurous sulfate reference electrode) at 25°. Figures on the abscissas give the time in minutes; the galvanometer zero and the current before the addition of periodate and after addition at $t = 0$ are marked. Full-scale sensitivity, 0.8 $\mu$A. [From P. Zuman, Advan. Phys. Org. Chem. 5, 8 (1967).]

For a reaction following first-order kinetics, it is sufficient to plot the logarithm of $i_0/i_p$ where $i_0$ is the limiting current measured at time $t = 0$ and $i_p$ at a selected interval $t = p$ against time and to determine the slope of the linear plot. The numerical value of the slope is equal in this case to $k/2.3$. Even simpler is the evaluation using an analog computer. Usually the output of these computers can be recorded as a function of time using the same recorder as that used for limiting currents, and the evaluation consists of modeling and comparison of the tracings obtained with the computer and the polarographically studied reaction. For evaluation of rate constants the settings of computer potentiometers are calibrated using known concentrations of the electroactive species.

For faster reactions the fast mixing of the components becomes essential, if the reaction is not started by a photochemical[47] or an electrochemical[56] generation of the reactant. Mixing chambers similar to those used for flow methods proved useful. The choice of the technique for recording of the concentration change with time again depends on the half-time of the studied reaction. For half-times of the order of seconds, the measurement of peak heights[57] using the single-sweep technique proved advantageous. For even faster reactions, measurements of the change in the instantaneous current with time on a single drop have worked well,[47] the recording being ensured for half-times greater than 0.15 seconds by using a string galvanometer, for those greater than 0.0015 seconds by using an oscilloscope.

Also oscillographic polarography[58] proved useful for the study of reactions with half-times of the order of seconds or a few minutes. Certain complications in these cases are presented by the fact that generally in oscillographic polarography the measured quantity is not linearly proportional to concentration of the electroactive species over a wider concentration range, and calibration curves have to be constructed. But, on the other hand, some substances that are electroinactive in classical polarography are electroactive under the conditions of oscillographic polarography with alternating current.

In the study of fast reactions, polarographic methods can be advantageously combined with flow methods. Whereas the polarographic indication of the concentration changes in the continuous flow presents difficulties that result from the perturbation of the diffusion layer round the electrode by the flowing reaction mixture, the combination with a stopped-flow technique seems to be free of such complications. The evaluation can be

[56] B. Kastening, *Collection Czech. Chem. Commun.* **30**, 4033 (1965).
[57] F. C. Snowden and H. T. Page, *Anal. Chem.* **22**, 969 (1950).
[58] R. Kalvoda, "Techniques of Oscillographic Polarography." Elsevier, Amsterdam, 1965.

carried out either using the instantaneous current measurements[59] or the changes in the oscillographic $dE/dt = f/E$ curves.[60] In the latter case, with mixing achieved in a period shorter than 0.01 second using a streaming electrode and making a ciné film record of the oscillographic curves, a reaction with a half-time of 1 second was followed.

In the above discussion, it was assumed that only one wave changes with time, or that all the waves present decrease or increase without changing the ratio of their heights. When the time changes of two waves are important, it is possible by using two sets of polarographic equipment to record simultaneously the changes in the height of two waves.

## C. Examples

Applications of polarography in the study of enzymatic processes can be classified either according to the enzymatic system involved or according to the electroactive substance determined polarographically. The latter system of classification was adopted here.

In the study of biological reactions it is possible to follow polarographically concentration changes of the enzyme, of the substrates, or of products formed in the process.

Among the electroactive substances, the reduction of hemin[40] can be mentioned first. $\alpha$-Chloroprotohemin and $\alpha$-hematin gave reduction waves (Fig. 28), the half-wave potentials of which are shifted toward more negative values with increasing pH (Fig. 21). The slope of the linear dependence, 0.058 mV/pH, the shape of the polarographic curves, the good agreement of polarographic half-wave potentials with standard potentials determined potentiometrically, and the results of oscillographic polarography[40] indicate that the system oxidized ($Fe^{III}h$) and reduced ($Fe^{II}h$) form of hemin is reversible. The study was complicated by the association of the compound in the solution, to which was attributed the decrease of hemin waves at pH < 7.9. The shifts of half-wave potentials with pH and the decrease of hemin waves at pH > 10 indicate that only the protonized form of hemin undergoes reduction.

Cozymase[61] and cytochrome $c$ are among other electroactive biological substances which can be reduced.[62] Cytochrome $c$ can be also determined using a catalytic wave[63,64] in a solution containing 0.001 $M$ hexaminocobaltichloride, 1 $M$ ammonia, and 1 $M$ ammonium chloride. This catalytic wave,

[59] E. Bauer, *Abhandl. Deut. Akad. Wiss. Berlin. Kl. Chem. Geol. Biol.* **1**, 411 (1964).
[60] E. Bauer and H. Berg, *Chem. Zvesti* **18**, 454 (1964).
[61] F. Šorm and Z. Šormová, *Chem. Listy* **42**, 82 (1948).
[62] H. Theorell, *Biochem. Z.* **298**, 258 (1938).
[63] C. Carruthers, *J. Biol. Chem.* **171**, 641 (1947).
[64] C. Carruthers and V. Suntzeff, *Arch. Biochem. Biophys.* **17**, 261 (1948).

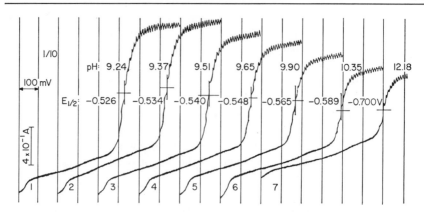

Fig. 28. Effect of pH on reduction waves of α-chlorohemin. Borate buffers, sodium hydroxide added, pH given on the polarogram, concentration of hemin: (1) and (2), $1 \times 10^{-3} M$; (3) $9 \times 10^{-4} M$; (4), $8 \times 10^{-4} M$; (5), $7.1 \times 10^{-4} M$; (6), $6.1 \times 10^{-4} M$; (7), $5.4 \times 10^{-4} M$. Curves starting at 0.0 V, 100 mV/absc., NCE, full-scale sensitivity, 2 μA.

the height of which is a function of cytochrome c concentration, corresponds to a catalytic hydrogen evolution. Similar curves can be used in the study of the reactions of papain[65] and ribonuclease.[66]

Nevertheless, more common is the polarographic determination of substrates and reaction products. Of these the simplest and most widely used is the polarographic determination of oxygen. In most cases it is possible to immerse the dropping electrode directly into the reaction mixture, in which the enzymatic process takes place. The limiting current of the first oxygen wave at $-0.5$ V is recorded as a function of time. When the effect of metallic mercury on the enzymatic process is feared, the electrode can be separated by a permeable membrane from the reaction mixture. The rate of diffusion of oxygen from the reaction compartment to the measurement compartment prevents the application for the fastest reactions. In many instances metallic mercury does not interfere with the enzymatic process, but care should be taken in preventing the mercury ions from mixing with the enzyme system. Consumption of oxygen was followed in the presence of various enzyme systems, both in the absence and in the presence of various substrates and added substances.[67] When the reaction of cytochrome oxidase with oxygen was followed,[68] 0.1 ml of

[65] J. D. Ivanov, "Polarography of Proteins, Enzymes and Aminoacids." Publ. House Acad. Sci. USSR, Moscow, 1961 (in Russian).

[66] H. Berg, Ric. Sci. 27 (Suppl. III), 184 (1957).

[67] M. Březina and P. Zuman, "Polarography in Medicine, Biochemistry, and Pharmacy," p. 152. Wiley (Interscience), New York, 1958.

[68] O. Pihar and L. Dupalová, Chem. Listy 48, 265 (1954).

the enzyme preparation was mixed with 0.8 ml of a 0.54% solution of
$p$-phenylenediamine in a 0.1 $N$ phosphate buffer pH 7.3 and with 0.1 ml of
0.002 $M$ cytochrome $c$. In the study of the reaction of succinyl oxidase, the
limiting current of oxygen was measured at $-1.1$ V in a solution containing
0.1 ml of the enzyme preparation, mixed with 0.1 ml of 0.5 $M$ sodium
succinate of pH 7.3, 0.05 ml of 0.002 $M$ cytochrome $c$, and 0.75 ml of a 2 $M$
phosphate buffer pH 7.3. The respiration of yeast suspensions and the
effects of various factors on this process has been studied.[69-71]

The decrease of hydrogen peroxide concentration was used to elucidate
kinetics of its reaction with catalase in the homogeneous solution.[72] Catalase activity of various bacteria was determined in this way. In a similar
way the reaction of peroxidase was studied[73] as well as the reaction of
peroxidase from milk, which catalyzes oxidation of uric acid by hydrogen
peroxide.[74] The increase in hydrogen peroxide concentration was followed
in the reaction of hypoxanthine with oxygen in the presence of xanthine
oxidase from milk.[75] The decrease of waves of $p$-benzoquinone and of
salicylaldehyde was followed[75] in the presence of aldehyde oxidase from
milk, and also the decrease of wave of Methylene blue in the presence of
xanthine oxidase in solutions containing hypoxanthine.[75]

In the evaluation of malease isolated from corn,[76] 1.0 ml of the suspension of the enzyme, 1 ml of 0.1% DPN, 1.0 ml of 0.5% triphenyltetrazolium
chloride, 1.0 ml of 0.1 $M$ sodium maleate and 1.0 ml of phosphate buffer
pH 7.5 were maintained under conditions required for enzymatic reaction.
The reaction was quenched by the addition of 50 ml of an ammonia-
ammonium chloride buffer, pH 7.5, diluted to 100 ml. Fumaric acid
content was determined.

Waves of the quinoid compound formed by the action of tyrosinase on
tyrsoine was used in the study of the enzymatic oxidation of tyrosine.[77]
The effect of carbonic acid anhydrase was followed by measuring the
shifts of potentials of the anodic and cathodic wave of quinhydrone in
poorly buffered carbonic acid–bicarbonate solutions.[78]

Anodic waves of hydroquinone, pyrogallol, and ascorbic acid were used

[69] J. P. Baumberger and F. W. Fales, *Federation Proc.* **9**, 9 (1950).

[70] L. Selzer and J. P. Baumberger, *J. Cell. Comp. Physiol.* **19**, 281 (1942).

[71] R. J. Winzler, *J. Cell. Comp. Physiol.* **17**, 263 (1941); **21**, 229 (1943).

[72] J. Molland, *Acta Pathol. Microbiol. Scand. Suppl.* **66**, 165 (1947).

[73] B. Chance, *Arch. Biochem.* **24**, 389, 410 (1949).

[74] E. Knobloch, *Sb. Lekar.* **46**, 12 (1944).

[75] E. Knobloch, *Collection Czech. Chem. Commun.* **12**, 407 (1947).

[76] W. Sacks and C. O. Jensen, *J. Biol. Chem.* **192**, 231 (1951).

[77] K. Wiesner, *Biochem. Z.* **314**, 214 (1943).

[78] O. H. Müller, *Federation Proc.* 783 (1948); *Proc. 1st Intern. Polarography Congr.,
Prague 1951*, Vol. I, p. 159. Přírodověd. Nakl., Prague, 1951.

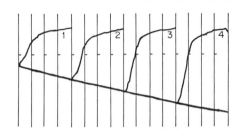

FIG. 29. Polarographic waves of thiocholine in the course of hydrolysis of acetylthio-choline in the presence of acetylcholinesterase. Phosphate buffer pH 8, 10% acetyl-cholinesterase solution (prepared from erythrocytes), and acetylthiocholine are added to make a 0.2% solution. Anodic waves recorded after: (1), 15; (2), 150; (3), 285; (4), 420 seconds. Curves were recorded starting at −0.8 V toward more positive potentials, NCE, 200 mV/absc. Continuous record corresponds to the time dependence of the thiocholine wave at 0.0 V (NCE) under the same conditions. Full-scale sensitivity, 7 μA.

for the study of their oxidation by hydrogen peroxide in the presence of peroxidase.[79] Anodic waves of hydroquinone and catechol were followed in the oxidation of these compounds by oxygen in the presence of tyrosinase.[80]

Anodic waves of thiocholine, corresponding to a mercury salt formation, were used in the study of the kinetics of the enzymatic hydrolysis of acetylthiocholine in the presence of acetylcholinesterase and cholinester-ase.[81] To a buffered acetylthiocholine solution was added a preparation of acetylcholinesterase or of cholinesterase. The increase of the anodic wave was continuously recorded (Fig. 29). The method was used to evaluate the activity of cholinesterase.

Finally, various types of proteolytic enzymes were followed by measure-ment of the catalytic wave of hydrogen evolution observed in ammoniacal cobalt solutions in the presence of proteins.[82] In particular, polarography was used in the evaluation of activity of pepsin[83,84] and trypsin.[85]

## D. Conclusions

Polarography offers for the studies of reaction kinetics a simple and easily accessible experimental technique that makes it possible to study reactions in dilute solutions and in small volumes and has therefore limited

[79] J. Doskočil, *Proc. 1st Intern. Polarography Congr., Prague 1951*, Vol. III, p. 65. Přírodověd. Nakl., Prague, 1952.

[80] J. Doskočil, *Collection Czech. Chem. Commun.* **15**, 614, 780 (1950).

[81] V. Fišerová-Bergerová, *Collection Czech. Chem. Commun.* **28**, 3311 (1963).

[82] Footnote 67, p. 650–655, where references are given.

[83] R. Brdička and J. Klumpar, *Casopis Lekaru Ceskych* **17**, 234 (1937).

[84] O. Hrdý, *Chem. Listy* **44**, 109 (1950).

[85] S. Štokrová, *Chem. Listy* **48**, 1083 (1954).

requirements as to the amount of the substances studied. The half-times accessible by classical polarography are of the order of seconds, and they can be reduced to tenths of seconds when oscillographic techniques are used. Times needed for mixing of solutions prevents the following of faster reactions, hence high dilutions of one component are to be used for the study of fastest reactions. Flow techniques can be adapted to polarographic indications, in particular the stop-flow method, when it is ensured that the motion of the solution does not affect conditions of the surface of the working electrode.

The main limitation of these polarographic techniques is that at least one of the components of the reaction mixture must be polarographically active. But this restriction is not so severe as it would appear, because most of the substances that can be studied spectrophotometrically are electro-active as well. Polarography offers the opportunity to follow the reaction using a sensor placed directly in the reaction mixture. In the study of enzymatic reactions, fear is sometimes expressed about the toxic effects of mercury. It should be kept in mind that usually only the mercury ions, not the metallic mercury, possess the toxic properties. When the dropping mercury electrode is properly handled, and a separate reference electrode is used, the concentration of mercury ions present in the solution is usually negligible for all except the most sensitive systems. If necessary the separation of the dropping electrode by a membrane is possible. Polarography can be readily applied in the study of light-sensitive systems, and polarographic vessels allow a simple temperature control over a wide temperature range.

# Section III

# Rapid Mixing

## [6] Rapid Mixing: Stopped Flow

*By* Quentin H. Gibson

## I. Introduction

A rapid chemical reaction which has been initiated in solution by mixing two reagents together may, in principle, be followed by two main ways. In the first, and historically the older of these, the mixed reagents are observed at a point downstream from the mixing chamber, where, so long as flow continues, the composition of the reaction mixture is invariant with time. This is the continuous-flow method; and, though calling for a considerable expenditure of reagents, it does not require the use of rapid recording methods for following the reaction. The second method is to examine a stationary portion of the mixture of the reactants, hence the "stopped-flow" method, using a recording procedure with a response time rapid enough to follow the reaction as it proceeds in the sample of fluid. It is obvious that this definition embraces any reaction followed, for example, in a conventional spectrophotometer; the discussion will, however, be limited to the case where a reaction mixture is generated by a specially constructed apparatus allowing observation to begin in times of the order of a few milliseconds and to proceed with a time resolution of the order of a millisecond or so.

Many of the general conditions that apply to the design of continuous-flow apparatus and which are dealt with in detail in Chapter [7], devoted to such apparatus, apply also to stopped-flow equipment, and no attempt will therefore be made to recapitulate them here. It is sufficient to refer to Chance,[1] who has shown how the fastest irreversible second-order reaction that can be followed in a flow apparatus is related to the response time of the recording system, the optical aperture of the monochromator, the sensitivity of the photomultiplier, and the extinction coefficient of the reactants. Although his analysis was primarily directed toward continuous-flow systems, much of the reasoning is equally applicable to the stopped-flow method. There is, however, one important difference: It is assumed that the precision with which the composition of a reacting mixture can be determined is limited ultimately by the numbers of quanta available for analysis during the time of observation. In the continuous-flow method the relevant quantity is the product of numbers of quanta per second times duration of flow, and it is possible to compensate for a light source of low intensity by the use of large volumes of solution with observation extended over a prolonged period. No such option is available, however, in the stopped-flow method where the response time of the system as a whole must be matched to the intrinsic rate of the reaction under observation; and indeed practical considerations suggest that the overall response time of the stopped-flow system is best held to about 0.1 of the rate constant being measured. If the overall response time is longer than this, and if a first-order reaction is being measured, formally it is necessary to replace the first-order rate equation with a rate equation for two consecutive first-order reactions, the velocity constant of the second of these being equal to the response time expressed as a rate constant for the recording system. In such circumstances, even though the value of one of the rate constants is known, the determination of the value of the other is somewhat time consuming. It is doubtful, also, if greater accuracy is achieved by using a longer time constant for the recording system since, although a record with less noise will be obtained, the form of the curve must be known more precisely to allow the evaluation of the rate constant for the chemical reaction. The practical effect is that the light sources used for stopped-flow work should be of the greatest possible surface brightness, monochromators should have large $f$ numbers, and the detectors have a high quantum yield.

The two designs chosen for detailed description here have the slightly unusual feature in common that the line of sight during observation is along the length of the observation tube rather than transverse to it. With this experimental arrangement the correct value of the velocity constant will be obtained for any first-order reaction, irrespective of the length of the

[1] B. Chance, *Discussions Faraday Soc.* **17**, 120 (1955).

observation tube employed in the stopped-flow method, although this is not true of second-order rate constants. Errors in the latter may be estimated by applying the "block" method of Roughton and Milikan,[2] dividing the length of the observation tube into an arbitrary number of blocks within each of which reaction is supposed to have proceeded to a calculated extent and summing the results for the various blocks.

## II. Stopped-Flow Apparatus of Gibson and Milnes

This apparatus has been chosen for description, first, because detailed constructional drawings have appeared and, second, because a slightly improved version is at present available commercially; for the author it has the additional advantage that its weak points are as well known as the stronger ones. It is probable, too, that more examples of this apparatus (the total is at present about 150) have been constructed than of any other single model. The apparatus in its original form as described by Gibson and Milnes[3] is shown in side elevation and plan in Figs. 1 and 2, which give the general layout of the parts of the apparatus.

The principle of operation may be followed by reference to Figs. 1 and 2. The reagents are placed in 25-ml reservoir syringes (marked $A$ in Figs. 1 and 2) from which they are transferred by proper setting of the valves to the 2-ml driving syringes ($B$ in Figs. 1 and 2). To make a determination, the driving syringes are forced in together by the syringe pushing block ($N$ in Figs. 1 and 2) by using a hydraulic unit ($O$ in Figs. 1 and 2). The reactants pass through the mixer ($C$ in Fig. 1) and then flow past the observation window ($E$ in Fig. 1) and on into the 2-ml stop syringe ($F$ in Figs. 1 and 2) driving its plunger upward until the handle strikes the stop ($G$ in Figs. 1 and 2). When this happens, flow is stopped suddenly without having been slowed down previously. The progress of the reaction is followed by the change in absorbance recorded by photomultiplier ($H$ in Figs. 1 and 2). To repeat the observation, the port valve ($I$ in Figs. 1 and 2) is opened by using the key ($P$ in Fig. 1) which allows the spent reaction mixture in the stopping syringe to drain away through the drain port ($J$ in Figs. 1 and 2) into a beaker. The port valve ($I$ in Fig. 1) is then closed and the cycle is repeated by operating the hydraulic unit ($O$ in Figs. 1 and 2).

## A. Construction Details

Most of the many stopped-flow apparatuses which have been described from time to time have been designed with a specific objective in view and

[2] F. J. W. Roughton and F. A. Milikan, *Proc. Roy. Soc. (London)* **A155**, 258 (1936).
[3] Q. H. Gibson and L. Milnes, *Biochem. J.* **91**, 161 (1964).

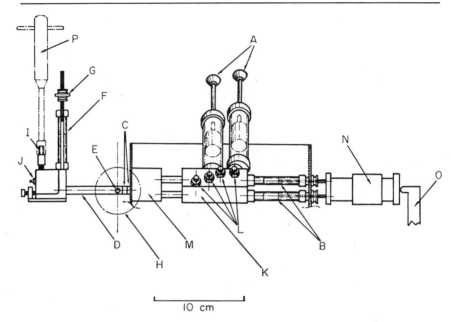

10 cm

Fig. 1. Side elevation of flow apparatus. *A*, Reservoir syringes (25 ml); *B*, driving syringes (2 ml); *C*, two mixers; *D*, observation chamber; *E*, observation window; *F*, stop syringe (2 ml); *G*, stop; *H*, position of photomultiplier; *I*, port valve; *J*, drain port; *K*, valve block; *L*, valves; *M*, delivery block; *N*, syringe-pushing block; *O*, lever to Armstrong R.C. 8 hydraulic actuator; *P*, position of key for operating stop valve. From Gibson and Milnes, footnote 3.

have therefore incorporated features that made them especially valuable for the study of a particular problem or group of problems. Thus, Chance[4,5] sought to obtain a high degree of fluid economy for his pioneer studies in the field of enzyme kinetics, and accepted in return for this high fluid economy a relatively short optical path length and the consequent need for great precision in photometric measurements and in stabilization of the light source; on the other hand, Allen, Brook, and Caldin[6] were concerned with nonaqueous systems and with making measurements at extremely low temperatures. Their stopped-flow apparatus is unique in using solutions driven by gas pressure rather than mechanically and is constructed throughout from materials immune to attack by ordinary chemical reagents. The instrument described here was designed with the kinetics of reactions of

[4] B. Chance, *J. Franklin Inst.* **229**, 455, 613, 737 (1940).
[5] B. Chance, *Advan. Enzymol.* **12**, 153 (1951).
[6] C. R. Allen, A. J. W. Brook, and E. F. Caldin, *Trans. Faraday Soc.* **56**, 788 (1960).

hemoglobin with gases in mind, and was subsequently modified somewhat to be more suitable for the study of enzyme intermediates in solution. One of the prime requirements of the design was good reproducibility and the measurement of moderately large absorbance changes with high precision.

In stopped-flow operation, it is necessary that the reagents be thoroughly mixed before the start of observation, as otherwise the records will be marred by apparent variations in absorbance caused by the alternation of materials of high and low absorbance in the field of vision of the photomultiplier. To reduce this noise to the minimum and also to eliminate the other artifacts that must arise if mixing is not completed rapidly, a relatively elaborate system of mixers was employed, the details of which are shown in Fig. 3. This figure shows the mixing units in cross section, in longitudinal section, and in a rear view. The reagents enter the mixing chamber, as shown in the center drawing, from the right-hand side and

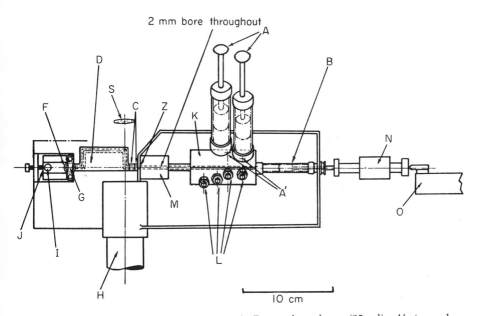

FIG. 2. Plan view of flow apparatus. $A$, Reservoir syringes (25 ml); $A'$, tapered ports for reservoir syringes (1-in-10 record taper); $B$, driving syringe (2 ml); $C$, two mixers; $D$, observation chamber; $F$, stop syringe (2 ml); $G$, stop; $H$, photomultiplier housing; $I$, port valve; $J$, drain port; $K$, valve block (stainless steel); $L$, valves; $M$, delivery block (stainless steel); $N$, syringe-pushing block; $O$, Armstrong R.C. 8 hydraulic actuator; $S$, biconvex quartz lens (25 mm diameter × 50 mm focal length); $Z$, baffle. The crosses on the valve block ($K$) show the positions of the valves ($L$) and of the tapered ports ($A'$) for the reservoir syringes ($A$). From Gibson and Milnes, footnote 3.

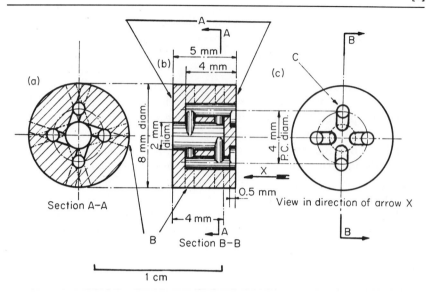

Fig. 3. End view (*a*), longitudinal section (*b*), and cross section (*c*) of Perspex mixers. *A*, Both faces lapped flat and parallel, *B*, holes drilled 0.5 mm or 0.75 mm in diameter, plugged with Perspex where indicated by the broken lines; *C*, four holes drilled 1 mm in diameter. (From Gibson and Milnes, footnote 3.)

are then driven into the central tube through 8 jets of 0.5 mm diameter which open tangentially into a 2-mm central hole. The jets are so placed that the two series attempt to spiral the liquid in opposite directions. The aim is to produce maximum turbulence while avoiding a violent spinning motion of liquid that might favor the onset of cavitation. In the original design two such mixers were used in series; but later experience showed that this was perhaps excessive, and appreciable "dead time" could be saved if only one mixer were used. So far, the mixers have been constructed only from Lucite; but there is no reason to doubt that other plastics could be employed if this were necessary. However, in the study of hemoglobin reactions as well as in most biochemical reactions, the solvent is water and Lucite is a suitable material.

The observation tube, shown in Fig. 4 was designed, first, to give a long pathlength and, second, to permit adequate temperature control. The material is stainless steel, partly because this gives a reasonable compromise between thermal conductivity and mechanical and chemical properties, and partly because polishing the internal wall of the tube allowed the 2-cm light path to be transversed without undue loss of light or need for thorough collimation.

The most recent designs of observation tubes differ somewhat from that figured. The Lucite windows shown have been replaced by silica, and the four right angles through which the fluid must turn have been reduced to two, with a significant decrease in back pressure. The original design had the advantage of allowing the observation tube to be mounted by pressure applied from the end $B$ in Fig. 4 in line with the seating for the mixer. In the more recent designs the observation tube is set in an accurately milled keyway, to absorb the sideways thrust due to the out-of-line positioning of the mixing chamber and stopping syringe.

A further constructional feature is the valves used for controlling the admission of reagents to the driving syringes. Details are shown in Fig. 5. They are needle valves with 60° PTFE tips (Teflon) which seat in corresponding 60° depressions machined in the valve block ($K$ in Figs. 1 and 2). Four valves are required, two to control the connection of the reservoir syringes to the 4-ml syringes and two more to control the admission of fluid to the mixing chamber. This rather elaborate arrangement is used in order to withstand quite high hydrostatic pressures. When the hydraulic drive is used, it is estimated that pressures of the order of 150 psi are generated and these could drive out the barrels of conventional ground glass taps. The chief disadvantage of the valves is that, when the PTFE

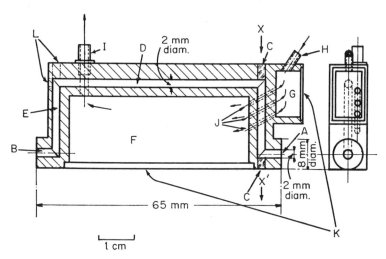

Fig. 4. Plan of observation chamber with a 20-mm optical-path observation tube. $A$ and $B$, Inlet and outlet, respectively, for reagent mixture; $C$, Perspex plugs polished at both ends to form windows; $D$ and $E$, 2 mm diameter drillings; $F$, main chamber; $G$, inlet chamber for temperature-controlled water; $H$ and $I$, inlet and outlet port, respectively, for temperature-controlled water; $J$, drillings between chambers $F$ and $G$; $K$, chamber lids pressed into recesses after milling; $L$, stainless-steel plugs pressed in after drilling; $X$-$X'$, light path. (From Gibson and Milnes, footnote 3.)

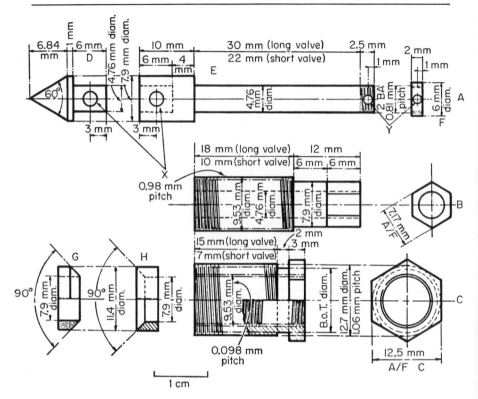

FIG. 5. Details of valves for controlling liquid flow. A, Central valve spindle (composed of three pieces: D, PTFE 60° tip; E, stainless steel shank; F, stainless steel nut); B, stainless steel collar; C, double-threaded stainless steel collar; G, polythene washer; H, dished stainless steel washer. The holes X are drilled to take a 2.38 mm diameter stainless steel pin, and the holes Y to take a 1.58 mm diameter stainless steel pin. (From Gibson and Milnes, footnote 3.)

tips are moved up and down by the action of the screw threads, there is an appreciable displacement of fluid which may come from the mixing chamber side or from the driving syringe side, depending upon whether or not the stopping syringe is full and where the greatest resistance to motion happens to be. As a result, the first operation of the apparatus after refilling the driving syringes must be regarded as unreliable and the result discarded. This is true even if the materials used for filling the syringes are the same as those previously employed.

*Temperature Control*

The arrangements for temperature control of the apparatus are thorough, not only because temperature control is in itself desirable, but because, with a long optical path, close temperature control is necessary to avoid optical artifacts due to thermal effects in the observation tube. The general character of these artifacts is illustrated in an exaggerated form by the experiment illustrated in Fig. 6. This figure shows the apparent changes in absorbance observed at 530 mμ (though the wavelength has little effect on the size of the absorbance change), when the temperature of the liquids entering the observation tube is widely different from the temperature of the tube itself. These changes have nothing to do with changes in absorbance as ordinarily understood in biochemistry, and are pure optical artifacts since water is the only liquid present in the stopped-flow apparatus. The upper half of the record shows the apparent changes in absorbance when ice cold water was injected into the observation tube at room temperature, while the lower half of the record shows the effect of introducing warm water into an observation tube cooled to 0°. The difference in temperature of 20° was the same in both cases. The two traces are almost mirror images of one another, the apparent change in absorbance reaching a maximum in a little over 0.1 second and disappearing again in the course of 2 or 3 seconds. The explanation offered is that, when the steel is warm and a cold solution is injected, a temperature gradient is quickly established across the liquid after stopping. Correspondingly, there is a gradient in refractive index with an effect equivalent to that of a positive lens. In the circumstances of the optical system used in the stopped-flow apparatus, the establishment of a positive lens within the observation tube gives a

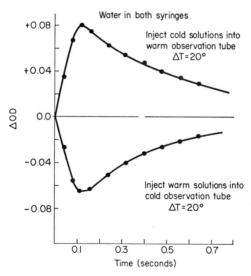

Fig. 6. Temperature difference artifact in stopped-flow apparatus; 2 cm path stainless steel; 2 mm observation tube.

change in light transmission corresponding to an apparent increase in absorbance. Naturally, the positive lens effect does not endure, since heat flows from the steel into the liquid and the gradient across the observation tube disappears, and with it the lens effect, in the course of a few seconds. The overall change observed is thus a rapid rise as the gradient is established, and a slower decrease as the temperature difference between the observation tube and the liquid which it contains disappears. Although in ordinary use great differences in temperature of the kind used in obtaining Fig. 4 would rarely be met with, small differences in temperature between the different parts of the apparatus are likely to be found; and, if the apparatus is employed without circulation of temperature-controlled water, even if all the solutions and the apparatus are supposed to be in equilibrium at room temperature, small changes in absorbance of the order of 0.001 to 0.002 are frequently seen to occur with a time scale similar to that shown in Fig. 4. Thus, even if determinations at room temperature are required, temperature-controlled water should always be circulated through the apparatus.

As the 2-ml driving syringes and the stainless steel valve block are immersed in an open water bath, it is necessary to arrange that a fixed volume of water be available for circulation. The scheme employed is shown in Fig. 7, where the bath water passes through a heat exchanger consisting of about 4 ft of 0.5-inch copper tubing coiled in a spiral that dips into a large water bath maintained at the desired temperature. The output from the pump is divided at a T-junction so that a part of the liquid circulates into the main water bath and a part passes through the observation tube block before being returned to the main bath. The return from the observation tube may be arranged so that the flow is visible, since it is important to be quite sure that adequate circulation in the observation tube block is taking place. If the liquid in the large temperature-controlled bath is briskly stirred, the difference between the temperature of the large bath and of the water bath in the stopped-flow apparatus will amount to about 1° for a difference of 20° between room temperature and the bath liquid. The actual temperature in the water bath of the stopped-flow apparatus is recorded with an immersion thermometer.

The arrangement described is not, of course, the only possible one, since temperature control units are currently available that might be applied to regulate the temperature of the water bath directly. The scheme described is, however, simple and effective, particularly when a large series of experiments at a single temperature is to be performed. When determinations are to be made at a series of temperatures, the large thermal capacity of the main water bath constitutes something of an embarrassment since most of the time of the experiment is spent in waiting for the

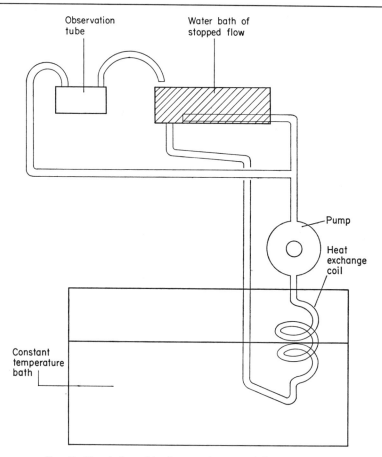

Fig. 7. Circulation of bath water in stopped-flow apparatus.

temperature of the water bath to change to a new value. An alternative is to maintain the large water bath at the lowest temperature that will be required and to put heaters in the water bath of the stopped-flow apparatus of a capacity sufficiently great to raise the temperature of the circulating water to the desired value. This arrangement requires a thermostat and also a variable voltage transformer so that the heat output can be adjusted reasonably closely to the total load required, in order to avoid excessive temperature swings in the water bath of the stopped-flow apparatus.

*Light Source*

By far the simplest source is a tungsten ribbon filament lamp supplied from a commercial stabilized current supply as, for example, Kepco or

Sorenson Nobatron. With such a source, the limiting factor in recording a relatively slow absorbance change (i.e., so that mechanical disturbance and the movement of scattering material in the observation tube does not come into play) appears to be set by the occurrence of slow oscillations in the behavior of the filament within the tungsten bulb. These oscillations have typically a period of 2–5 seconds and give rise to apparent absorbance changes of the order of 0.0002. Their extent varies markedly for different lamps and is also to some extent a function of the current through the lamp. While the cause is not known, it seems possible that an effect analogous to that in an automobile turn-indicator flasher may be responsible since they are almost certainly not associated with changes in the current, and are seen whether the current is supplied from a stabilized electronic source or from a large-capacity storage battery. The remedy is to have a supply of lamps and to select one that shows only a small low frequency component.

A worthwhile extension into the ultraviolet region can be obtained by substituting a quartz-iodine-tungsten lamp for a conventional ribbon filament lamp. These lamps have a coiled filament and are quite well adapted to use with ordinary monochromators. Significant output is available from these lamps down to about 280 m$\mu$. For observations at still shorter wavelengths, it is necessary to use a xenon discharge lamp supplied from a direct current source. These lamps have been improved significantly in recent years, but there is still some tendency for the arc to jump within the lamp from time to time. For this reason they are not well suited for use in a single-ended instrument unless the absorbance change to be measured is of the order of 0.1 or more and if long-term stability is not required, which is indeed often the case with the stopped-flow apparatus. The xenon lamp has the advantage of an intrinsic brightness which is perhaps some 50 times that of a ribbon filament and so is capable of giving a lower shot noise effect than the other sources, but this advantage is generally more than counterbalanced by the long-term stability problems. Since the arc usually moves laterally within the bulb, some improvement may be obtained by mounting the monochromator with the slits horizontal.

### Monochromators

The use of a monochromator is by no means essential, but the convenience and flexibility which it provides are such that it is recommended wherever at all possible. The most desirable quality in a monochromator to be used with the stopped-flow apparatus is a large $f$ number. The monochromator need not be of great physical size, however, since the entrance pupil of the observation tube is small and the larger amount of light obtainable from a monochromator with large grating surface or large prism cannot be effectively used. Perhaps the most important question to be decided is whether to use a grating or a prism monochromator. The disadvantage of the grating monochromator is that it tends to have an inherently higher level of stray light than a prism instrument of corresponding quality and, furthermore, when used at short wavelengths, filters must be used to remove second-order light. This presents serious difficulties at wavelengths shorter than about 250 m$\mu$ since filters which will transmit in that region while being opaque to longer wavelengths are not readily available. For this purpose a prism instrument has considerable advantages since a stray light filter is not required. For most general work, however, a combination consisting of a tungsten light source and a grating monochromator with a blaze in the middle ultraviolet at about 330 m$\mu$ is very satisfactory. The blaze of the monochromator together with the multiplier characteristic and the lamp characteristic work together to give an approximately uniform sensitivity between about 380 m$\mu$ and 750 m$\mu$, if a multiplier with extended red sensitivity is employed. In this same range, if a prism instrument is used, it will be necessary to readjust the slits on a number of occasions to maintain an approximately uniform bandwidth, whereas with the grating monochromator, the bandwidth is independent of wavelength. Furthermore, if it is intended to use the instrument for fluorescence recording, the prism monochromator offers an appreciable disadvantage in that the output available falls off because of the decreasing bandwidth at any given slit opening at the same time that the output from the exciting source is falling off. The result may well be that it is difficult to obtain sufficient excitation energy at, say, 280 m$\mu$; and much more satisfactory performance may be secured from a grating instrument.

If filters are used, it will be necessary to prepare a calibration curve if accurate results are to be obtained; and this may be true in some cases even when interference filters rather than glass filters are employed. Calculations using known photomultiplier response curves, filter transmission curves, and light source characteristics show that in the case of compounds having sharp absorption bands, such as hemoproteins, significant deviations from linearity between concentration and absorbance are to be expected even if the total absorbance change is limited to a value as small as 0.05.

*Photomultiplier*

The application of photomultipliers in stopped-flow apparatuses employed for following absorbance changes does not place any serious demands upon them, and any photomultiplier may be used with satisfactory results. The only quality of importance is the photocathode sensitivity since this will determine the photomultiplier contribution toward the overall noise level. In all circumstances there is an abundance of light, and high multiplication factors are quite unnecessary. The choice of spectral response characteristic depends upon the work to be carried out, but perhaps the most satisfactory multiplier is an extended red sensitivity type with appreciable response up to 800 m$\mu$ and with the cathode deposited on a silica surface. Among end window types, the EMI type 9558Q is expensive but satisfactory.

In typical biochemical applications where a small absorbance change is to be measured, the linearity of response of the photomultiplier is not important since deviations from linearity over a range of, say, 5% are likely to be small under almost any conditions. However, where linearity of response is important, consideration should be given to strapping together sufficient of the dynodes to allow a reasonable voltage per stage to be applied to the multiplier. The alternative approach of reducing the light input is unsatisfactory because of the effect on the noise level. Although it is convenient to employ a cathode follower or emitter follower, this is in no way necessary; and, provided that the anode load of the multiplier is not made greater than about 100 K, an oscilloscope may be operated directly. The stray capacitance associated with a normal length of lead is such that the time constant of the resulting combination will be of the order of 1 msec or less. When a follower is employed, it will be necessary to supply loading capacitances to bring down the time response to a convenient value; this is best done by providing a series of capacitances and a wafer switch so that a time constant may be selected as dictated by the reaction under study. It is also convenient to be able to switch from one time constant to another during the course of a recording when, for example, a rapid change is succeeded by a slow one. The oscilloscope may be triggered for a second sweep still using the same solution in the stopped-flow observation tube, and the time constant for recording be changed at the same time that the time base of the oscilloscope is switched. On the whole, an emitter follower is preferable to a cathode follower since fewer voltages must be supplied and microphony is not a problem in the case of the emitter follower. It is scarcely feasible, however, to give circuits and circuit values since these must depend upon the design in individual cases.

### Photomultiplier Supplies

A well-regulated PM supply (e.g., Fluke) is necessary since the output of the photomultiplier is sensitive to small changes in the supply voltage. Any good commercial supply will be found satisfactory in this respect, however, if it has regulation to 0.1% or better. Since very short time constants are generally inapplicable in stopped-flow work, it is sufficient to have a chain current of the order of 0.5–1 mA, which will not impose limitations due to the rate of charging of stray capacitances associated with the dynodes and their wiring. A relatively high resistance chain also avoids the need for air cooling of the chain and artifacts due to "warming up" effects.

Some commercial PM supplies have appreciable external magnetic fields to which photomultipliers are very sensitive. These fields will show themselves as a modulation of the voltage output with a steady light source, usually at 120 cycles. The trouble can be cured by reorienting the power supply or placing it at a greater distance from the photomultiplier.

## B. Recording Results

For the most rapid reaction that can be studied by the stopped-flow method, the use of a cathode ray oscilloscope and camera is almost obligatory. Among oscilloscopes by far the most suitable is a storage oscilloscope since, with it, all the information that must be recorded for a single complete kinetic run may be assembled on the screen and inspected before a photograph is taken. The only disadvantage of the storage oscilloscope is that contrast and photographic quality are not so good as with a conventional photographic oscilloscope operated under ideal conditions.

The oscilloscope may be triggered either by the absorbance changes that occur when a fresh portion of reactants is delivered into the observation tube, or by causing the stopping syringe to provide a signal applied to the external trigger of the oscilloscope time base. The second method is preferable since irregular operation may occur occasionally with the first when the signal level is low as compared with the noise. The least satisfactory is a velocity-operated trigger where, with high noise levels, triggering may become quite irregular. The signal-operated trigger, whether velocity or DC level, has the disadvantage that the triggering necessarily occurs early in the flow cycle, and at the highest speeds a large proportion of the trace may be devoted to recording changes in absorbance during flow.

Photography of the oscilloscope screen may be with a Polaroid oscilloscope camera or a 35 mm camera. With the Polaroid camera, a record is available immediately and is usually of the same size as the original image on the oscilloscope screen. Recording on 35 mm film is economical, and in addition, all the exposures made in a day's work are permanently fixed on the film in an invariable order, while each oscilloscope photograph obtained by the Polaroid method must be identified immediately upon preparation.

In general the simplest possible camera is the most serviceable in this application, particularly where a photographic oscilloscope rather than a storage oscilloscope must be used. In these circumstances, a camera with automatic film advance is a liability, since in order to obtain a complete frame with (i) a reaction record, (ii) a line corresponding to the absorbance when the reaction is completed, and (iii) the necessary reference voltages, it is necessary to open the shutter 4 times; and these openings should not be accompanied by film transport. Furthermore, no elaborate shutter mechanism is needed since brief time exposures are required with slow film to photograph storage oscilloscope traces; and, where a photographic oscilloscope is employed, the camera shutter must be opened before the oscilloscope time base is triggered.

To evaluate the 35 mm film records, the film is put into a conventional photographic enlarger and the record is traced by hand at a considerable magnification. The tracing may then be measured using an ordinary ruler. It is important to use the simplest means of measuring the record, since by far the slowest process in the whole series of operations is reduction of the data from voltage excursions to changes in absorbance.

The steps involved in data reduction are illustrated in Fig. 8. In the upper part of this figure are shown the data of a typical reaction record. At the left is the voltage excursion produced on a relatively insensitive range of the oscilloscope when the light beam is interrupted. This record, which is regularly obtained a few seconds after the reaction record is run, gives a measure of the voltage output for the fully reacted mixture and will be abbreviated VFR. At the right-hand side is the reaction record made on a scale ten times more sensitive, and with a rapid time base sweep. There is also a record of the transmission of the fully reacted mixture which corresponds to the VFR value as recorded at the left-hand side of the figure. In Figure 8B is a diagram showing the voltage changes as they would appear if the reaction record and the reference voltage had been recorded with uniform oscilloscope sensitivity. It follows from the definition of absorbance as $\log_{10} I_0/I$ that the difference in absorbance between the fully reacted mixture and the mixture at any earlier time is given by $\log_{10} \text{VFR}/V_t$, where $V$ is the voltage output at any time $t$. The necessary figures might be obtained by direct measurement of the record B, but may be more precisely determined by measuring the excursion from VFR at any time $t$ on an expanded record of the kind shown in panel A and subtracting the voltage so obtained from the reference voltage recorded on the less sensitive scale of the oscilloscope. The example shown in Fig. 8 is that of a reaction in which the change measured is associated with an increase in the amount of light transmitted. If the change measured is associated with a decrease in the amount of light transmitted, that is, if the reaction record of panel A is inverted, then $\Delta$ absorbance is given by $\log_{10} V_t/\text{VFR}$.

Where the total absorbance change is of the order of 0.05, the error in taking $\log x$ as proportional to $x$ is less than 2%, so that it is sufficient to make a measurement of the reaction record, obtain the total absorbance change observed, and equate this with the total size of the excursion as

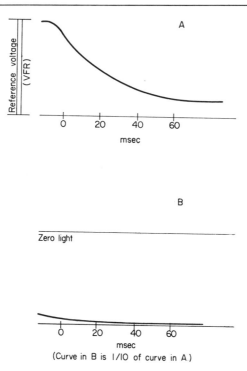

Fig. 8. Example showing method of recording and evaluating stopped-flow reaction record. Panel A shows the record as regularly obtained with VFR and trace. Panel B shows the appearance that would be seen if the oscilloscope sensitivity used in recording the trace was the same as that used in measuring VFR.

measured in millimeters on the record and to express the values at successive time intervals by supposing that the absorbance is proportional to the excursion on the reaction record.

## C. Performance Figures

Two useful indices of the performance of a flow apparatus, are: (1) The greatest first-order rate constant which can be measured, and (2) the greatest second-order rate constant that can be determined in an irreversible second-order reaction. The first of these indices depends solely on the efficiency of mixing and on the speed with which the reaction mixture can be introduced into the light beam; the second depends in addition on the extinction coefficients of the reaction, the characteristics of the light source, and the minimum absorbance change that can be measured with a given precision. For the apparatus described here, the greatest rate constant that would normally be measured with a 2-cm path is of the order of

200–250 per second, while with the 2-mm path observation cell it is about 500 per second. The greatest second-order rate constant that could be measured in the case of a reaction with a difference in specific molar absorptivity of $10^5$ as the result of the reaction (values of this order are reached in some hemoprotein reactions) would be of the order of $5 \times 10^8$ $M^{-1}$ $sec^{-1}$ to $1 \times 10^9$ $M^{-1}$ $sec^{-1}$ for a 2-cm tube, and about one-fifth as great for a 2 mm depth of solution. The gain in the first-order rate constant which can be measured with the shorter path length cell is due to the increase in flow velocity which can be obtained with the straight through flow system used in the latter case, since the observed value of first-order constant is independent of path lengths in the case where the observation beam passes along the length of the observation tube.

The greatest rate constant, either first or second order, that can be measured in a stopped-flow apparatus is determined effectively by the "dead time" of the apparatus, which is a measure of the time which is required for reaction mixture to flow from the mixing chamber to the point of observation. Where observation is made at right angles to the direction of flow and the volume observed is small compared with the total volume between the mixing chamber and the point of observation (i.e., when the observation is made at a point along the observation tube), then the dead time may be determined equally well by measuring the volume of the system between the mixing chamber and the point of observation and the flow velocity in the apparatus; or, alternatively, by using a chemical reaction whose rate may readily be varied and determining the extent of the observed reaction for two widely different rates. A convenient reaction for this purpose is the combination of reduced myoglobin with carbon monoxide, whose rate is proportional to carbon monoxide concentration. If one of the syringes contains buffer equilibrated with 1 atm of carbon monoxide, the observed rate of combination of a dilute solution of myoglobin will be about 250 per second; and proportionately less with lower partial pressures of carbon monoxide. It is not, of course, necessary to measure the carbon monoxide concentration accurately, but only to measure the rate constants for the two concentrations of carbon monoxide, together with the total change in absorbance in each case. Then, writing the absorbance change seen for the first concentration of carbon monoxide as $A_1$, and that with the second concentration of carbon monoxide as $A_2$, and the total absorbance change which would have been observed had the stopped-flow apparatus operated instantaneously as $A_0$, we have $A_1 = A_0 \exp(-k_1 t)$ where $k_1$ is the rate constant measured with carbon monoxide concentration 1 and $A_2 = A_0 \exp(-k_2 t)$ where $k_2$ is the rate measured with the second carbon monoxide concentration and $t$ in each case is the dead time to be determined. Dividing the equations by one another and passing from numbers to logarithm, gives: $\ln(A_1/A_2) = (k_2 - k_1)t$. When the observation tube is viewed longitudinally, a different set of relations pertains. If the rate of the reaction is such that the transit of an element of reaction mixture through the length of the observation tube occupies several half-lives of the reaction, then the parameter that will determine whether or not some reaction is observed is the volume between the point of mixing and the point at which reaction mixture enters the first part of the observation tube included in the observing optical system. On the other hand, if the rate of the reaction is relatively slow so that the time taken for an element of fluid to traverse the observed volume is short compared with the half-time of the reaction, then the chemically determined dead time will approximate to the time required to fill one-half of the observed volume.

To give concreteness to the discussion, if it is assumed that the reaction under study is first order and the flow down the tube is of the mass type, then writing the total change in absorbance observed as $A_{obs}$, $k$ as the rate constant for the reaction, $A_0$ as the absorbance change which would be observed in the case where the dead time was zero, and $t_1$ the time taken by an element of fluid to travel from the point of mixing to the beginning of the observed volume, and $t_2$ the time for an element to travel from the point of mixing to the end of the observed volume, then it may readily be shown that

$$A_{obs} = A_0[\exp(-kt_1) - \exp(-kt_2)]/k$$

For the practical case which would occur in the case of the stopped-flow apparatus under discussion with a volume flow of 30 ml/sec, $t_1 = 1.2$ msec, $t_2 = 3.3$ msec, the geometric dead time, i.e., the time taken for an element of fluid to travel from the point of mixing to the center of the observation tube, is 2.2 msec. If the chemical dead time were determined with a reaction proceeding at 250 per second, and any low rate for the second measurement, the dead time would come out to 1.98 msec, while if the more rapid of the reactions studied were raised to 1000 per second, the apparent dead time would drop to 1.31 msec.

## D. Noise Level

The shot noise exhibited by the apparatus depends, of course, upon the surface brightness of the source, bandwidth of the monochromator, cathode sensitivity of the photomultiplier, and geometric setup in the optical path, as well as on the response time of the electronic circuits. The following figures apply to a Durrum-Gibson apparatus using a 108 W ribbon-filament tungsten lamp operated at rated current with a Bausch and Lomb small-grating monochromator and EMI 9558Q photomultiplier, with overall response time 1 msec and nominal bandwidth 2 m$\mu$ at all wave-lengths. The results were 340 m$\mu$, 3.3%; 400 m$\mu$, 0.9%; 530 m$\mu$, 0.9%; 600 m$\mu$, 1%; 750 m$\mu$, 1.5%. All these readings refer to determinations with water in the observation tube. Where a more strongly absorbing solution is employed, the figures should be multiplied by the square root of the ratio $I_{water}/I_{solution}$, where $I_{solution}$ is expressed in terms of percentage transmission. The noise level can, of course, be reduced by using a longer time constant or a greater bandwidth. In general the improvement in signal-to-noise level will be proportional to the nominal bandwidth employed at the monochromator, and in proportion to the square root of the increase in the time constant.

The signal-to-noise ratio together with the precision with which the rate constants are required determine the minimal quantities of material and the minimal absorbance changes which must be used. In general, the absorbance change measured may profitably be made of the order of 0.05 to 0.5 in the solutions after mixing (i.e., in a 2-cm pathlength in the instrument under discussion). The upper limit is set by the difficulty in making precise measurements of the small percentage transmissions that will be encountered if the absorbance change is greater than this. The total amount of an enzyme preparation required to make the determination is set not only by the absorbance requirements but by the volume of solution that must be delivered in the operation of the apparatus. To some extent, this volume is arbitrary, and satisfactory results may be obtained if a total volume of the order of 0.3 ml or so is delivered in obtaining each stopped-flow record; but, if sufficient solution is available, it is more satisfactory to use a volume of the order of 0.6 ml which washes out the observed volume with about a 7-fold excess of new reaction mixture. The least volume of solution which can be used is further influenced by the necessity of filling the working syringes, and here, the least volume is of the order of 2 ml. For enzyme solutions, the most satisfactory procedure is to fill the apparatus with the buffer which will be used, to put the enzyme solution in a storage syringe of 2-ml capacity, and then to transfer the solutions to and fro between the working and storage syringes until there is a uniform mixture within the apparatus. Thereafter, two blank squirts of the apparatus must follow in order to sweep out the drillings in the steel blocks and the approaches to the mixing chamber and fill them with reactants and reaction mixture. The apparatus is then adjusted to the sensitivity ranges and speed ranges dictated by the individual reaction, and two or three reaction records are obtained. Where the same concentration of an enzyme solution is to be studied with variable concentrations of a substrate, say, the requirement for enzyme can be scaled-down below the limit of 2 ml of solution per unit of information and satisfactory results may regularly be obtained with 1–1.5 ml.

## E. Problems and Difficulties of the Apparatus

### Air Bubbles

It is absolutely essential to have the apparatus completely free from air bubbles when operation is attempted; even a very small bubble in the observation tube will cause serious changes in absorbance, while larger amounts of air, if trapped in the stopping syringe, impart an elasticity to the system causing an oscillation of the fluid in the observation tube at the end of the driving period. The apparatus must initially be purged of bubbles by pushing liquid as rapidly as possible from the working syringes into the reservoir syringes, discarding the air bubbles as they are dislodged. Small gas bubbles may be removed by solution in gas-free water or buffer; and at the end of a day's work, after the apparatus has been washed out with clean water, it should be filled with gas-free water to discourage the separation of the bubbles in the event of possible changes in the environmental temperature. Furthermore, it is advantageous to carry out all transfers of solutions under positive pressures; that is to say, when filling the working syringe from a reservoir syringe, the preferred procedure is to push on the reservoir syringe rather than to pull on the driving syringe. Once the apparatus has been freed from air bubbles, however, there is no reason why, with proper care, new ones should be introduced.

### Leakage

There should be no difficulty in preventing leakage from the valve stems or past the valves themselves. The syringes, unfortunately, are another matter. Here there is no remedy but selection of individual syringes having good performance characteristics. As the plungers are of more closely uniform dimensions than the barrels, it is not necessary, in carrying out selection, to examine all possible combinations; but the syringes should be tested by measuring for leakage when exposed to a constant high pressure with water inside.

Some improvement in syringe performance may be obtained by lubricating with a high-viscosity mineral oil or a suitable silicone fluid. The chief reason for continuing to use these syringes, which have a ceramic plunger and glass barrel, is that the coefficients of expansion of the material of the plunger and of the barrel are the same; and the syringes will operate equally well over the range of temperature from 0 to 40°. Attempts to combine materials such as plastics and glass or plastics and stainless steel have been unsuccessful because of the wide difference in temperature coefficients of the materials concerned. It is possible that satisfactory results could be obtained using plastic disposable syringes with rubber pistons since these appear to have very good low pressure leakage characteristics, but they have not yet been tried out at the high operating pressures required in the stopped-flow apparatus.

### Diffusion from One Working Syringe to Another

If a long interval is allowed to elapse between successive ejections of fluid from the working syringes, some diffusion from one side of the apparatus to the other will occur; and, in cases where the reagents differ widely in concentration, this diffusion may reach proportions sufficient to influence the absorbance change observed. For example, when glucose oxidase was allowed to react with glucose under anaerobic conditions, the concentration of glucose was sometimes 10,000 times that of the enzyme; and, under these circumstances, if the apparatus were allowed to stand for more than 2 or 3 minutes between successive activations of the driving syringe, the excursion observed was reduced, presumably due to diffusion of glucose into the enzyme side of the apparatus. Problems of this kind are particularly likely to be met with in the case of enzyme reactions when the concentrations of the reagents show wide disparity. There does not appear to be a remedy for this problem other than making several squirts in close succession and an increase in the volume delivered at each individual squirt.

### Syringe Breakage

There is a sharp water hammer effect at the stoppage of flow which can clearly be heard in the form of an audible "ping" as flow stops. The pressures used in the apparatus are such that the glass syringe used for stopping flow is stressed to near the limit of its endurance; and an exceptionally violent pull on the operating lever of the hydraulic apparatus may break the stopping syringe. This is particularly true of the Durrum-Gibson apparatus which uses a more effective hydraulic unit than that available for the earlier Gibson-Milnes apparatus.

### Breaking Valve Tips

The drawings of the valve mechanism shown in Fig. 3 call for the provision of Teflon tips whose shanks have a diameter of $3/16$ inch and pierced by a hole of $1/16$ inch for a locating pin. These tips are somewhat fragile, and if the valves are screwed home too firmly they will fracture at the level of the retaining pin. Although it requires only a few minutes to turn up a new insert from a piece of Teflon rod, the loss of time and perhaps of enzyme may be irritating. It is recommended that, when tightening the valves, the pressure applied should not exceed that available from the fingers alone when applied to a tommy bar thrust through the nut driver used for operating the valve. The power available from the wrist and arm is not required to give a fluid-tight seal.

The valves should be removed at intervals of a few months and inspected for damage to the seating surfaces. The high pressures can lead to cutting of the Teflon tip by the valve seatings, and they should be removed and refaced from time to time. When carrying out this operation, it should be borne in mind that the tip of the valve is able to rotate independently of the body; otherwise some difficulty may be experienced in chucking the work.

### Anaerobic Operations

It is impossible to transfer fluid into the stopped-flow apparatus without some contamination by atmospheric oxygen; and, even if the apparatus is first flushed out with thoroughly degassed water, and then the working solution is introduced directly from a glass tonometer utilizing a short length of capillary plastic tubing as a connector in place of a reservoir syringe, some oxygen is inevitably entrained. With ordinary careful operation, however, the concentration of oxygen can be reduced to about $2 \times 10^{-7} M$. To obtain a significant improvement on this figure, it would seem necessary to enclose the apparatus as a whole within a nitrogen tent of plastic material. It should be noted that because of the rather large number of angles and crannies, especially in the valve block and in the syringes themselves, a few small gas bubbles may be trapped; and these can give rise to problems within the time of experimentation with a system that is particularly sensitive to small quantities of oxygen. If the concentration of residual oxygen is greater than the limit just mentioned, the remedy is to fill the apparatus with degassed buffer and allow it to stand for several hours. On repeating the experiment, satisfactory results will usually be obtained.

### Contact with Stainless Steel

Although stainless steel has been employed in construction because of its high resistance to corrosion, examples have been met with where enzymes and substrates have been influenced by contact with the interior of the apparatus. It was found, for example, that quite rapid hydrolysis of orthocarboxyphenylphosphate occurred when it was left in contact with the interior of the valve block, but did not take place in the glass syringes. It was possible to reduce the effect greatly by drying the apparatus and coating it with a liquid silicone. It is noteworthy that even large concentrations of EDTA did not have any perceptible effect in reducing this hydrolysis. Although such effects have occurred rarely, the possibility should always be borne in mind when anomalous results are obtained.

### III. Microstopped-Flow Apparatus of Strittmatter[7]

This apparatus was developed especially to work with small volumes of solutions and so is particularly interesting in biochemical applications where the supply of material often determines whether or not stopped-flow spectrophotometry can be employed. The apparatus owes its fluid economy to a combination of the functions of a mixing block and observation tube within a single unit, on the principle shown in Fig. 9. Before operation begins the plunger containing the mixer rests at the bottom of the cuvette as in drawing A. After operation, the plunger has moved up because some fluid has been expelled into the microcell, thus opening up a channel for light to pass from the monochromator to the photomultiplier and providing an observation tube of 1 cm path length for an expenditure of 30 mm³ of fluid. To repeat the determination, the plunger is pressed gently back to

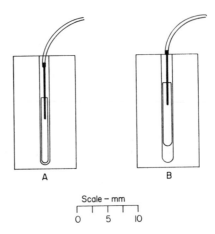

A                    B

Scale – mm

0        5      10

Fig. 9. Plunger positions in the microcell before (A) and after (B) mixing.

[7] P. Strittmatter, *in* "Rapid Mixing and Sampling Techniques in Biochemistry" (B. Chance, R. Eisenhardt, Q. Gibson, and K. Lonberg-Holm, eds.), p. 76. Academic Press, New York, 1964.

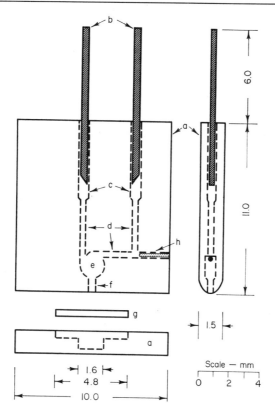

F1G. 10. The construction details of the plunger containing the mixing chamber. $a$, Plexiglas plunger; $b$, sections of number 22 stainless steel needles; $c$, 0.53-mm diameter channels; $d$, 0.38-mm diameter channels; $e$, mixing chamber; $f$, 0.50-mm diameter exit channel; $g$, Plexiglas plug; $h$, tapered glass rod. From Strittmatter, footnote 7.

the bottom of the cuvette and a fresh portion of solution is delivered. This arrangement appears to offer something like the ultimate in terms of fluid economy since the total volume of liquid delivered from the syringes is utilized for observation. As the author points out, however, this economy of fluid is obtained at the cost of the carry-over of 3–5% of fully reacted solution from the previous operation of the apparatus into the new reaction mixture, since, of course, no wash-through of the observation cuvette is attempted. Further details of the construction of this apparatus are shown in Figs. 10 and 11. As shown in Fig. 10, the mixing chamber is built into the lower part of the plunger and delivers through a short channel, $f$, into the cuvette. The design calls for a mixing chamber with a volume of approximately 1 $\mu$l in the plunger, but there is perhaps some

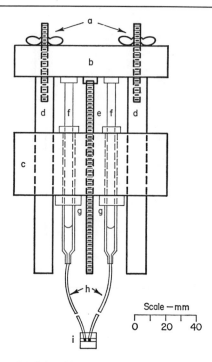

Fig. 11. Syringes and driving block. *a*, Threaded rod; *b*, aluminum driving block; *c*, aluminum syringe block 4.0 × 3.0 × 10.0 cm; *d*, 1.3 cm aluminum rods; *e*, threaded rod; *f*, 1.0-ml tuberculin syringes; *g*, rubber sleeves; *h*, polyethylene tube, 0.58 mm internal diameter; *i*, microcell and plunger. From Strittmatter, see footnote 7.

doubt as to whether this is really necessary since mixing of fluid will continue within the cuvette after the liquids have passed through jet F. There is also a period of the order of 2–3 msec which is required for the movement of the plunger from its position of rest to its working position, so this time is also available for the completion of mixing.

As the plunger must move to deliver the sample into the observation tube, there must be a flexible connection between the syringes and the plunger. This connection, as illustrated in Fig. 11, is made through fine-bore polyethylene tubing; the reagents are contained in 1-ml tuberculin syringes mounted in an appropriate frame whose movement is controlled by the threaded rod (*e*) shown in the drawing. In the form described, monochromatic light comes from the optics of a Beckman DU spectrophotometer, and changes in absorbance are followed with a 1P28 photomultiplier working into a 100 K resistor driving a Tektronix oscilloscope directly. The time constant of the combination is not given, but, depending on the stray capacity and length of leads, it may be of the order of 1 msec. Because of the small angle embraced by the light path, it was necessary to use rather large slits and, instead of computing absorbance changes directly, calibration curves were prepared by injecting samples of known composition into the microcell and reading their apparent absorbance with the bandwidth actually employed in the experiment.

Temperature regulation was provided by the normal thermospacer for the Model DU spectrophotometer cell housing and an aluminum shield surrounding the syringes and polyethylene tubing outside the cell housing. Fluid from a constant-temperature bath was circulated through the system.

Among other attractive features of the design is the availability of a triggering mechanism, depending upon the fact that the Lucite plunger in which the mixing chamber and delivery channel are machined cuts off the light more strongly than most solutions likely to be used with the instrument. Thus, there is a large transient in optical absorbance at the time when the plunger moves out of the light path of the instrument, and this is utilized to trigger the time base of the storage oscilloscope.

The design has other advantages: (1) A micromixing arrangement within the moving plunger which is separated from the reactant syringes by flexible tubing is quite versatile. It can be adapted to many observation vessel shapes so that it provides a general method for following rapid reactions by spectrophotometric, fluorescence, or other physical methods. (2) The mixing plunger itself can be constructed with common laboratory materials and equipment. (3) Customary laboratory equipment can be adapted to provide a light source and detector system.

The development of this design is not yet complete, and insufficient performance figures are available to allow a detailed comparison between it and the older equipment described in detail earlier in the article. Some of the disadvantages entailed by the small fluid volume are very fairly pointed out by the author in his article. It appears, in addition, that the apparatus could not readily be modified to work with anaerobic preparations, partly because of the difficulty of filling the syringes without contact of the solutions with air and partly because of the very high permeability of polyethylene to dissolved gases that would rapidly soak through the connection tubing and contaminate the portion of solution due to be used in a given squirt of the apparatus. Last, there is the question how far the economy of fluid usage is real, since, although the volume of solution employed to obtain a single record is very small, being only about $\frac{1}{20}$ of that used in the Gibson apparatus, this is not really the relevant quantity for a comparison. As has already been said, a minimum of 2 ml of solution of appropriate strength is required to obtain a unit of kinetic information using the Gibson apparatus, and it may be that, allowing for filling the syringes and the manipulations involved in setting up, a quantity of the order of 0.3–0.4 ml would be required in Strittmatter's apparatus. On account of the longer path length used in the macro apparatus, the solutions need only be about half as concentrated as those required for the micro apparatus, so that the real gain in fluid economy may be between 2 and 5 times, rather than 20 times as would be suggested by the relative volume of fluid delivered.

Although it is perhaps too early to attempt a final assessment of this equipment, there can be no doubt that it is among the most ingenious and novel designs to appear in recent years; and, as such, worthy of mention, especially as it can be constructed using comparatively simple machine tools.

### IV. Observation of Rapid Fluorescence Changes by the Stopped-Flow Method

Apparatus for following changes both in intensity of fluorescence and in polarization of fluorescence has been developed and is described by Gibson, Hastings, Greenwood, Massa, and Weber.[8] The same equipment may be used for both purposes, and a diagram showing construction of the observation cell and the layout for following polarization is shown in Fig. 12.

The observation tube for following changes in fluorescence intensity alone can readily be fitted into an ordinary stopped-flow apparatus as described in the first section of this article. For following changes in fluorescence polarization, however, it is necessary to employ two photomultipliers, and a specially constructed stopped-flow apparatus is needed. In the case of the apparatus figured here, the modifications include the substitution of a xenon lamp for a tungsten filament and a lens and mirror system capable of focusing the exciting light upon a window placed on top of the observation tube. The observation chamber itself is constructed by drilling two holes of 3 mm diameter through a block of stainless steel in directions perpendicular to one another and to the direction of flow of the fluid from the mixing chamber. Of the four holes now present in the block, three are closed with windows, while the fourth is sealed with a polished plug with a rhodium-plated end which is placed opposite the window used for admitting the exciting light, and which serves to reflect the light from the bottom of the observation tube and so increase the intensity of the fluorescence available. The windows are fixed in place with epoxy cement which must, however, be liberally filled with lampblack in order to reduce its fluorescence. The volume of liquid contained in the observation chamber

[8] Q. H. Gibson, J. W. Hastings, C. Greenwood, J. Massa, and G. Weber, *Biochem. Biophys. Acta*, in press (1965).

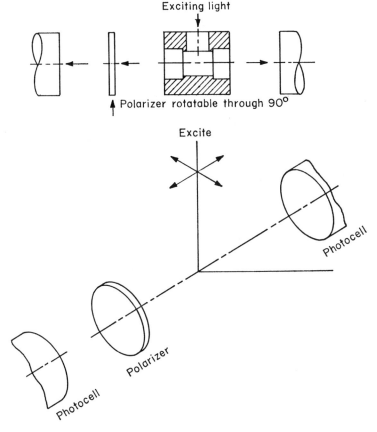

FIG. 12. Diagrammatic layout of apparatus for following changes in fluorescence polarization by the stopped-flow method. [From Q. H. Gibson, J. W. Hastings, G. Weber, W. Duane, and J. Massa, *in* "Flavins and Flavoproteins" (E. C. Slater), p. 360. Elsevier, Amsterdam, 1966.]

is of the order of 65 mm³, and so is closely comparable with that employed in the absorbance stopped-flow equipment.

## A. Performance

While in theory the performance of fluorescence equipment can be increased almost indefinitely by increasing the intensity of the exciting light, in practice, and especially with the stopped-flow equipment, the limit is set by the appearance of background fluorescence. An idea of the sensitivity that can conveniently be obtained may be given by saying that with 0.05 mg of protein per milliliter before mixing, using BSA and a time constant of 1 msec, the noise represents 3% of the total light intensity observed when fluorescence is excited at 285 m$\mu$, using a nominal bandwidth of 12 m$\mu$ in the monochromator. In the more favorable case of FMN, a comparable noise level is reached at about $5 \times 10^{-8} M$. The greatest second-order constant so far measured is $8 \times 10^8 M^{-1}$ sec$^{-1}$ for the binding of 1-anilinonaphthalene-8-sulfonate by BSA.[8a]

## B. Measurement of Fluorescence Polarization

A basic decision is whether to excite fluorescence with natural or with polarized light. The choice made in the present case is to use natural light for excitation since in this way a greater intensity of light is available for measurements of fluorescence alone using the photomultiplier which does not have a polarizer interposed between it and the source of fluorescence. With natural light excitation, the multiplier without a polarizer sees constantly the sum of $I_{\parallel} + I_{\perp}$. The multiplier with the polarizer sees alternatively $2I_{\perp}$ when it is set to pass light polarized vertically and $I_{\parallel} + I_{\perp}$ when it is set horizontally. In use, the outputs from the two photomultipliers are applied to a difference amplifier, and with the polarizer set in the horizontal position, the outputs of the two photomultipliers are adjusted to be equal by varying the anode load used in the multiplier circuit on the side without a polarizer. (This allows equal PM voltages to be applied to the two multipliers, which is important in securing uniformity of response over a wide range of intensities; R. DeSa, personal communication.) The polarizer is now rotated through 90 degrees, when the net output from the difference amplifier becomes $I_{\parallel} - I_{\perp}$. To obtain the polarization, it is only necessary to divide by $I_{\parallel} + I_{\perp}$ which is equal to the total intensity seen by the multiplier on the side without a polarizer. The polarization so obtained is numerically the same as that observed using polarized excitation if the intensities are measured by rotating an analyzer before a single photomultiplier. An example of the application of this apparatus to an enzyme problem is shown in Fig. 13.

## V. Stopped-Flow Measurements of Electrical Conductivity

Suitable apparatus has been described by Sirs[9] and Prince.[10] The apparatus of Sirs was constructed principally from Lucite and used 2-ml syringes that were filled from reservoir syringes by means of three-way glass stopcocks. The observation tube was 3 mm in diameter, and the electrodes were 1.2 mm diameter stainless steel rods located 12 mm from the mixing chamber. With a hand-driven apparatus, this geometry will yield a dead time of about 10 msec in stopped-flow operation. For measurement purposes the electrodes form one arm of an AC bridge fed with

[8a] G. Weber and Q. H. Gibson, unpublished.

[9] J. A. Sirs, *Trans. Faraday Soc.* **54**, 201 (1958).

[10] R. H. Prince, *Trans. Faraday Soc.* **58**, 838 (1958).

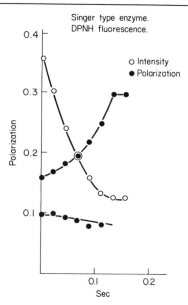

Fig. 13. Example of the application of the fluorescence polarization stopped-flow apparatus. Changes in fluorescence measured with a stopped-flow apparatus on mixing a DPNH dehydrogenase (2.5 mg/ml after mixing) containing $1.2 \times 10^{-4} M$ potassium ferricyanide with $5 \times 10^{-5} M$ DPNH under aerobic conditions. The expected initial level of fluorescence shown was higher than that observed. Fluorescence excitation was carried out using a xenon lamp and Bausch and Lomb small-grating monochromator at 340 m$\mu$. Fluorescence was observed through Wratten 2B filter. The trace was recorded photographically using an oscilloscope and camera. The open circles show the change in fluorescence intensity on an arbitrary scale. The filled symbols show changes in polarization of fluorescence during the reaction. The upper curve shows changes in polarization of the total fluorescence with time; the lower curve shows the polarization of the variable part of the fluorescence calculated on the assumption that an intensity and polarization of fluorescence equal to the final values is present throughout the reaction as a background. [From Q. H. Gibson, J. W. Hastings, G. Weber, W. Duane, and J. Massa, in "Flavins and Flavoproteins" (E. C. Slater, ed.), p. 360. Elsevier, Amsterdam, 1966.]

alternating current at 50 kc. The bridge was balanced with fully reacted mixture in the observation tube. Then, on driving a sample of reactants into the observation tube, an out-of-balance voltage appeared across the bridge; this voltage was amplified, rectified, and presented on a cathode ray oscilloscope using a time constant in line with the dead time of the apparatus as a whole. The application of the apparatus has so far been restricted to situations in which the change in conductivity was of the order of 10% or more, and it has not yet found extensive biochemical application.

The apparatus of Prince is chiefly interesting because it was intended for operation with nonaqueous media and duplicated many of the features of Sirs' apparatus without using Lucite. In stopped-flow applications with comparatively slow reactions, difficulty was encountered due to diffusion of reagents from the tubes leading to the mixing chamber to the point of observation. This difficulty was met by incorporating an additional glass tap in the observation tube between the mixing chamber and the point of observation. For a slow reaction, the syringes were actuated and the stopcocks turned, thus isolating the portion of the solution under study from the working syringes of the stopped-flow apparatus.

## VI. Measurement of H⁺ Concentration Using the Glass Electrode

The most recent description of stopped-flow apparatus using a glass electrode is given by Sirs.[11] The chief limitation of the use of the stopped-flow method for following rapid reactions by means of pH changes is the rather sluggish response of the electrode system, which has a very high internal resistance that cannot conveniently be decreased by increasing the area of electrode surface, since this would lead to an unacceptably large increase in the volume of solution and so produce limitation due to the volume delivery from the flow system. In the practical apparatus set up by Sirs, the response time following a sudden change in pH was of the order of 50 msec with the best compromise that he achieved. There is a considerable field for the development of pH recording in biochemical applications, particularly as an adjunct to other methods for following rapid reactions.

## VII. Polarographic Recording

Chance[12] has used a bare platinum microelectrode for following changes in oxygen concentration in stopped-flow apparatus. The method is quite satisfactory if observation is not required during or soon after the flow period, since during the flow period a set of diffusion relations pertains that is very different from those found some time after flow has stopped; the transition period from the flow to the stopped-flow condition requires several seconds. For these reasons the use of platinum electrodes is more satisfactory in continuous flow systems, though even here appreciable difficulties in relation to the thickness of the stationary film may arise. If slow reactions are to be followed, coated electrodes of the type developed by Clark[13] may be useful; these have response times of the order of 2–10 sec so that reactions which can be studied by the stopped-flow method scarcely fall within the category of rapid reactions, as the term is usually understood, since the reactions might well be initiated in ordinary laboratory glassware without the use of special mixing equipment.

[11] J. A. Sirs, *Trans. Faraday Soc.* **54,** 207 (1958).
[12] B. Chance, *Biochem. J.* **46,** 387 (1950).
[13] L. C. Clark, *Trans. Am. Soc. Artificial Internal Organs* **11,** 41 (1956).

## VIII. Thermal Stopped-Flow Apparatus

A thermal stopped-flow apparatus has been described by Berger and Stoddart.[14] The thermal method for following rapid reactions in continuous flow systems has been developed primarily by Roughton.[15] The application of stopped-flow procedures to thermal measurements is no less demanding in sensitivity (the temperature rise that could be expected in an aqueous solution on mixing millimolar solutions in which a reaction liberating 20 kcal/mole is only 0.01°), and it has turned out that the engineering problems involved are severe. Furthermore, it is necessary to measure these small changes in temperature with a time resolution compatible with utilization of the stopped-flow method. In the apparatus described the drive of the solutions is by gas pressure and flow is controlled by electrically operated solenoid valves. It has proved difficult to bring the solutions exactly to the same temperatures in spite of immersion in a water bath and the surrounding of the apparatus with a thermal shield, and temperature "jumps" on starting flow of the order of a few thousandths of a degree have been common, even after prolonged periods of equilibration. These jumps are believed to arise from nonsimultaneity of opening the solenoid valves. It is necessary also to use highly efficient mixing; for this purpose, a very highly developed ten-jet mixer has been produced which is illustrated on p. 34 (loc. cit.). The apparatus of Berger and Stoddart includes also provision for monitoring the reaction simultaneously by optical methods using light guides to allow immersion of the whole apparatus in a water bath. Although this apparatus has not yet found extensive application to biochemical problems, and although its development is not complete, many of the technical ideas are novel and might profitably be applied in the design of other flow systems.

## IX. Stopped-Flow Detection of EPR Changes

Although flow apparatus have been developed for use with EPR apparatus, for example Hartridge and Roughton,[16] Borg,[16a] Piette,[16b] and Berger and Stoddart,[17] no design has yet appeared specifically intended for carrying out stopped-flow work with aqueous solutions. Both articles cited, however, contain some comments on the possibilities and difficulties of applying the stopped-flow method to EPR recording.

[14] R. L. Berger and L. C. Stoddart, in "Rapid Mixing and Sampling Techniques in Biochemistry" (B. Chance, R. Eisenhardt, Q. Gibson, and K. Lonberg-Holm, eds.), p. 105. Academic Press, New York, 1964.

[15] F. J. W. Roughton, in "Techniques in Organic Chemistry" (A. Weissberger, E. S. Lewis, and S. L. Friess, eds.), Vol. 8, Part II, p. 758. Wiley (Interscience), New York, 1963.

[16] H. Hartridge and F. J. W. Roughton, Proc. Roy. Soc. London **A104**, 376 (1923).

## X. Dual Wavelength Recording

The very important procedures, which have been developed principally by Chance, for minimizing the changes which are caused by turbidity in suspensions of cells and other biological materials have chiefly been applied with apparatus with inherent time response limitations of the order of 0.1 second. Furthermore, comparatively large optical paths have been used so that the flow systems employed have either had to deliver large volumes of fluid or have required long flow times. Some of these problems have been circumvented by ingenious flow equipment utilizing the regenerative and time differential principles. The problems inherent in dual wavelength spectrophotometry can, however, be attacked using stopped-flow apparatus of the conventional kind described in this article. The two requirements for the extension of dual wavelength methods to orthodox stopped-flow equipment are, first, that the frequency of interchange of the two observing wavelengths be made much higher than can be achieved by an ordinary vibrating mirror and, second, that the volume of sample observed be reduced to permit both fluid economy and a short dead time. The principle of the dual wavelength stopped-flow apparatus may be followed by reference to Fig. 14. The heart of the system is an oscillator operating at 5 kc which serves to determine the frequency of modulation of the two light sources and also to control the gating circuits necessary to separate the pulses from the photomultiplier. Starting from the light sources, $L_1$ and $L_2$, which are xenon arcs modulated at 5 kc to give square pulses, the light passes through monochromators 1 and 2 whose output at the two wavelengths is mixed. The mixed beam is divided into two parts, one of which passes through the cuvette containing the sample under observation to a photomultiplier $P_2$. The other one passes directly on to a monitoring photomultiplier labeled $P_1$. The output of both photomultipliers contains signals derived from both wavelengths, and these are separated by the gating circuits represented by $G_1$ through $G_4$, so that four signals are available to give information about the behavior of the two light sources with and without modification due to the sample in the cuvette. A portion of these signals is used to provide a feedback control to the power supplies for the two lamps, $E_1$ and $E_2$, while the remaining

[16a] D. C. Borg, *in* "Rapid Mixing and Sampling Techniques in Biochemistry" (B. Chance, R. Eisenhardt, Q. Gibson, and K. Lonberg-Holm, eds.), p. 135. Academic Press, New York, 1964.

[16b] L. H. Piette, *in* "Rapid Mixing and Sampling Techniques in Biochemistry" (B. Chance, R. Eisenhardt, Q. Gibson, and K. Lonberg-Holm, eds.), p. 131. Academic Press, New York, 1964.

[17] R. L. Berger and L. C. Stoddart, *in* "Rapid Mixing and Sampling Techniques in Biochemistry" (B. Chance, R. Eisenhardt, Q. Gibson, and K. Lonberg-Holm, eds.), p. 105. Academic Press, New York, 1964.

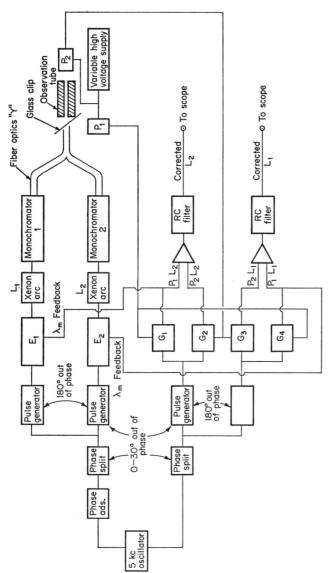

Fig. 14. Schematic diagram of stopped-flow dual-wavelength apparatus. For fuller explanation see text. (From DeSa and Gibson, footnote 17a.)

signals are applied to difference amplifiers which provide the final readout information as a difference between the effects of the sample on the two wavelengths used, as is customary in dual wavelength technique. One of these wavelengths would be chosen to be isosbestic while the other, as near as possible to the first, would be a wavelength where significant absorbance changes occur during the reaction being studied. The apparatus operates with an overall time constant of approximately 1 msec for the recording circuits, and is capable of registering satisfactorily changes of less than 1% in transmission. The equipment can be used not only for experiments with scattering materials but also for enzyme kinetic experiments in which only small amounts of enzyme are available. Since the design gives first-order correction for variations in lamp intensity, enzyme concentrations of the order of one-fourth to one-fifth of those required with the ordinary stopped-flow apparatus can be used, while the dual wavelength feature can be employed under these circumstances to obtain reaction records at two wavelengths simultaneously. This is a requirement which is frequently met in enzyme kinetic studies, and it is particularly valuable, quite apart from the question of economy of materials, to have information at more than one wavelength recorded simultaneously from a single sample of the preparation, since spurious effects due to changes in the enzyme solution between experiments is thereby eliminated. The design is made possible by the high surface brightness of modern xenon lamps, which permit the use of small fluid samples and narrow bandwidths. It should be admitted, however, that the instrument is somewhat simpler in principle than in practice; and the original publication must be consulted for details for its construction and operation.[17a]

### XI. Combination of Stopped-Flow with Other Reaction Kinetic Methods

By far the most important limitation of the stopped-flow method, and indeed of liquid flow methods in general, is the relatively long dead time imposed by the necessity of bringing about adequate mixing of the fluids. This restricts its use to what are, from a chemical point of view, relatively slow reactions with activation energies typically in the range of 7 kcal or more. The advantage of the method, on the other hand, is that it can be applied to a wide range of reactions; the only requirements are solubility, a degree of chemical inertness sufficient to allow the construction of the apparatus, and the possibility of following the reaction by an appropriate physical method. Many ingenious and powerful methods for following reactions with higher rates than those accessible to the ordinary stopped-flow technique have been described and some are dealt with in other chapters of this volume. They are usually applicable only to the equilibrium

[17a] R. DeSa and Q. H. Gibson, *Rev. Sci. Instr.* **37,** 900 (1966).

systems. Perturbation and relaxation methods consist of rapidly changing
an external parameter such as pressure, temperature, or electrical field
density which perturbs the equilibrium at a rate faster than the chemical
system under study can reequilibrate. The relaxation of the system to the
new equilibrium position is then followed by appropriate recording methods.
In terms of dead time, these procedures may be of the order of 100–1000
times more powerful than flow techniques; and new developments in
methods for producing the perturbations offer promise of even further
increases in speed. This speed, however, is purchased at the expense of two
serious difficulties. The first is that only equilibrium systems can be
studied, while the second limitation is that, save in exceptional circumstances, only small departures from equilibrium can be produced by the
application of the perturbation. There is the corresponding advantage that
all processes under these conditions may be treated as first order, with
corresponding simplification of the interpretation of the results.

One exception to the difficulty that only a small percentage displacement can be obtained is found in the application of flash photolysis to
appropriate equilibrium systems, where with systems of high quantum
yield a high proportion of the total reactant initially present can be converted to a new form by a flash. Although nothing can be done to increase
the size of excursion in the systems with low temperature or photosensitivity, the equilibrium requirement can sometimes be evaded by the combination of flow and perturbation techniques. The range of applicability of

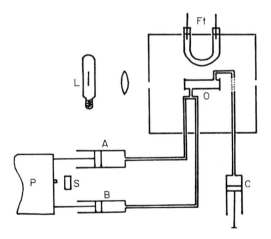

FIG. 15. Block diagram of flow-flash apparatus. $A$, Large syringe to contain the CO
compound of the reduced enzyme; $B$, small syringe for $O_2$ solution; $C$, receiving syringe;
$P$, pushing block; $S$, switch of flash-photolysis apparatus; $Ft$, flash tube; $L$, light source
for recording the reaction; $O$, 5-cm observation tube. [From Q. H. Gibson and C.
Greenwood, $Biochem.\ J.$ **86,** 541 (1963).]

these methods is very small indeed, and special apparatus and procedures must be introduced to meet each specific case.

By way of example, a brief description of the application of combinations of stopped-flow and photochemical methods in the study of cytochrome oxidase will be given. The reaction of cytochrome oxidase with oxygen is too rapid to be studied conveniently by the stopped-flow method over any appreciable range of concentration of oxygen since the second-order velocity constant is about $5 \times 10^7\ M^{-1}\ \text{sec}^{-1}$ at room temperature, and even with equivalent quantities of reactants and concentrations of enzyme that fall conveniently in the detection range of the stopped-flow apparatus, a considerable part of the reaction necessarily occurs during the dead time. This limitation can be substantially removed by combining stopped-flow with flash photolysis if the carbon monoxide compound of the cytochrome oxidase is mixed with oxygen instead of the reduced cytochrome oxidase. This mixture is made in the stopped-flow apparatus and has a half-life of the order of 20 seconds or so at room temperature. If, however, a photolysis flash is fired at any time after mixing has taken place, the remaining carbon monoxide can be removed in a time that is short as compared with the fluid mixing time. Using ordinary gas-filled flash tubes, the carbon monoxide can be removed with a half-time of the order of 3 μsec, and calculations suggest that an appropriate laser flash

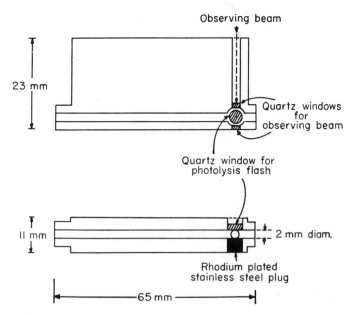

FIG. 16. Sketch of cell for photolysis experiment. [From Q. H. Gibson, C. Greenwood, D. C. Wharton, and G. Palmer, *J. Biol. Chem.* **240**, 888 (1965).]

could do very much better still. The reaction between the free reduced cytochrome oxidase liberated by the flash then proceeds in solution at a rate dictated by the concentrations of the reactants and is no longer limited by the mixing times of the flow equipment. A block diagram of the apparatus used for this purpose is shown in Fig. 15.

Another application of stopped-flow and photochemical methods in connection with cytochrome oxidase has been made in studying the reduction of cytochrome oxidase by cytochrome $c$. Here the spectrophotometric evidence suggested that cytochrome $a$ became reduced quite rapidly, but cytochrome $a_3$ did not. This was confirmed by carrying out the reaction in the presence of carbon monoxide, which can combine with reduced cytochrome $a_3$ but not with reduced cytochrome $a$. Then, after the reaction had been initiated by mixing reduced cytochrome $c$ with the oxidized cytochrome oxidase in the presence of carbon monoxide, a photolysis flash was fired. The formation of a photosensitive compound was taken as evidence that cytochrome $a_3$ had been reduced by cytochrome $c$, and by firing the flash at different intervals after the initiation of the reaction in the stopped-flow apparatus, the time course of the formation of the photosensitive compound could be obtained. The apparatus used for doing this is shown in Fig. 16 and an experiment showing the use of the apparatus is given in Fig. 17.

Another combination of the stopped-flow with a perturbation method, which has been discussed, is the temperature-jump with stopped-flow method (see Chap. [1]). Again, the advantage would lie in extending the range of application of the method. This advantage might be especially true in the study of enzyme kinetic systems, where steady state might be induced by mixing in the flow apparatus. In this case, the steady state persists for a few milliseconds or tens of milliseconds during which time temperature perturbation could be carried out. The difficulties that remain, however, continue to be serious since such a system is likely to be exceedingly complex; and the overall rate may well depend on the interaction of a considerable number of independent rate constants, each of which will have its own temperature dependence. It seems that the most profitable application of such a piece of apparatus may be to verify the results of analysis of an enzyme kinetic reaction by conventional methods, using the results of the conventional analysis to predict the effect of the temperature-jump procedure and comparing the observed and predicted courses of the reaction, after jumping the temperature.

## XII. Future Developments of the Stopped-Flow Method

The limitations of the stopped-flow method at the present time are largely those of the hydraulic system. The developments in design of the apparatus and of the recording systems have been brought to the point

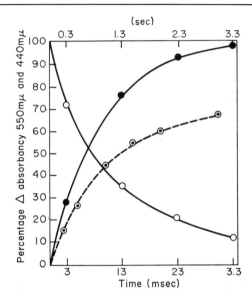

FIG. 17. The rate of appearance of photosensitivity in $2.5 \times 10^{-5} M$ cytochrome oxidase during reduction by $5 \times 10^{-5} M$ cytochrome $c$ in the presence of $5 \times 10^{-4} M$ carbon monoxide. The figure shows the absorbance changes at 445 m$\mu$ (●) and 550 m$\mu$ (○) on the millisecond time scale, and the appearance of photosensitivity (◉) on the time scale in seconds. The buffer was $0.1 M$ phosphate, pH 7.6, containing 0.1% deoxycholate; path, 4 mm; temperature, 20°. [From Q. H. Gibson, C. Greenwood, D. C. Wharton, and G. Palmer, *J. Biol. Chem.* **240**, 888 (1965).]

where little important change in fluid economy can be expected, though past gains have been spectacular indeed. The fluid economy of the apparatus described by Chance[4] shows a gain of the order of 250,000 times over the original flow apparatus of Hartridge and Roughton.[16] The picture is far different, however, when the rate constant which can be measured is considered, Hartridge and Roughton having, in a single stride, realized the full potential of the method, so that the design has yet to appear giving a significant improvement in time resolution over their apparatus.

So far as the stopped-flow apparatus itself is concerned, serious difficulties arise if attempts are made to drive the reagents through much faster than is done in current equipment. One of these is due to the inertia of the moving fluid which, to be usefully employed in stopped flow, must be brought to rest suddenly, but which, because of its arrest, will generate large forces within the observation tube and stopping device. This point has been discussed recently by Berger and Stoddart.[14]

The limit set by these forces is already approached in the current apparatus using glass syringes, but it seems that an improvement by a factor of 2 or 3 in time resolution might be obtained if stainless steel syringes were substituted. If this were done, however, a new set of problems would undoubtedly appear; and in particular it might be found that cavitation would give rise to significant troubles, though the experience of Chance in the use of pulsed flow methods[18] suggests that problems due to cavitation may not be too severe in the case of an apparatus that delivers into a stopping syringe that offers appreciable resistance to flow.

Further development beyond a factor of 2 or 3 times over the existing apparatus calls for the application of very high pressures indeed and the redesign of apparatus to deal with them. With high pressures and high flow velocities the question of acceleration would become significant and with it the problems of fluid economy would reappear, since quite substantial volumes of liquid would have to be delivered in order to attain maximal flow velocity with these high pressures and high speeds. Sudden stopping would be impossible and a pulsed-flow apparatus rather than a stopped-flow apparatus would be the end result. While apparatus using extreme pressures may well be developed for specialized use, it may be that future developments will be by way of revolution rather than evolution and that the stopped-flow apparatus in the general biochemical laboratory may continue to be of the present form, at least in the near future.

## Note Added in Proof

Since this chapter was written, a number of modifications have been made in the apparatus described. These have been prompted by the application of automatic data-handling procedures as a substitute for photographic recording of the photomultiplier output.

The modifications to the stopped-flow apparatus are:

*i.* Substitution of compressed air pistons (Airco Inc., Angola, Ind.) for the hydraulic activating levers. The advantages of compressed air over the manually operated hydraulic system are those of reproducibility and control. It has also been found possible to use substantially higher driving pressures, up to 250 lb/in.$^2$ in the syringe. The resultant increase in flow rate has reduced the dead time to 1.6 msec for a 2-cm light path, and 500 $\mu$sec for a 2-mm path.

*ii.* The higher pressures resulted in excessive leakage past the ceramic syringe plungers. These were replaced by stainless steel with Teflon rings held in an adjustable screw seating. Although leakage may effectively be eliminated with such a plunger, it requires readjustment if the temperature is varied by more than about 5°, because of the large temperature coefficient of expansion of the Teflon.

*iii.* The shock of flow-stopping, increased by the higher pressures, required resilient mounting of the monochromator.

The data-collection system consisted of a PDP8/S computer and analog-to-digital converter, together with a home-built timing unit. The general principle of operation is as follows. Just before flow stops the stopping syringe closes a pair of contacts which were formerly used to trigger the oscilloscope sweep. These are now used to set a mono-

[18] B. Chance, *in* "Rapid Mixing and Sampling Techniques in Biochemistry" (B. Chance, R. Eisenhardt, Q. Gibson, and K. Lonberg-Holm, eds.), p. 125. Academic Press, New York, 1964.

stable timing circuit whose duration is adjusted to correspond with the remaining period of flow time, ordinarily about 2 msec. When the monostable timer resets, the pulse–timing clock is turned on, and a stream of pulses follows, each one of which causes the voltage output from the photomultiplier to be sampled, digitized by the analog-to-digital converter, and the binary number deposited in the computer memory. Usually, 400 such samples have been taken to represent a reaction record. A further sample of 256 points is collected to represent the final reaction voltage. The 400 samples are then graduated by fitting a second degree curve to groups of 20 points by least squares, and using the polynomial to provide an estimate of the voltage at the beginning of each time interval corresponding to the collection of 20 samples. The 20 smoothed values so obtained are then used to calculate absorbance values with, optionally, first-order rate constants. These values are typed out together with the corresponding times by the teletypewriter of the computer. The whole operation requires about 40 sec using the 8/S computer.

After about one year's experience of using the method, it is the opinion of the author that it is greatly to be preferred to photographic recording. First, the result is immediately available in its final form, and graphs showing the progress of an experiment may be prepared as each new stopped-flow run is obtained, allowing repetition of doubtful determinations. Next, since the labor of evaluation is slight, replicates may be used regularly, and the computer program allows for the collection of successive runs with output of mean values either with or without individual results. The precision with which the results are obtained is greater than that of measurement of photographs, and it now appears that under the best conditions the standard error of determining a point is about 0.001 or 0.5% of the change in absorbance measured, whichever is the greater.

It is my further opinion that a small general-purpose computer is much to be preferred to a special-purpose wired-program machine in this application. It is readily possible to write programs, which must be in assembler language, to perform other operations than those described. For example, fluorescence polarization may be measured using two photomultipliers and a multiplexer for the analog-to-digital converter, together with appropriate program modifications, and dual wavelength spectrophotometry is possible without electronic balancing circuits, again with a multiplexer and program modifications.

The range of times accessible to the equipment is limited, at the high-speed end, primarily by the computer. If data points are taken in by conventional means with the computer waiting in a loop for the application of an external signal to the skip bus, eight further instructions are required to acquire, deposit, and count a point for a minimal time of about 500 $\mu$sec, as compared with the conversion time of 37 $\mu$sec for the converter used. A large increase in rate may be obtained, at the cost of complicating the timing hardware, by depositing the required instructions in linear array. In such a scheme, only two instructions need be executed to acquire a data point, and these require only 80 $\mu$sec. It is probably desirable, therefore, to use a computer faster than the PDP8/S; several suitable small machines by various manufacturers are currently available.

The external timing clock, which runs from a 100 kc crystal, permits patchcord selection of 495 different sampling rates, together with changeover facilities allowing the sampling rate to be changed part way through a given data collection sequence, i.e., after the collection of, for example, 120 points at one rate, the rate could be changed for the collection of the remaining 280 points. This facility has proved valuable in dealing with the biphasic reactions, which are common in biochemical practice, both because material is saved and because both slow and fast phases are recorded on a single sample of reaction mixture, allowing better correlation of results.

All indications are that lower prices and higher performance for the small laboratory computer lie ahead; it is recommended that such a computer be considered very seriously as an integral part of a rapid kinetic system.

## [7] Rapid Mixing: Continuous Flow

### By H. Gutfreund

### I. Historical Survey of Principles

It is a sobering thought that for 40 years after Hartridge and Roughton[1] first measured rapid reactions in solutions by flow techniques, the time resolution of approximately 1 msec was not significantly improved. There is now a great deal of effort directed toward the design of equipment to enable the making of observations within a fraction of a millisecond. The effort has just begun to yield data on the progress of reactions with half-times in the 100 $\mu$sec range. This lag is due not to shortage of ideas, but to limited engineering facilities. A study of the literature on rapid-flow techniques reveals that many useful and promising ideas and suggestions for technical improvements, which would increase considerably the range of applications, have not been taken up in practice. For this reason some techniques will be described here that may not yet have been tried. Quite apart from the use of rapid-flow methods for the study of general chemical problems, it is only in the last few years that their application to the elucidation of steps in enzyme catalysis and in the reactions of other biological macromolecules has been appreciated by more than half a dozen investigators. Papers presented at a recent symposium[2] showed an increasing number of designs of flow equipment in operation and under construction. Also a variety of commercial instruments are being announced. This article is intended to convey enough of the principles involved to help in design, testing, and modification for specialized purposes. There are, however, so many variables that only personal experience, trial and error, can help to perfect the technique.

The scope of this chapter is the discussion of the theory, design, and

---

[1] H. Hartridge and F. J. W. Roughton, *Proc. Roy. Soc.* (*London*) **A104**, 37b (1923); *Proc. Cambridge Phil. Soc.* **22**, 426 (1924); *ibid.* **23**, 450.

[2] B. Chance *et al.* (eds.), "Rapid Mixing and Sampling Techniques." Academic Press, New York, 1964.

application of methods involving the initiation of reaction by rapid mixing of two fluids and observations during the flow of the resulting reaction mixture. There are a number of possible modes of flow and observation. The main groups are continuous flow and stopped flow. In the present chapter we are primarily concerned with continuous flow. The stopped-flow technique, which involves observation of the reaction mixture after flow has stopped, is discussed by Gibson (this volume [6]). Here we shall refer to this procedure only for comparative purposes.

The essential feature of any procedure for rate measurements is that one must be able to get observations of the extent of the reaction as a function of time, i.e., at different ages of the reaction mixture. This can be achieved in flow systems in three ways. First, one can vary the age of the reaction mixture at the point of observation by varying the rate of delivery (ml/sec) from the mixing chamber. Second, one can vary the diameter of the flow tube and with that the volume of fluid between point of mixing and point of observation. Third, one can move the point of observation along the flow tube. The age of the reaction mixture at the point of observation is given by volume between point of mixing and point of observation divided by the flow rate [ml/(ml/sec) = sec]. The volume is, of course, a function of the diameter of the flow tube, which is usually circular, and the distance traveled.

Later consideration of problems involved in mixing, and the necessity of maintaining turbulent flow in the tube, will show that more precise and reproducible data are obtained if the flow rate and tube diameter are kept constant. The convenience of using change in rate of flow does make this an attractive procedure as long as it is used with care. Methods that involve changing flow tube diameters may sometimes be useful to increase the sensitivity of optical observations and when relatively long delays between points of mixing and observation are required, but such procedures are very uneconomical in terms of reactants. The classical work of Hartridge and Roughton[1] on the reactions of hemoglobin with various ligands was performed by optical or thermal observation of the progress of the reaction at different points at an increasing distance from the mixing chamber. With the reaction mixture flowing through the tube at 10 meters/sec and observations taken at 1-cm intervals along the tube, concentration changes can be measured at 1-msec intervals. Although one could make measurements at shorter intervals than 1 cm, this would give an illusory accuracy because of two inherent limitations: The age of the reaction mixture depends on the distance between the point of mixing and the point of observation, and neither of these are true points; mixing as well as the "point" of observation must cover a finite distance along the axis of flow (see Sections II and IV, respectively).

Studies of optical and thermal changes by Hartridge and Roughton involved fluid flow at the steady-state velocity for the period while visual readings with the reversion spectrophotometer, or of galvanometer deflections produced by thermocouples, were taken at various points along the flow tube. Flow tubes with diameters of 5 mm were used, and through such tubes 196.25 ml of reaction mixture are discharged per second at a flow rate of 10 meters/second. The collection of data necessary for the determination of the rate constant of a reaction under a particular condition usually involved the expenditure of about 3 liters of solution. The problems of balancing fluid economy against resolution with respect to time and concentration changes were first discussed by Roughton and Millikan.[3]

The advent of electronic detection, amplification, and recording techniques has contributed to economies in reactants both by reducing the time required for observation and the diameter of the flow tube. For example, if an accurate record of the reactant concentration in a 1-mm diameter tube can be obtained on an oscilloscope during 0.1-second flow at 10 meters per second, only 0.785 ml of reaction mixture is expended. These two factors are discussed in Section IV. From such pulsed-flow methods Chance was led to the accelerated stopped-flow technique, and Gibson to the stopped-flow technique. The latter provides the ultimate in fluid economy and simplicity of operation, but its time resolution is not as good as that attainable in continuous-flow devices. In the economical continuous-flow apparatus of Millikan,[4] in Chance's[5] accelerated stopped-flow and Gibson's[6] stopped-flow techniques, the two reactants are delivered from syringes driven by a motor-, pneumatic- or hand-activated barrier.

During accelerated and stopped flow, observations are made continuously, and flow velocity, which determines the age of the reaction mixture at the observation point, is recorded in parallel with the progress of the reaction. Gibson's contribution to this volume gives detail of the stopped-flow method, and accelerated flow is discussed further in Section III of this chapter.

Clearly much of the equipment described by other contributors to this volume is suitable for the construction and operation of every kind of flow apparatus. Many of the components are interchangeable between the different methods described. For those who want to develop new applications, hybrid versions of components and monitoring methods may turn out to be most suitable. As an example, one should point out that the valve

[3] F. J. W. Roughton and G. A. Millikan, *Proc. Roy. Soc. (London)* **A155,** 258 (1936).

[4] G. A. Millikan, *Proc. Roy. Soc. (London)* **A155,** 277 (1936).

[5] B. Chance, *in* "Techniques in Organic Chemistry" (S. L. Friess, E. S. Lewis, and A. Weissberger, eds.), 2nd ed., Part II, Vol. VIII, p. 733. Wiley (Interscience), New York, 1963.

[6] Q. H. Gibson, this volume [6].

block and mixing chamber described by Gibson can be used directly for various continuous-flow devices.

In this connection it must also be made clear that no attempt is made to go over all the ground so well covered by Roughton and Chance[7] in their survey of the state of the art up to 1962 and by the various contributors to the International Symposium on Rapid Mixing and Sampling Techniques[2] held at the Johnson Foundation in 1964. Any research worker intending to construct new equipment for the study of rapid reactions by one of the continuous-flow methods, is likely to get all the pertinent information and references only if he studies the chapters of this volume and also the two other volumes referred to above.

## II. Mixing

In continuous and pulsed-flow techniques, the ultimate limitation in time resolution is almost certainly the time required to mix two solutions and to bring them to the first point of observation. Reactions that are faster than this limiting process can be measured only by one of the techniques involving the application of some physical perturbation to initiate the reaction in the chemically complete mixture. The perturbation and relaxation techniques, discussed in other chapters, are complementary to the flow techniques, both in time resolution and in the chemical information they provide.

The extent of mixing has to be correlated with one of three interrelated quantities: (1) *the volume* required to fill the space between the first point of contact of the two solutions and the point at which the extent of mixing is measured; (2) *the time* taken for the space between these two points to be filled, i.e., the time taken to deliver the volume referred to in (1) to be delivered through the mixing jets; (3) *the distance* between point of contact and point at which extent of mixing is measured.

In practice, one is interested in determining how close to the initial point of contact one obtains adequate mixing; we shall have to define later what is adequate. It is probable that in most mixing devices a good deal of the mixing occurs at the beginning of the flow tube. The function of the mixing chamber is to start the distribution of the two solutions and to create turbulence; mixing of the turbulent fluid is then continued in the flow tube. In testing a new experimental arrangement with fixed dimensions, one examines the percentage of mixing achieved at a number of points along the flow tube at each of several flow rates. One obtains in this way

[7] F. J. W. Roughton and B. Chance, *in* "Techniques in Organic Chemistry" (S. L. Friess, E. S. Lewis, and A. Weissberger, eds.), 2nd ed., Part II, Vol. VIII, Chap. XIV. Wiley (Interscience), New York, 1963.

FIG. 1. Two-jet mixing chamber leading into an observation post. The clamp below can be used to fix a capillary tube in position against an O-ring in the mixing block.

the dependence of "mixing distance" on flow velocity and the age of the adequately mixed solution at different flow velocities. It is essential that this age be low compared with the half-life of the reaction to be investigated. It is also important to provide a rate of flow fast enough to obtain turbulent flow; laminar flow is readily observed by some optical test for schlieren patterns or double refraction. Flow rates of 5–20 meters/second through tubes of about 1 mm diameter are the likely range to avoid laminar flow or cavitation. The latter problem, which arises during fast flow rates, is discussed in the next section.

The simplest mixing chamber is illustrated in Fig. 1. This consists of two jets, one for each of the two solutions, which are offset slightly to produce a turbulent effect at the point of contact. This device can be constructed in any workshop from a variety of plastic materials which make it easy to cement in plugs to seal the holes produced by drilling into the block from the sides. More complex mixing chambers consisting of a larger number of jets or constructed to include additional barriers to increase turbulence have been designed and tested in a number of different laboratories. Only representative examples of these will be described here in conjunction with methods to test their efficiency.

To improve mixing efficiency and to reduce the frictional resistance to the rapid delivery of a lot of fluid into a small mixing chamber, a variety of arrangements with 4, 6, 8, or infinite numbers of jets have been constructed (see footnotes 1, 2, 7). A mixing chamber with a total number of 8 jets was found very satisfactory in the author's laboratory. Four jets from each reactant are produced by dividing the input twice, the second division occurring very near the mixing chamber. Pairs of jets from each solution are alternatively placed around the mixing circle, slightly offset from each other in the direction of flow to the observation point.

Different principles of mixing are described by Gibson in this volume [6]. Dr. R. L. Berger of the National Heart Institute, Bethesda, Maryland, has carried out some interesting experiments with a different type of mixer. One reactant solution comes out through holes in the vertical baffle, and the other through a jacket from the horizontal channels (Fig. 2). A related design was described by Eigen and De Maeyer[2] in connection with their combined-flow and temperature-jump apparatus. In this device eight channels leading into the mixing chamber are supplied by eight syringes and the semispherical electrode, used for the heating discharge, acts as a baffle. This arrangement adds considerable versatility since the eight syringes allow the mixing of a greater number of different solutions as well as different proportions of two solutions.

The variety of mixing devices described should not confuse the potential user. Equal volumes of aqueous solutions of reactants with similar physical properties can be mixed efficiently on a 2-msec time scale by the simplest mixer, as long as the rate of flow is sufficient to maintain turbulence. A more detailed and personal examination with specific tests for each individual problem is necessary if a better time resolution is required or more complex mixing is to be achieved. Special problems arise when unequal volumes of the two reactants are to be mixed.

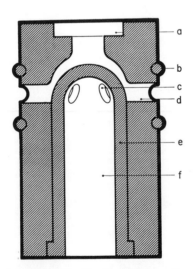

Fig. 2. Mixing chamber designed by Dr. R. L. Berger: One reagent flows from tube (f) through holes (c), while the other reagent comes through holes (d) in the main block to flow mixture to (a). O rings at (b) seal the mixing chamber in the syringe block.

It is not possible to state categorically any specific value for percentage of mixing that is adequate for the point (in space or time) of start of observation. If the heat of mixing is large or if there is a considerable optical density difference, a much higher degree of mixing is required at a satisfactory starting point for thermal or optical density records, respectively, than if there is a negligible change in the monitored property during the mixing process. It must also be pointed out that large differences in other physical quantities such as viscosity, density, and refractive index, will affect the performance of any mixing device. It is, therefore, essential to examine the efficiency and reproducibility of mixing with any new system one wishes to study.

The most widely applicable operational procedures for testing mixing and time resolution of flow devices depend on optical or thermal measurements during reactions that are essentially instantaneous on the time scale under consideration. The extent to which such reactions have progressed at the point of observation is a measure of the degree of mixing achieved by the time the fluid has reached that point. Neither optical changes of the protonation of an ionized dye nor heat changes during some suitable ionic reaction give a perfect picture of the degree of mixing. Each method has its own shortcomings in the integration of the chemical change in a still heterogeneous observation volume in the flow tube. A more empirical approach was used by Barman and Gutfreund[2] to test the overall accuracy of using the distance between the center of the mixing chamber and the point of observation for the calculation of the age of the solution at the latter point. Pseudo first-order reactions, such as the hydrolysis of 2,4-dinitrophenyl acetate by NaOH, can be examined over a wide range of rates by increasing the concentration of the excess reagent. First-order plots of the progress of the reaction can be examined for deviations of the extrapolated straight line from the coincidence of zero reaction and zero time.

Other systems suitable for optical testing of the total dead time—the earliest possible time for correct information about a reaction—are worth mentioning. There is a great deal to be said for the operational test of time resolution with a pseudo first-order reaction. This procedure exposes any unsuspected or unknown causes of errors in results obtained from newly constructed equipment. The hydration of pyruvate to form the corresponding diol and the reduction of 2,4-dichlorophenolindophenol by ascorbate have been found in the author's laboratory to be particularly suitable for tests. The disappearance of the absorption at 317 m$\mu$ (which can still be followed at 340 m$\mu$) due to the conversion of the ketone to the diol form of pyruvate is a convenient test. A first-order reaction is observed when pyruvate dissolved in phosphate buffer pH 2 is mixed with NaOH solutions to bring the reaction mixture to an alkaline pH. At 25° and pH 11.5 the reaction has a half-time of 7 msec, at pH 7.6, $t_{1/2} = 1$ sec.

The reaction used most extensively for testing flow equipment, other than quenching devices, is the reduction of indophenol observed in the 600 m$u$ region. The dependence of the pseudo first-order reaction on pH and ascorbate concentration provides a wide range of half times. At pH 5.5 the half-time is 3 msec when ascorbate is 30 m$M$ and 30 msec when ascorbate is 3 m$M$. The second-order rate constant 7700 $M^{-1}$ sec$^{-1}$ decreases with increasing pH. A plot of log $k$ against pH has the slope $-0.83$ when the reaction is carried out in a standard medium containing 0.5 $M$ phosphate and 0.1 $M$ citrate. The indophenol concentration is conveniently held at about 0.05 m$M$.

It should be emphasized here that it is of utmost importance to test how far back toward zero time first-order kinetics are observed with one of these test reactions. In many cases rapid flow techniques are used to detect initial transients, deviations from single exponentials. Dead times calculated from flow rates and mixing efficiencies measured on instantaneous reactions, are not sufficient evidence for a correct readout of information at short periods after mixing. This is especially true, and often overlooked, in attempts to improve the time resolution of stopped-flow equipment.

## III. Modes of Flow and Design of Apparatus

All flow methods for the measurement of reaction rates are variations based either on continuous flow or on stopped flow. Here we are principally concerned with observations during flow but need not neglect that further kinetic data can be obtained after flow has stopped. The principle of these methods depends on the establishment of a steady state, or a number of approximate steady states, resulting in reaction mixture of a defined age at any one point in the flow system, and the equipment is designed in such a way that one can sample either at one point or at different points. If one can sample only at one point, one has to vary the age of the reaction mixture at the point by a change in flow rate or a change in volume between point of mixing and point of observation.

Some special devices involving continuous flow should be mentioned here: the regenerative flow machine of Chance and the stirred flow reactors of Denbigh and Hammett. Although these have proved very useful for particular applications, a discussion of the problems involved would take us too far from our main theme.

The classical arrangement of Hartridge and Roughton with solutions of the reactants stored in bottles and being driven by compressed gas into the mixing chamber and observation tube, is still usefully applied to systems where fluid economy is not a serious consideration. In many cases, however, it is convenient to use the more economical, compact, and self-contained arrangement of motor-driven syringes first developed by Millikan. In such a device, steady-state flow rates can be achieved rapidly and can be accurately controlled and reproduced.

The apparatus shown in Fig. 3 is one of many possible arrangements that can be used for such automatically controlled flow experiments. It was designed for pulsed flow (continuous flow of short duration) of the type required for rapid quenching experiments (Section V), but the apparatus can be adapted with readily exchanged units to permit observation by a variety of physical techniques. An important consideration in pulsed-flow experiments is a precise definition of the period for which the flow rate is constant. In quenching experiments, where all the output during the time flow is collected, the sharpness of the front of constant flow velocity and of deceleration is also of importance. Either one can have a record of the flow velocity by introducing a potentiometer with a moving arm driven directly or indirectly by the barrier that moves the syringe plungers, or one can apply the following operational tests: (1) One can record the optical density or temperature at a set point in the flow tube during the pulsed flow of a suitable reaction mixture. An optical record, with its rapid response time, will give the best picture of the sharpness of ascent to steady state. (2) One can estimate the volume delivered during different flow times and obtain a value for the deviation from the ideal of an infinitely fast acceleration and deceleration by extrapolation to zero flow time. The volumes delivered can be determined very accurately either by weighing or analysis of some marker substance dissolved in the solution.

The design described by the legend of Fig. 3 has as yet been tested only for the rapid sampling involved in the chemical quenching method, and as shown in Section V an overall accuracy of 1 msec must include any errors due to acceleration and deceleration delays. A contributory factor to the rapid acceleration and deceleration is likely to be the fact that the motor runs continuously and start and stop are controlled by a firmly engaging magnetic clutch and a magnetic brake, which in turn are controlled by microswitches operated by a lever that moves with the syringe barrier. The use of motors with variable speed or changeable gear ratios would permit variation in the flow rate, which is desirable for optical observations. For thermal measurements or quenching, this can be combined with a facility to vary the distance between mixing and sensing element or quenching.

Motor-driven devices can be obtained with accurately adjustable variable speeds controlled by a feedback circuit to give satisfactorily constant movement. The precise speed of movement can also be recorded by attaching the arm of a potentiometer to the syringe drive. In England

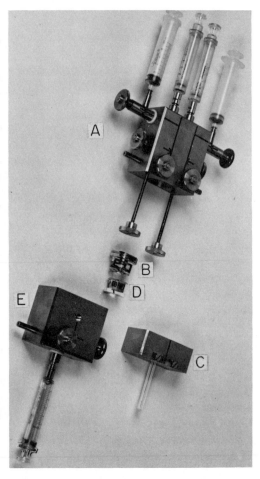

(A)                              (B)

FIG. 3A. Pulsed-flow apparatus comprising (from top to bottom) a 700 rpm motor geared to shaft, magnetic clutch, and magnetic brake (90 V, direct current), arm to microswitch controlling clutch and brake, syringe drive, two storage syringes (on the outside), and two driving syringes (in center) block as illustrated in Fig. 3B.

FIG. 3B. Components made of stainless steel are: syringe block with needle valves (A), stopping block (E), and clamp for capillary (C). For stopped-flow observation, mixing chamber (B) and observation chamber (D), which are both made of a suitable plastic (Lucite), are inserted into (A) and (E), respectively, and the two blocks are screwed together. O rings inserted in the plastic components serve as seals. For pulsed-flow or quenching operation, the mixing chamber (B) is inserted into (A) and (C) is clamped onto (A).

suitable equipment is manufactured by Lancashire Dynamo Electronic Products Ltd. under the name of Stardrive thyristor-controlled adjustable-speed drives. In the United States similar equipment is available as Pacific Industrial Control type E-16 with push-button reversing and dynamic breaking. In either case, $\frac{1}{4}$-horsepower motors should be suitable for the operation.

The advantages of the continuous- and pulsed-flow operation for the accurate observation of the rates of rapid chemical reactions are discussed in the next section. Some discussion is appropriate here of an automatically controlled flow apparatus for mixed modes of operation, where one observes a succession of quasi-steady states. In this latter type of operation a programmed rate of acceleration must be introduced into the syringe drive mechanism. This can be done mechanically with appropriate gears, with the electrical supply to the motor with equipment mentioned above, or with hydraulic devices. The program must provide a suitable range of speeds: the precise speed at any time can be measured and recorded. The lowest speed permissible for reliable observations depends on the achievement of turbulent flow, which is essential for mixing and maintenance of a homogeneous cross section. The fastest speed will depend on the force that can be applied without the occurrence of cavitation and that the apparatus will stand without breakage or leakage. Observation during a 5- to 10-fold change of flow rate is a realistic aim, and greater ranges can be achieved.

If the drive of the continuous-flow apparatus is programmed in such a way that the flow rate increases linearly during 1 sec from 0 to 15 msec through a tube of 2-mm diameter, one will use 28.5 ml of the combined reaction mixture. A useful record will be obtained during the period of 3 msec to 15 msec flow rate. The advantage of this procedure over the stopped-flow technique is illustrated by the following example. If we consider a reaction with a half-time of 2 msec, a stopped-flow experiment would require prohibitive corrections if an instrumental time constant of more than 0.2 msec is introduced. If we observe the accelerated flow mode 3 cm from the point of mixing, we obtain data over the range of 1–5 half-lives of the reaction. A response time of 10 msec for the detector and other components could easily be accommodated, and one could correct for even longer response time. The advantages of methods that do not require short response times is discussed in the next section. Observation at 0.5 msec after mixing should just be possible with this mode of observation. Apart from one not well documented claim, it is usually found that stopped-flow observations result in ill-defined records for the first millisecond after the fluid is arrested.

Chance *et al.*[8] have shown recently that rates of reactions with half-times of about 100 $\mu$sec can be measured in flow devices. The principle which he employs is to prevent cavitation, which is likely to occur at flow rates above 15 meters/sec, by applying back pressure. The rate of flow (up to 35 meters/second) is controlled by a balance of hydraulic pressure applied to the driving and to the cushioning piston. Chance[9] has reported observations on reaction mixtures 50 $\mu$sec old. For precise optical observation, tubes several millimeters in diameter are used and the volumes of reaction mixture required become large. This is compensated for by the amount of information obtained from each experiment and by the use of the regenerative flow technique. The latter involves the mixing of 1 volume of substrate solution with 80 volumes of enzyme solution and recycling of the solution after the substrate is used up.

Other combined methods discussed in detail in other chapters are the combination of flow with relaxation or photochemical techniques. If the formation and decomposition of some reaction intermediate is much too fast for observation by flow techniques it may still be possible to establish an appreciable steady-state concentration in a flow tube. It is then often possible to displace this steady state by, for example, a rapid temperature change and to observe the rates of attainment of the new steady state, this volume [1] (see also Eigen and De Maeyer[2]). The relaxation to the new steady state must be rapid compared with replacement of the fluid under observation in the flow tube.

## IV. Methods for the Observation of Rapid Reactions

Most of the procedures for the study of rapid reactions, which are described in this volume, make use of the same principal methods for the observation of the progress of reactions. The special features of the observation of reactions during continuous and pulsed flow are as follows: first, the prolonged observation of one point in a steady state, or quasi-steady state, allows the introduction of long time constants; this increases the signal-to-noise ratio in the case of every monitoring device and permits the use of some devices with slow response times (glass and oxygen electrodes, thermocouples, and thermistors). Second, these modes of flow permit the chemical analysis of changes as described in Section V.

The resolution of rapid reaction methods involving observation during flow, as distinct from those involving observation after flow has stopped,

---

[8] B. Chance *et al.*, "Fast Reactions and Primary Processes in Chemical Kinetics" (S. Claesson, ed.), 5th Nobel Symp., Stockholm, p. 437. Wiley (Interscience), New York, 1967.

[9] B. Chance, *Discussions Faraday Soc.* **17**, 120 (1955); G. Czerlinski and A. Weiss, *Appl. Opt.* **4**, 59 (1965).

is superior both with respect to time and reactant concentration. These factors are involved in the limitation of the largest second-order rate constants that can be determined. The limitation of the determination of first-order rate constants, which are independent of reactant concentration, lies entirely in the time resolution of the technique. The time resolution of flow techniques is superior not only because it eliminates the time constants introduced by some measurements, but also because more rapid flow is possible when rapid stopping is not essential.

Within certain practical and theoretical limitations, the accuracy of the determination of a change in reactant concentration is a function of the magnitude of that change and of the time spent in observing it. This relation between the resolution with respect to concentration and the period of observation can be expressed in a variety of ways for different forms of observation. The general treatment of this subject in any depth requires an analysis in terms of communication theory and/or information theory. Unfortunately, a discussion of this interesting field would take us too far from the normal practical needs during the design of rapid reaction equipment. Some of the aspects of the time dependence of the precision of observation will be apparent from the discussions of individual monitoring devices. Many problems of the accuracy of observations are also discussed by Gibson [6] (see also footnote 8).

## A. Optical Observations

The principal optical method for the observation of the progress of a reaction is the measurement of optical density changes at fixed wavelengths. In their original studies of the kinetics of the reactions of hemoglobin with various ligands, Hartridge and Roughton used the reversion spectroscope for visual observation of changes of light absorption at a series of points along the flow tube. Another method proposed by Hartridge and Roughton in the pre-electronic era was the photographic recording of the light absorption in a flow tube. Surprisingly enough this simple and promising method does not appear to have been tried in practice. Apart from these methods, all the devices that can be used for optical density measurements in continuous flow systems are described by Gibson in connection with stopped-flow equipment. Although the precision of optical density records and the time resolution of the continuous-flow method can be superior to that of stopped-flow procedures, the more laborious operation of the former has prevented it from being used by more than a few experts. Since no continuous-flow device is commercially available, this relatively limited application is also true for the modes of observation described below, which cannot be used in conjunction with stopped-flow techniques.

## B. Calorimetric Measurements

Kinetic measurements with a continuous-flow thermal apparatus were among the earliest rapid reaction methods used by Hartridge and Roughton; an excellent survey of this field is given by Roughton,[7] and only a more limited account is given here. For rapid calorimetry the continuous-flow method has a number of advantages. Although some experiments have been carried out with stopped-flow calorimeters (see, e.g., Berger and Stoddart[10]) original results on reactions of proteins or enzyme catalysis have, up to now, resulted only from continuous flow devices. The sensing devices (thermocouples or thermistor beads) have response times at least as large as the half-times of many of the reactions that one usually wishes to measure with rapid reaction methods. On the other hand, in the continuous-flow tube temperature measurements can be carried out for a required period at any one point along the tube (i.e., at any constant age of the reaction mixture). The movement of a thermocouple in the flow tube to different positions is relatively easy and can in principle be automated.

While commercial amplifiers with a frequency response in the fractionally msec range have maximum sensitivities of about 10 $\mu$V, for measurements over periods of the order of a second it is possible to get equipment with sensitivities in the region of $10^{-9}$ V. To translate this into sensitivities in terms of temperature changes and chemical reactions, the following considerations may be worthwhile. A single junction pair of thermocouples (copper-constantan) gives a change of approximately 40 $\mu$V per degree temperature change. Assuming a measuring accuracy of 0.01 $\mu$V one obtains an accuracy of $0.25 \times 10^{-3}$ degree, and this means that one can measure a reaction with a heat of 10 kcal with an accuracy of 25 $\mu M$ change. Roughton gives $4 \times 10^{-5}$ degree as the limit for the differential method (see below).

In Roughton's and in Berger's laboratory the classical galvanometer amplifier of A. V. Hill is still used with great success, but commercial equipment with nanovolt sensitivity is now available. A great convenience of the continuous-flow technique for rapid heat measurements is the possibility of using a differential technique to avoid the difficulties of measuring small differences between large quantities when reactions are preceded by large heats of mixing. If the reference junction is placed at a point past the mixing chamber where mixing is just completed, or, at a point downstream, the flow tube where the reaction is complete and the measuring junction is moved along the tube, suitable temperature difference reaching between reference point and measured point can be obtained.

[10] R. L. Berger and L. C. Stoddart, *Rev. Sci. Inst.* **36,** 78 (1965); R. L. Berger and N. Davids, *ibid.* **36,** 88 (1965); L. C. Stoddart and R. L. Berger, *ibid.* **36,** 85 (1965).

If a pair of suitably spaced thermocouples is moved down the tube together, keeping the distance between them constant, data can be obtained that are suitable for Guggenheim plots for first-order reactions. The testing of mixing efficiency, heat losses of the system, and calibration of thermocouples and amplification to the final record can all be combined. The neutralization of HCl by excess of NaOH results in a rise in temperature ($\Delta H = -13.4$ kcal/mole), which is instantaneous on the time scale of flow experiments. If the two reactants are used in a continuous-flow apparatus any temperature changes observed in the flow tube are due to incomplete mixing and to loss of heat to the surroundings. Control experiments for the measurement of heat changes during the flow of each of the reactants on its own have to be carried out to correct for frictional and other nonspecific effects.

Recent studies by Chipperfield[11] of the kinetics of temperature changes during the reactions of $CO_2$ with various amino acids, illustrate that valuable data could be obtained with relatively simple equipment. In this report it is also illustrated how the overall heat of a rapid reaction can be determined in a flow system. Clearly this quantity has to be known for the interpretation of kinetic data. Although Chipperfield uses the galvanometer amplifier which is a specialty of a few laboratories (Roughton[7]), commercial equipment of suitable sensitivity is also available.

Flow equipment with mechanical drives operating syringes for controlled short periods will be very useful for thermal rapid reaction kinetics, but such devices have not yet been adequately tested. The motor-driven flow apparatus described in Section III may well prove to be suitable for thermal measurements.

## C. Electrode Measurements

Measurements of rapid changes in hydrogen ion, $CO_2$, or $O_2$ concentrations are very valuable for the study of the mechanisms of many biochemical reactions. The response times of the specific electrodes, and of the high impedance amplifiers required for the detection of their signals, are too long for the measurement of rapid reactions in continuous-flow devices. Since the time of Roughton's[7] survey of electrometric methods for monitoring rapid reactions, there has been very limited progress. Roughton and Rossi-Bernardi[12] have used $CO_2$ electrodes for measurements in a flow cell, and Rossi-Bernardi and Berger[13] have succeeded in carrying out pH measurements to an accuracy of 0.001 pH unit on a millisecond time

[11] J. R. Chipperfield, *Proc. Roy. Soc.* (*London*) **B164,** 401 (1966).

[12] F. J. W. Roughton and L. Rossi-Bernardi, *Proc. Roy. Soc.* (*London*) **B164,** 381 (1966).

[13] L. Rossi-Bernardi and R. L. Berger, personal communication.

scale. No special equipment is described in the literature, but there is no difficulty in principle in using a specific electrode in any adaptation of existing flow systems and recording signals with one of a number of available commercial amplifiers. The main limitation is likely to be the relation between size of sensitive electrode, volume of "observation cell" and precision of the definition of the age of the solution in the "observation volume."

## V. Quenching and Chemical Sampling[13a]

The study of a number of chemical reactions requires that more than two different solutions be mixed rapidly and at measured short intervals. The following typical situations for which such arrangements are required may serve to illustrate a wide range of applications. First let us consider the two reactions

$$A + B \xrightarrow{k_1} C \xrightarrow{k_2} X + Y$$
$$C + D \xrightarrow{k_3} E$$

Under a number of conditions, depending on the magnitudes and ratios of the three rate constants, it is most convenient to study the individual steps by first mixing A and B and then mixing D with the first reaction mixture after a chosen time interval. Arrangements for the observatoin of reactions of this type are illustrated by Hurwitz and Kustin.[14] Although one can imagine a great variety of similar devices, not many have been tried in practice.

Another purpose for a double mixing arrangement arises if one wishes to stop a reaction rapidly by the addition of a quenching reagent. Roughton[7] surveyed the rapid quenching techniques used up to 1963 and concluded that the two main drawbacks of the method are the relative rarity of the cases in which a suitable quencher can be found, and the uncertainty as to its applicability for reaction times of less than 0.01 sec. During the last three years both these objections have been removed, and detailed consideration of the method and its application to enzyme reactions is now pertinent.

Although one would obviously always prefer to observe the rate of formation or decomposition of a reaction intermediate by recording the change in some physical quantity that specifically reflects the concentration of this intermediate, often this is not possible. Methods involving rapid quenching of a reaction at different time intervals from its start and

[13a] This section was written in collaboration with Dr. T. E. Barman.
[14] P. Hurwitz and K. Kustin, *Trans. Faraday Soc.* **62**, 427 (1966).

subsequent chemical analysis of the composition of the samples thus taken, can add considerably to the scope of rapid flow techniques. Enzyme reactions are critically dependent on the state of ionization of one or more of the groups on the protein molecule. The arresting of an enzyme reaction by addition of acid or base involves ionic processes that are rapid compared with any steps suitable for investigation by flow techniques. In some cases the effect of acid or base is reversible, and in others the enzyme is irreversibly inactivated. The subsequent treatment of the samples will depend on this, as is illustrated by some of the systems described as examples. It turned out to be remarkably simple to quench enzyme reactions with acids and to obtain samples of reaction mixtures with an age defined to within 1 msec.

The pulsed-flow apparatus described in Section III has been used with interchangeable capillary tubes of different lengths. The quenching method involves mixing of equal or known ratios of volumes of the solutions of the two reagents in the mixing chamber whence the resulting reaction mixture is made to flow through the capillary tube, the end of which is immersed in the quenching solution. Since the rate of flow of the reaction mixture is known, its age on being quenched is determined by the length of capillary tube. With the apparatus used by the authors, reaction times from 3 msec were used with a flow rate of 8.4 m sec$^{-1}$ through the capillary tube (diameter = 0.1 cm).

The quenching process raises several problems. An ideal quenching agent must "freeze" the reaction mixture rapidly and without in any other way affecting the reaction under study. Any sluggishness in arresting the reaction would give rise to a "quenching time"—when small this can readily be corrected for. The quenching time turned out to be sufficiently short when the end of the capillary tube carrying the reaction mixture was immersed in a beaker containing the quenching agent (equal in volume to the reaction mixture, usually about 3 ml) to within 0.2 cm of the bottom of the vessel. Additional stirring of the medium with a magnetic stirrer causes an increase in the quenching time, presumably due to vortex formation. The method is illustrated here with three enzyme reactions: the trypsin-catalyzed hydrolysis of N-benzoyl L-arginine ethyl ester, the chymotrypsin-catalyzed hydrolysis of N-β-(2-furyl) acryloyl-L-tyrosine ethyl ester (quenched in 0.36 M sulfuric acid) (Barman and Gutfreund[15]), and the alkaline decomposition of phosphoryl-alkaline phosphatase (quenched in 7 N perchloric acid) (Barman and Gutfreund[16,17]). A quench-

[15] T. E. Barman and H. Gutfreund, *Biochem. J.* **101,** 411 (1966).
[16] W. N. Aldridge, T. E. Barman, and H. Gutfreund, *Biochem. J.* **92,** 23c (1964).
[17] T. E. Barman and H. Gutfreund, *Biochem. J.* **101,** 460 (1966).

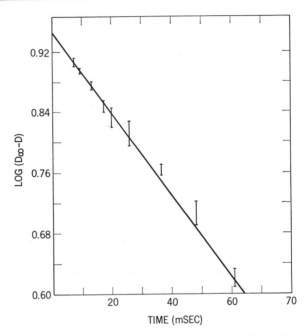

Fɪɢ. 4. First-order plot of the rate of hydrolysis of 2,4-dinitrophenyl acetate in NaOH. Reaction mixture (about 3.5 ml) = 0.625 m$M$ 2,4-dinitrophenyl acetate + 0.237 $M$ NaOH, quenched in 3 ml of 4 $M$ HCl. The mixture was adjusted immediately to pH 4.0 by the addition of 3 ml of 15 $M$ potassium acetate buffer, pH 4.5, and the optical density was read at 360 m$\mu$. The lines for each reaction time represents the standard deviation of five samples.

ing method involving freezing of the reaction mixture in an organic solvent is discussed by Bray;[2] crystals of reaction mixture can, however, be studied only in very special cases.

A suitable method to test the performance of quenched flow methods is the alkaline hydrolysis of 2,4-dinitrophenyl acetate (Barman and Gutfreund[2]). One of the products of the reaction, 2,4-dinitrophenol, has a molar absorption at 360 m$\mu$ of 8.8 $\times$ 10$^3$ $M^{-1}$ cm$^{-1}$ at pH 4.0. The conditions of the experiment are described in Fig. 4. The results yielded a second-order constant for the hydrolysis $k$ = 55 $M^{-1}$ sec$^{-1}$, a value in close agreement with comparative measurements carried out with the optical stopped-flow apparatus. In this experiment—involving the quenching of a strongly alkaline reaction mixture—the quenching time was about 3 msec, but the quenching time of enzyme systems in dilute buffers was considerably shorter, as can be seen from the extrapolation to zero time (see Fig. 4).

Enzyme Reactions Studied by the Rapid-Flow Quenching Technique

*Dephosphorylation of Phosphoryl-Alkaline Phosphatase.* It is known that alkaline phosphatases from a number of sources are phosphorylated when incubated with orthophosphate under slightly acid conditions; at alkaline pH the phosphoryl enzyme is rapidly dephosphorylated. The kinetics of the dephosphorylation of the *Escherichia coli* phosphoryl alkaline phosphatase have been investigated by the rapid-flow quenching method (Aldridge, Barman, and Gutfreund[16]); the conditions of the experiment and results obtained are summarized in Fig. 5. At pH 8.4, the half-life of the dephosphorylation was 6 msec corresponding to a rate constant of 115 sec$^{-1}$. The quenching time was <0.5 msec.

*The Chymotrypsin-Catalyzed Hydrolysis of N-β-(2-Furyl) Acryloyl-L-tyrosine Ethyl Ester (FATEE).* Barman and Gutfreund[17] studied the kinetics of the liberation of ethanol in the system chymotrypsin-FATEE. The reaction mixtures were quenched in equal volumes of 0.36 *M* sulfuric acid (this concentration of acid neither precipitated the enzyme protein nor hydrolyzed the substrate), and the ethanol was recovered by vacuum distillation and estimated by the alcohol dehydrogenase method of

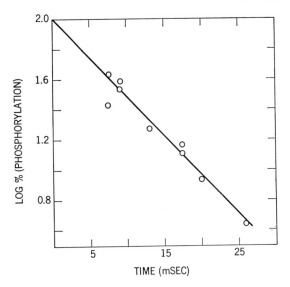

Fig. 5. First-order plot of the dephosphorylation of phosphoryl phosphatase. The reaction mixture (about 3.5 ml) was obtained by mixing equal volumes of phosphoryl enzyme solution (0.33 mg of alkaline phosphatase per milliliter + 33 m*M* magnesium acetate + 1 m*M* orthophosphate-$^{32}$P + 0.05 *M* sodium acetate buffer, pH 5.4) and 0.1 *M* sodium carbonate-bicarbonate buffer containing 6 mg of serum albumin per milliliter to give a final pH of 8.4. The reaction mixture was quenched in 7 *N* perchloric acid.

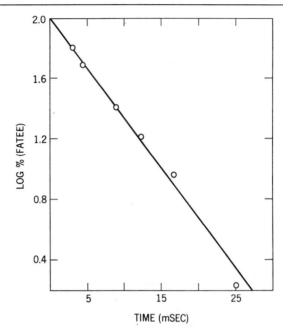

FIG. 6. Chymotrypsin: FATEE. First-order plot of the rate of liberation of ethanol. Reaction mixture (about 3.5 ml): 0.270 m$M$ chymotrypsin + 0.224 m$M$ FATEE + 0.08 $M$ phosphate buffer, pH 6.6; quenched in 0.36 $M$ H$_2$SO$_4$.

Lunquist. It is clear from Fig. 6 that this relatively mild quenching agent very effectively arrested the reaction (quenching time <0.5 msec). The half-time of the reaction was 4.8 msec, corresponding to a rate constant of 147 sec$^{-1}$.

*The Trypsin-Catalyzed Hydrolysis of N-Benzoyl-L-arginine Ethyl Ester (BAEE)*. In experiments with the system trypsin–BAEE (in 0.08 $M$ phosphate buffer), it was found that 0.15 $M$ hydrochloric acid was as effective in arresting the reaction as 0.36 $M$ sulfuric acid of 7 $M$ perchloric acid (Barman and Gutfreund, unpublished observations).

It is clear that the rapid-flow quenching technique is of wide application for the study of enzyme reactions. Its limitations are given by the efficiency and possible side effects of the quenching agent and by the availability of methods of analyzing the quenched reaction mixtures. It must be emphasized that *quenched* reaction mixtures are analyzed rather than the reaction mixtures themselves: a careful choice of a number of different quenching agents should be made with each system under study so as to reduce potential side effects to the minimum. It is doubtful whether or not first-order reactions with half-lives of less than 1 msec can be studied by the method, but it should be possible to study the kinetics of very fast second-order reactions by the use of large volumes of dilute solutions of reactants. The quenched solutions of products can be concentrated prior to analysis.

The study of the rates of quenching of enzyme reactions by various reagents can also give useful information and can be carried out by the methods described here.

### Acknowledgments

I wish to thank my colleagues T. E. Barman, D. R. Trentham, C. H. McMurray, and D. W. Yates for considerable help in discussing the procedures for testing the limitation of flow devices. The financial support of the Agricultural Research Council and the U.S. Public Health Service is also acknowledged.

# Section IV

# Irradiation

# [8] Photochemical Reactions in Nucleic Acids

## *By* H. E. JOHNS

## I. Introduction

Ionizing radiations have been extensively used to perturb biological systems in the hope that the observed effects could be used to increase our understanding of these systems. When ionizing radiations interact with a cell, each ionizing event involves enough energy to disrupt any chemical bond within a macromolecule, and it is usually impossible to pinpoint the nature of the damage. In contrast, when ultraviolet light (UV) in the range 200–300 millimicrons (m$\mu$) interacts with a molecule, energy tends to be absorbed in specific bonds within a molecule to produce a number of specific photoproducts. These photoproducts in turn give rise to alterations in the biological activity of the cell. Because of the importance of RNA and DNA in controlling the metabolism of the cell, and since these nucleic acids strongly absorb UV, the nature of UV damage to nucleic acids and their components has been studied extensively. In this chapter, we will be concerned with photochemical reactions in UV-irradiated nucleic acid components.

## A. UV Photoproducts in Pyrimidines

It has generally been observed that the pyrimidines are much more sensitive to UV than the purines[1] and may be altered in two important ways, as illustrated in Fig. 1. The two main photoproducts are the dimer,[2] in which two pyrimidines are linked together in a cyclobutane ring, and the hydrate,[3,4] in which water is added across the 5–6 double bond.

The presence of the 5–6 double bond in cytosine, thymine, and uracil gives these compounds a peak absorption at a particular wavelength, $\lambda_{max}$, which at neutral pH values occurs in the range 260–270 m$\mu$. Since the hydrate and dimer lack the 5–6 double bond, they have very little absorption at the $\lambda_{max}$ where the parent absorbs, so that irradiation leading to either hydrate or dimer brings about a drop in the absorption at $\lambda_{max}$. Detailed discussions on dimers and hydrates may be found in recent review articles by McLaren and Shugar,[1] Setlow,[5] Wacker et al.,[6] Johns et al.,[7] Smith,[8] and Setlow.[9] Here we will content ourselves with a very brief discussion of a few of their properties, because these properties, in part, determine the experimental techniques which have been developed for dealing with them.

## 1. *Dimers*

The dimer may be formed between thymine rings when thymine is irradiated in frozen solution,[2] but not in liquid solutions.[10] Presumably the

[1] A. D. McLaren and D. Shugar, "Photochemistry of Proteins and Nucleic Acids." Macmillan, New York, 1964.

[2] R. Beukers and W. Berends, *Biochim. Biophys. Acta* **49**, 181 (1961).

[3] R. L. Sinsheimer, *Radiation Res.* **1**, 505 (1954).

[4] A. M. Moore, *Can. J. Chem.* **36**, 281 (1958).

[5] R. B. Setlow, *J. Cell. Comp. Physiol.* **64** (Suppl. 1), 51 (1964).

[6] A. Wacker, *Progr. Nucleic Acid Res.* **1**, 369 (1963).

[7] H. E. Johns, M. L. Pearson, C. W. Helleiner, and D. M. Logan, *18th Ann. Symp. Fundamental Cancer Res. Univ. Texas, M. D. Anderson Tumor Inst. 1964.* p. 29. Williams & Wilkins, Baltimore, Maryland, 1965.

[8] K. C. Smith, *in* "Photophysiology" (A. C. Giese, ed.), Vol. 2, p. 329. Academic Press, New York, 1964.

[9] J. K. Setlow, *in* "Current Topics in Radiation Research" (M. Ebert and A. Howard, eds.), Vol. 2, p. 195. North Holland Publ., Amsterdam, 1966.

[10] Dimers of thymine, uracil, orotic acid, and their derivatives can also be formed in aqueous solutions if $O_2$ is removed from the solution during the irradiation,[11,12] but the yield of dimers is less than when the solutions are frozen. The fact that the yield in solution is reduced by the presence of $O_2$ suggests that the dimer comes from the excited triplet state since $O_2$ is a known quencher of such states.[13,14] Additional evidence for the involvement of the triplet state comes from the fact that dimers can be produced using a sensitizer which absorbs the radiation energy and transfers it to the triplet state of the pyrimidine.[15–20] At the moment it appears as though both

absorption of a photon can lead to the dimer only if the thymine rings have the correct orientation in space to favor dimerization.[20a,21] Dimers are also formed in frozen solutions of uracil[22] and cytosine,[23] and mixed dimers[22] are produced between any two pyrimidines in frozen solutions containing thymine, uracil, and cytosine. Irradiation of *dinucleotides* of the pyrimidines in *liquid* solution produces dimers[24–30] but with a smaller yield, $\frac{1}{20}$ to $\frac{1}{100}$, of that in frozen solutions. As one might expect, dimers are also formed between adjacent pyrimidines in irradiated DNA and RNA,[31,32] since local regions of these molecules have the same structure as the dinucleotides.

All the dimers discussed above have the important property that they can be uncoupled by radiation. This property enables one to carry out an easy test for the dimer. If, for example, a frozen solution of thymine is irradiated, dimers are formed and the absorption of the thawed solution at $\lambda_{max}$ is less than it was originally. If now the thawed solution is irradiated as a liquid solution the absorption at $\lambda_{max}$ will increase as dimers are

---

singlet and triplet states may be involved—the singlet for dimers produced in ice and the triplet for dimers in solution. The involvement of the triplet state will not be discussed further in this chapter.

[11] E. Sztumpf-Kulikowska, D. Shugar, and J. W. Boag, *Photochem. Photobiol.* **6,** 41 (1967).

[12] C. L. Greenstock, I. H. Brown, J. W. Hunt, and H. E. Johns, *Biochem. Biophys. Res. Commun.* **27,** 431 (1967).

[13] I. H. Brown and H. E. Johns, *Photochem. Photobiol.* **8,** 273 (1968).

[14] D. Whillans and H. E. Johns, *Photochem. Photobiol.* **9,** 323 (1969).

[15] A. A. Lamola and T. Yamane, *Proc. Natl. Acad. Sci. U.S.* **58,** 443 (1967).

[16] I. Von Wilucki, M. Matthaus, and C. M. Krauch, *Photochem. Photobiol.* **6,** 497 (1967).

[17] M. Charlier and C. Hélène, *Photochem. Photobiol.* **6,** 501 (1967).

[18] D. Elad, C. Krüger, and G. M. J. Schmidt, *Photochem. Photobiol.* **6,** 495 (1967).

[19] A. A. Lamola and J. P. Mittal, *Science* **154,** 1560 (1966).

[20] C. L. Greenstock and H. E. Johns, *Biochem. Biophys. Res. Commun.* **30,** 21 (1968).

[20a] S. Y. Wang, *Nature* **200,** 879 (1963).

[21] C. Hélène, *Biochem. Biophys. Res. Commun.* **22,** 237 (1966).

[22] A. Wacker, D. Weinblum, L. Träger, and Z. H. Moustafa, *J. Mol. Biol.* **3,** 790 (1961).

[23] D. Weinblum, Ph. D. Thesis, Technische Universität, Berlin, 1961.

[24] H. E. Johns, M. L. Pearson, J. C. LeBlanc, and C. W. Helleiner, *J. Mol. Biol.* **9,** 503 (1964).

[25] R. A. Deering and R. B. Setlow, *Biochim. Biophys. Acta* **68,** 526 (1963).

[26] A. Wacker, H. Dellweg, and E. Lodemann, *Angew. Chem.* **73,** 64 (1961).

[27] H. E. Johns, S. A. Rapaport, and M. Delbrück, *J. Mol. Biol.* **4,** 104 (1962).

[28] C. W. Helleiner, M. L. Pearson, and H. E. Johns, *Proc. Natl. Acad. Sci. U.S.* **50,** 761 (1963).

[29] I. H. Brown, K. B. Freeman, and H. E. Johns, *J. Mol. Biol.* **15,** 640 (1966).

[30] K. B. Freeman, P. V. Hariharan, and H. E. Johns, *J. Mol. Biol.* **13,** 833 (1965).

[31] H. Dellweg and A. Wacker, *Z. Naturforsch.* **17b,** 827 (1962).

[32] R. B. Setlow, W. L. Carrier, and F. J. Bollum, *Proc. Natl. Acad. Sci. U.S.* **53,** 1111 (1965).

uncoupled. When a dinucleotide such as TpT[32a] is irradiated[24,27] the situation is more complicated, for now dimerization and reversal (uncoupling of dimers) proceeds simultaneously and after long exposures equilibrium is established between the forward reaction leading to the dimer, represented by T͡pT, and the reverse reaction leading back to the parent, according to Eq. (1).[32b]

$$\text{TpT} \overset{h\nu}{\underset{h\nu}{\rightleftarrows}} \widehat{\text{TpT}} \qquad (1)$$

The equilibrium concentrations of dimer and parent depend upon the wavelength of the irradiation, the forward reaction being favored at long wavelengths and the reverse reactions at short wavelengths. If one irradiates first at long wavelengths to a sufficient dose to produce dimers and then irradiates the mixture of parent and dimer with short wavelength radiation, dimers may be uncoupled and the absorbance may increase with dose. The increase in absorbance with short wavelength irradiation following long wavelength irradiation is the simplest test for the presence of a dimer. Conditions under which this test will detect dimers are discussed by Brown and Johns[33] and Setlow and Setlow.[34]

### 2. *Hydrates*

The hydrate can be formed in uracil[3,4,28,29,35,36] or in cytosine[30,37] but has not been detected in thymine, although Fahr[38] has observed a short-lived irradiation product of thymine which may be the hydrate. The hydrate is unstable and on heating reverses to the parent compound. The rate of reversal is pH dependent, the half-life varying from a few minutes to days. In general, cytosine hydrate is much less stable than uracil hydrate. The simplest single test for the presence of a hydrate is the return of absorbance at $\lambda_{\max}$ on heating.

### 3. *Deamination*

Cytosine dimers and hydrates, in addition to wavelength and heat reversal, may also undergo deamination[30,37,38] leading to the corresponding

[32a] TpT, thymidylyl-(3′,5′)-thymidine.
[32b] This reaction scheme is a first approximation to the complete reaction described in Eq. (51).
[33] I. H. Brown and H. E. Johns, *Photochem. Photobiol.* **6**, 469 (1967).
[34] R. B. Setlow and J. K. Setlow, *Photochem. Photobiol.* **4**, 939 (1965).
[35] P. A. Swenson and R. B. Setlow, *Photochem. Photobiol.* **2**, 419 (1963).
[36] D. M. Logan and G. F. Whitmore, *Photochem. Photobiol.* **5**, 143 (1966).
[37] H. E. Johns, J. C. LeBlanc, and K. B. Freeman, *J. Mol. Biol.* **13**, 849 (1965).
[38] E. Fahr, R. Kleber, and E. Boebinger, *Z. Naturforsch.* **21b**, 219 (1966).

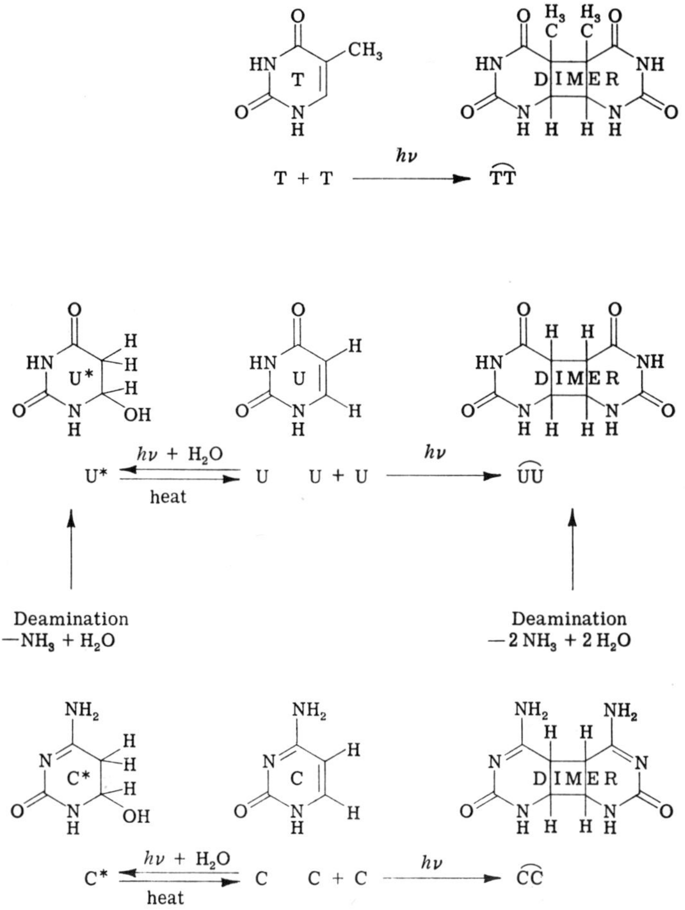

FIG. 1. Structure of thymine (T) and its photoproduct the dimer T̂T; uracil (U) and its photoproducts the dimer ÛU and the hydrate U*; cytosine (C) and its photoproduct the dimer ĈC, and the hydrate C*. The diagram shows how the cytosine photoproducts can deaminate to the corresponding uracil compounds by the addition of $H_2O$ and the loss of $NH_3$.

uracil compound. The deamination reactions involving the loss of $NH_3$ and the gain of $H_2O$ are shown in Fig. 1.

### 4. Biological Importance of Dimers and Hydrates

A number of investigators have attempted to determine in detail the effects of dimers and hydrates on various biological systems. Much of this work has been handicapped by lack of knowledge of the basic photochemistry of dimers and hydrates. However, there is good evidence that dimers of thymine, uracil, and cytosine are involved in the photoreactivation of bacteria and yeast[32,39-43] and that a repair process exists for excising such dimers from DNA.[44,45] A number of investigations have been carried out on the effects of UV photoproducts on protein synthesis.[46-53] In the case of uracil, specific information concerning the coding ability of its photoproducts is now available. For example, when irradiated triplets of uracil are assayed for their ability to bind amino acids to ribosomes U*

behaves as C and operates with about 90% efficiency while $\widehat{UU}$ in the first position of the triplet can be read as either UU or GU with about 10% efficiency.[54,55]

A detailed discussion of the photochemistry of the nucleic acids would be out of place here. The above remarks have been included to give the reader some idea of the properties of dimers and hydrates. We shall use the production of hydrates and dimers to illustrate photochemical reactions. In the following sections we will discuss the basic concepts of cross sections and quantum yield, the production of monoenergetic radiation, the measurement of radiation, the detection of photoproducts, the measurement of the yield of photoproduct, and the mathematical interpretation of the results in terms of cross sections.

## II. Basic Concepts

### 1. *Photon Energy*

Photochemical reactions are produced by the absorption of individual photons or quanta. The energy $\mathcal{E}$ of one photon is related to its wavelength and frequency by the relation

[39] R. B. Setlow and J. K. Setlow, *Proc. Natl. Acad. Sci. U.S.* **48,** 1250 (1962).

[40] C. S. Rupert, *J. Gen. Physiol.* **43,** 573 (1960).

[41] D. L. Wulff and C. S. Rupert, *Biochem. Biophys. Res. Commun.* **7,** 237 (1962).

[42] J. K. Setlow, M. E. Boling, and F. J. Bollum, *Proc. Natl. Acad. Sci. U.S.* **53,** 1430 (1965).

[43] A. Muhammed, *J. Biol. Chem.* **241,** 516 (1966).

[44] R. P. Boyce and P. Howard-Flanders, *Proc. Natl. Acad. Sci. U.S.* **51,** 293 (1964).

[45] R. B. Setlow and W. L. Carrier, *Proc. Natl. Acad. Sci. U.S.* **51,** 226 (1964).

[46] D. M. Logan, Doctoral Dissertation, University of Toronto, Toronto, 1965.

[47] L. Grossman, *Proc. Natl. Acad. Sci. U.S.* **48,** 1609 (1962).

[48] L. Grossman, *Proc. Natl. Acad. Sci. U.S.* **50,** 657 (1963).

[49] P. D. Harriman and H. G. Zachau, *J. Mol. Biol.* **16,** 387 (1966).

[50] M. H. Buc and J. F. Scott, *Biochem. Biophys. Res. Commun.* **22,** 459 (1966).

[51] F. Fawaz-Estrup and R. B. Setlow, *Biochim. Biophys. Acta* **87,** 28 (1964).

[52] L. Grossman, J. Ono, and R. G. Wilson, *Federation Proc.* **24** (Suppl. 15), S-80 (1965).

[53] J. Ono, R. G. Wilson, and L. Grossman, *J. Mol. Biol.* **11,** 600 (1965).

[54] P. Ottensmeyer and G. F. Whitmore, *J. Mol. Biol.* **38,** 1 (1968).

[55] P. Ottensmeyer and G. F. Whitmore, *J. Mol. Biol.* **38,** 17 (1968).

$$\varepsilon = h\nu = \frac{hc}{\lambda} \tag{2}$$

where $h$ is Planck's constant ($6.62 \times 10^{-27}$ erg sec), $c$ the velocity of light ($3.00 \times 10^{10}$ cm/sec), $\nu$ the frequency of the light, and $\lambda$ the wavelength. If the energy $\varepsilon$ is expressed in electron volts (eV) and $\lambda$ in m$\mu$, then Eq. (2) becomes

$$\varepsilon(eV) = \frac{1240}{\lambda(m\mu)} \tag{3}$$

Thus photons with wavelengths ranging from 220 to 300 m$\mu$ (the region of interest in this chapter) have energies varying from 5.64 to 4.13 eV corresponding to 130 and 95 kcal per mole of photons, respectively (see left part of Table I). Photons in this wavelength range thus have sufficient energy to disrupt most chemical bonds and so produce chemical reactions.

TABLE I

RELATION BETWEEN PHOTON ENERGY, ENERGY FLUENCE, PHOTON FLUENCE, AND WAVELENGTH FOR THE PROMINENT LINES IN THE MERCURY SPECTRUM

| Wave-length ($\lambda$), m$\mu$ | Photon energy ($\varepsilon$) | | Relation between energy fluence and photon fluence | | |
|---|---|---|---|---|---|
| | | | | Photon fluence ($L$) | |
| | | | Energy fluence ($F$), ergs/mm² | | |
| | eV | kcal/mole | | $\mu$E/cm² | Photons/cm² ($\times 10^{-17}$) |
| 225 | 5.511 | 127.1 | $10^4$ | 0.1880 | 1.133 |
| 235 | 5.277 | 121.7 | $10^4$ | 0.1974 | 1.189 |
| 240 | 5.167 | 119.2 | $10^4$ | 0.2005 | 1.208 |
| 248 | 5.000 | 115.3 | $10^4$ | 0.2072 | 1.248 |
| 254 | 4.882 | 112.6 | $10^4$ | 0.2122 | 1.279 |
| 265 | 4.679 | 107.9 | $10^4$ | 0.2214 | 1.334 |
| 275 | 4.509 | 104.0 | $10^4$ | 0.2298 | 1.385 |
| 280 | 4.429 | 102.2 | $10^4$ | 0.2339 | 1.409 |
| 289 | 4.291 | 99.0 | $10^4$ | 0.2415 | 1.455 |
| 297 | 4.175 | 96.3 | $10^4$ | 0.2482 | 1.495 |
| 302 | 4.106 | 94.7 | $10^4$ | 0.2523 | 1.520 |
| 313 | 3.962 | 93.9 | $10^4$ | 0.2615 | 1.575 |

2. *Units for Photochemistry*

Photochemical literature contains expressions such as energy flux, intensity, flux density, dose, etc. Since these mean different quantities to different people, we choose to use the notation which has been developed for dealing with ionizing radiation.[56] We shall refer to the number of

[56] Intern. Comm. Radiological Units and Measurements (ICRU) Report 10a, Handbook 84, U.S. Natl. Bur. Stand., Washington, D.C. (1962).

photons that cross a unit area as the *photon fluence*, $L$, and the number of photons crossing a unit area per unit time as the *photon fluence rate*. When one is interested in the energy carried by these photons, the terms *energy fluence* and *energy fluence rate* may be used. These have dimensions energy per unit area and energy per unit area per unit time.

In photochemistry one is often interested in converting large quantities of a material to some photoproduct, and it is convenient to deal with a mole $(6.02 \times 10^{23})$ or micromole $(6.02 \times 10^{17})$ of photons, and these are referred to as an Einstein (E) and $\mu E$, respectively. Thus a photon fluence could be expressed in photons/cm$^2$, E/cm$^2$, or $\mu E$/cm$^2$. Similarly a photon fluence rate could be expressed as $\mu E$/cm$^2$ sec.

For ionizing radiations, dose is restricted to energy *absorbed* and the special unit in which dose is measured is the rad (1 rad = 100 ergs/gram). The word exposure has come to refer to the amount of radiation to which a system is exposed, and the special unit is the roentgen. Thus with these two concepts a distinction may be made between the amounts of energy *incident* on a system and the amount of energy *absorbed* by the system. The author feels that some uniformity of notation should be introduced into photochemistry and proposes that the same basic distinction be made between dose and exposure. We will therefore use the word *exposure* as an alternative term to *photon fluence* $(L)$. We will not use the expression dose since this should be confined to the *energy absorbed* by the system being *exposed* to the radiation. The *average photon fluence* represents the average fluence "seen" by *all* the molecules in the irradiated vessel and a convenient unit is the $\mu E$/cm$^2$. We could thus say, for example, that all the molecules in the vessel were given an average exposure of $\bar{L} = 1.3$ $\mu E$/cm$^2$ or an average photon fluence of 1.3 $\mu E$/cm$^2$ or $1.3 \times 6.02 \times 10^{17}$ photons/cm$^2$.

In many UV irradiations the beam of photons impinges on an unstirred solution, and "doses" are often expressed in ergs/cm$^2$. This is the *energy fluence* "seen" only by the surface molecules. An exposure expressed this way is meaningless. If the solution is vigorously stirred all molecules will "see" the same radiation and the term exposure has meaning. An *average energy fluence* or *average photon fluence* (exposure) can be calculated from the incident energy fluence provided the absorbance, thickness, and volume of the solution are given. Often, however, these parameters are *not* given so that meaningful calculations are difficult. Methods for calculating an average fluence are given in Section IV.

Of the two units *average photon fluence* and *average energy fluence*, the author favors the first since this focuses attention on the idea that it is the number of photons per unit area which is important rather than the energy carried by these photons. It will be realized that an interaction requires the absorption of a photon rather than a given amount of energy.

Thus identical energy fluences at 400 mμ and 200 mμ, respectively, correspond to photon fluences which differ by a factor of 2.0 (see Table I). Conversion from one unit to the other can be made using Table I.

### 3. Meaning of Cross Section

In describing photochemical reactions it is convenient to use the concept of the cross section to describe the probability that a given type of alteration will take place. The cross section may be thought of in two different ways and may have dimensions of area per photon or of area per molecule. These will now be considered.

*a. Cross Section in cm²/Photon.* Suppose a solution containing $n$ molecules/cm³ occupies a slab 1 cm $\times$ 1 cm and thickness $\Delta x$ as illustrated in Fig. 2a. The number of molecules per unit area will be $N = n\Delta x$. Suppose now this solution is given an increment of exposure $\Delta L$ photons/cm² and that in the process $\Delta N_B$ molecules are altered to form B. Then the cross section $\sigma_B$ for the production of photoproduct B is defined by

$$\Delta N_B = \sigma_B \cdot N \cdot \Delta L \qquad (4)$$

This equation merely states that the number $\Delta N_B$ is directly proportional to the increment of fluence $\Delta L$ and the number of molecules $N$ in the slab. The constant of proportionality is $\sigma_B$. Since $\Delta N_B$ and $N$ have the same dimension, $\sigma_B$ must have dimensions $1/\Delta L$. Thus, if $L$ is measured in photons/cm², $\sigma_B$ will be expressed as cm²/photon. Similarly if $\Delta L$ is expressed in μE/cm², $\sigma_B$ would be in cm²/μE.

The meaning of the cross section in cm²/photon is illustrated in Fig. 2a, where we imagine each photon as being a pencil of area $\sigma_B$. Clearly, the total area of *all* these pencils is $\sigma_B \cdot \Delta L$. If we now think of each molecule as being a point and that a molecule is altered to type B if one of the

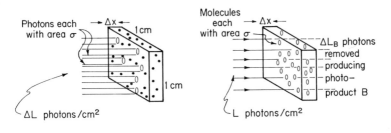

(A)  Cross section cm²/photon          (B)  Cross section cm²/molecule

Fig. 2. Meaning of cross section and relation between absorption cross section and absorbance. Solutions contain $n$ molecules/cm³, slab of solution of thickness $\Delta x$ contains $N = n\Delta x$ molecules/cm².

pencils intercepts it, then the number of interactions is simply $\sigma_B \cdot \Delta L \cdot N$ as given by Eq. (4).

b. *Cross Section in cm²/Molecule.* The alternative way of thinking of a cross section is illustrated in Fig. 2b, where we now imagine $L$ photons/cm² impinging on the same solution. Suppose $\Delta L_B$ of these photons interact with molecules in the slab to produce photoproduct B, then

$$\Delta L_B = -\sigma_B \cdot L \cdot N = -\sigma_B \cdot L \cdot n \cdot \Delta x \tag{5}$$

This equation merely states that the change in the fluence due to production of B will be proportional to the fluence $L$ and the number of molecules in the slab and that the constant of proportionality is $\sigma_B$. The negative sign indicates that $L$ decreases as one passes through the absorber.

Since $L$ and $\Delta L$ have the same dimensions, it follows that $\sigma_B$ has dimensions of $1/N$ or cm²/molecule. From a physical point of view we can imagine that each molecule has an area $\sigma_B$ and the total area of all the molecules is $\sigma_B \cdot N$. The number of photons that pass through this area and so interact with the molecules is $\sigma_B \cdot N \cdot L$ as given by Eq. (5).

From these two discussions it is evident that the cross section can be thought of either as an *area per photon* or an *area per molecule*. When we think of a photoproduct being produced in a *fixed target* by an *increment of fluence*, then the cross section should be expressed as cm²/photon according to Eq. (4) and Fig. 2a. On the other hand, when a target of *variable thickness* is thought of as being bombarded by a *fixed fluence* then we should express the cross section in cm²/molecule, according to Eq. 5 and Fig. 2b. It must be emphasized that both concepts are correct and either form may be used. Thus, *a cross section expressed in cm²/photon has the same numerical value as the cross section in cm²/molecule.* The difficulty of this concept partly arises from the mechanistic way the cross section is illustrated: in Fig. 2a, the photon is thought of as having an area while the molecule is a point, whereas in Fig. 2b the photon is thought of as a point while the target has area. Clearly, in the actual situation both entities have some "area" and *the chance of an interaction depends upon* the properties of *both the photon and the molecule.* For the same molecule the cross section depends upon the wavelength or energy of the photon, while for the same wavelength the cross section depends upon the structure of the absorbing molecule and the photoproduct considered.

### 4. *Absorption Cross Section*

In Fig. 2b, $\Delta L_B$ represents the number of photons that were removed from the beam *to produce photoproduct B.* In general there would be many other interactions leading to photoproducts C, D, E, etc. In addition, there would be interactions leading to excited states of the molecule which might or might not be detectable. Now each interaction, regardless of its type would either absorb a photon, or divert it from its original direction so that a measure of $\Delta L$ would be a measure of *all* the interactions and thus a measure of the absorption of the solution. The cross section for *absorption* will be represented by $\sigma_a$. It can be thought of as either an area per photon or as an area per molecule. It is usual to express it as $cm^2$/molecule or $cm^2$/mole. By analogy with the preceding section and in line with the above discussion it is defined by

$$\Delta L = -\sigma_a \cdot L \cdot N = -\sigma_a \cdot L \cdot n \cdot \Delta x \tag{6}$$

In this expression $\Delta L$ is the increment in fluence produced by *all* processes in the slab and no information is given concerning the types of interactions.

5. *Relation between Molar Absorptivity*[57] *(Extinction Coefficient) and Absorption Cross Section*

To measure an absorption cross section one usually measures the transmission of radiation through a thick layer of the solution and expresses this absorption as an absorbance $A$ defined by

$$A = \log_{10}(L_0/L) = \frac{1}{2.30} (\ln L_0 - \ln L) \tag{7}$$

where $L_0$ is the incident fluence and $L$ the transmitted fluence.[58a] Absorbance is therefore a function of wavelength, material, concentration, and path length.

This absorbance is in turn related to the molar absorptivity, $\epsilon$, defined by

$$A = \epsilon \cdot C \cdot b \tag{8}$$

where $C$ is the concentration of the solution in moles/liter $(M)$ and $b$ is the path length through the solution in centimeters.

Combining Eqs. (7) and (8) one obtains

$$\ln L_0 - \ln L = 2.30\epsilon \cdot C \cdot b \tag{9}$$

Now Eqs. (6) and (9) really describe the same phenomena, but Eq. (6) applies to an incremental layer of thickness $\Delta x$ while Eq. (9) applies to a layer of any thickness $x$. To compare them, one either integrates (6) or differentiates (9). Differentiating (9) with respect to $b$ one obtains

---

[57] Where possible the symbols and definitions recommended by the ASTM Committee E-13, on absorption spectroscopy will be followed.[58]

[58] "Manual on Recommended Practices in Spectrophotometry." Am. Soc. for Testing and Materials, Committee E-13 on Absorption Spectroscopy, Philadelphia, Pennsylvania (1966).

[58a] $L_0/L$ could equally well be the ratio of the incident and transmitted intensity.

$$\frac{dL}{L} = -2.30\epsilon \cdot C \cdot db \tag{10}$$

Comparing this with Eq. (6), it is clear that $\sigma_a$ and $\epsilon$ are related, thus

$$\sigma_a = 2.30 \frac{C}{n} \cdot \epsilon \tag{11}$$

Now $C$ and $n$ are both expressions for the concentration in different units. If we place $C = 1$ mole/liter then $n = 6.02 \times 10^{20}$ molecules/ml and

$$\sigma_a = \frac{2.30}{6.02 \times 10^{20}} \cdot \epsilon \frac{cm^2}{molecule} \tag{12}$$

Alternative expressions for $\sigma_a$ are

$$\sigma_a = 2.30 \times 10^3 \epsilon \frac{cm^2}{mole} = 2.30 \times 10^{-3} \epsilon \frac{cm^2}{\mu mole} \tag{13}$$

Equation (13) thus relates the cross section for absorption ($\sigma_a$) to the molar absorptivity ($\epsilon$). Another useful relation between $\sigma_a$ and the absorbance $A$ of a solution of thickness $b$ containing $n$ molecules/cm³ may be obtained by combining Eqs. (8) and (11) to give

$$\sigma_a = 2.30 \frac{A}{bn} \tag{14}$$

## 6. Quantum Yield

Suppose a solution has an absorption cross section $\sigma_a$ and a cross section $\sigma_B$ for the production of photoproduct B [defined by Eq. (4)], then the quantum yield is defined as

$$\Phi_B = \frac{\sigma_B}{\sigma_a} = \frac{\text{number of photons that convert molecules to type B}}{\text{number of photons absorbed}}$$

$$= \frac{\text{number of molecules converted to type B}}{\text{number of photons absorbed}} \tag{15}$$

The quantum yield is of theoretical importance in interpreting photochemical reactions since it relates the number of photons that give rise to a photoproduct to the number of photons absorbed. When dimers are formed by irradiating frozen solutions containing pyrimidines, the quantum yield is close to 1.0,[59] suggesting that the ring structures must be stacked in such a way that once the ring is excited by the absorption of energy, the energy cannot escape but gives rise to the dimer. In contrast irradiation of the pyrimidine dinucleotides (TpT, CpC, UpU) in solution produces dimers with a quantum yield of about 0.02.[7] Presumably now the orienta-

[59] W. Füchtbauer and P. Mazur, Photochem. Photobiol. **5**, 323 (1966).

tion of the rings is such that only one photon out of 50 which are absorbed can give rise to the dimer. Quantum yields of about 0.02 are also observed for the production of the hydrate.[7] The quantum yield for the reversal of dimers is close to 1.0 for all dimers so far investigated, indicating that nearly every photon that is absorbed in the dimer disrupts the cyclobutane ring.

A precise determination of a quantum yield is by no means easy since it requires a determination of a fluence $\Delta L$ and the number of photoproduct molecules formed from a known number of irradiated molecules. In addition one requires a knowledge of $\sigma_a$ which depends upon the molar absorptivity of the parent molecule. Since the cross sections for absorption and conversion are both functions of wavelength, one also requires a source of monochromatic light. In addition, this source of radiation should be intense in order to produce a sufficient amount of photoproduct in a reasonable time. In subsequent sections of this chapter we will discuss methods for obtaining monoenergetic radiation, for measuring the fluence, and for detecting the number and types of photoproducts.

## III. Irradiation Facility

### 1. Monochromator

We have developed a powerful UV monochromator for photochemical studies the details of which can be found elsewhere.[60,61] Briefly this monochromator uses two blazed gratings side by side with total area 27 × 20.8 cm²; identical collimator and telescope mirrors of diameter 25 cm and focal length 186 cm; entrance and exit slits of 15 cm height and up to 1.5 cm width; dispersion 4 mμ/cm at exit slit. The most useful UV source is a high-pressure Hg vapor lamp of either the water-cooled variety (Philips SP-500 W) or the air-cooled type (General Electric BH-6). By the use of a short focal length condensing mirror, the system can be operated with an f/1 aperture at the source. Figure 3 shows the number of photons per second which may be obtained at the exit slit as a function of wavelength when this monochromator is set to pass a wavelength band of 2.4 mμ. Figure 3 also shows that this instrument gives 10–20 times the intensity of two commercially available instruments.

### 2. Arrangement of Components at Exit Slit of Monochromator

The original papers[60,61] describe this instrument in detail. Since these were written certain improvements have been made in the irradiation

---

[60] H. E. Johns and A. M. Rauth, *Photochem. Photobiol.* **4**, 673 (1965).
[61] H. E. Johns and A. M. Rauth, *Photochem. Photobiol.* **4**, 693 (1965).

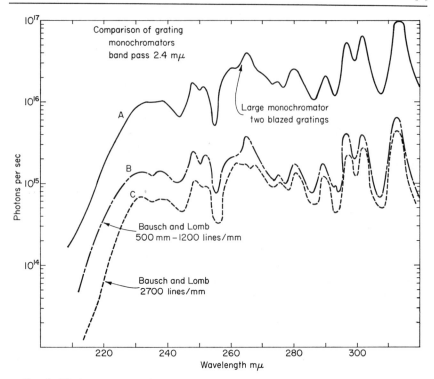

Fig. 3. Photons per second transmitted through three grating monochromators as a function of wavelength. Source SP-500 W Philips lamp operated in a water jacket. Curve $A$ was obtained with the large monochromator using two blazed gratings No. 33-53-27-03 with 1200 lines/mm; curve $B$ using the Bausch and Lomb 500 mm focal length instrument No. 33-86-45 with a grating of 1200 lines/mm; curve $C$ using Bausch and Lomb instrument No. 33-86-01 with a grating of 2700 lines/mm. Taken from H. E. Johns and A. M. Rauth [*Photochem. Photobiol.* **4**, 673, 693 (1965)].

facility at the exit slit, the details of which are now shown in Figs. 4 and 5. From the discussion on cross sections it was evident that the determination of a cross section required a knowledge of the fluence $L$ (photons/cm²) to which the molecules in the irradiation vessel are exposed. A precise determination of $L$ requires: (a) a device for monitoring, at frequent intervals, the photon flux (photons/unit time) being directed into the irradiation vessel; (b) a method for measuring the absorbance of the liquid being irradiated at suitably spaced intervals during the irradiation; (c) an absolute method for relating the monitor reading $M$ to the photon flux (photons/unit time) which enters the irradiation vessel. This absolute calibration can most readily be achieved using malachite green leucocyanide (MGL). An alternative method is to use a thermopile. The methods for obtaining an absolute calibration will be discussed later. Here we will describe in detail methods for achieving (a) and (b).

Beam monitoring and absorbance measurements can be achieved by coupling a Zeiss spectrophotometer and a monitor photocell to the exit optics of the UV monochromator as illustrated in Fig. 4. The quartz cuvette holding the solution is mounted so that it can be moved from point $P$ on the axis of the exit beam of the monochromator to point $Q$ on the axis of the beam of the Zeiss spectrophotometer.

The monitor photocell is rigidly connected to the cuvette holder so that when the cuvette is in the beam of the Zeiss spectrophotometer at $Q$ the photocell is in the beam of the monochromator at $P$. During the irradiation the cell is at $P$ and the photocell is out of the beam. After a suitable increment of fluence $\Delta L$, the cuvette is moved to position $Q$ and the absorbance of the cell measured with the Zeiss, and at the same time the beam from the monochromator is monitored with the photocell. The cell may then be moved back to position $P$ for the next increment of fluence. The details of the various components will now be described.

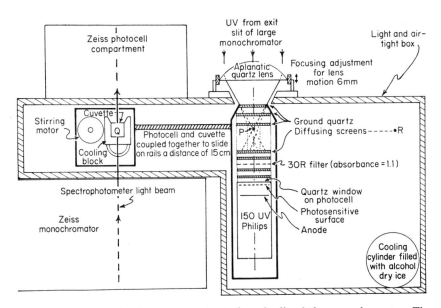

FIG. 4. Diagram showing components at the exit slit of the monochromator. The beam is focused at $P$ by the aplanatic lens. The quartz cuvette and photocell monitor are coupled together so that the photocell moves from $R$ to $P$ while the cuvette moves from $P$ to $Q$. With the cuvette at $P$ the cell is irradiated, while at $Q$ it intercepts the beam from the Zeiss spectrophotometer and may be measured.

### 3. *Photocell Monitor*

A convenient monitor is the Philips (Philips, Holland) photocell (150 UV) connected to an Eldorado photometer.[61a] To be useful as a monitor its calibration factor $K$ defined by Eq. (16) should meet certain requirements.

$$K(\lambda) = \frac{\text{number photons incident on solution in irradiation vessel in time } \Delta t}{\text{monitor reading } (M) \times \Delta t \text{ (sec)}}$$

(16)

(a) $K$ should be independent of how the entrance slit of the monochromator is illuminated. Provided the same number of photons emerge from the exit slit and are directed on the cell, the same reading on the monitor should be obtained regardless of their distribution over the length of the slit.

(b) $K$ should vary with wavelength in a smooth manner otherwise an accurate determination of $K$ will be difficult (see Section IV,B). If, for example, the monitor contained quartz with an absorption band at one wavelength, this would produce an unpredictable change in $K$ at this wavelength.

(c) $K$ should be independent of intensity so that an increase in the number of incident photons by any factor should increase the response of the photometer by the same factor.

The Philips (150 UV) photocell uses an end window with a circular sensitive surface of area 7.0 cm², and an Sb-Cs surface that has good sensitivity in the UV. The anode for the tube is behind the surface and so cannot intercept any of the beam, as is the case with the RCA type 935A. In order to meet the first two requirements for $K$, we have found it useful to arrange the photocell as shown in Fig. 4. Since the beam from the monochromator is very intense and would burn out the photocell surface if it were focused on it, the beam is passed through 7 diffusing screens and a neutral filter (No. 30R, $A = 1.1$).[61b] Diffusing screens of good quality quartz (with no absorption band at 248 m$\mu$)[61c] were ground on both sides, wet emery paper being used. The diffusers and filter are mounted in a light- and dust-tight aluminum tube. The diffusing screens serve to spread the beam over an area some two times as great as that of the sensitive surface and furthermore scatter the beam sufficiently so that the response of the sensitive surface is independent of the distribution of intensity at the exit slit. At any one wavelength, with this arrangement, the sensitive

[61a] Eldorado Electronics, 2821 Tenth Street, Berkeley, California.
[61b] Perforated Products Inc., 68 Harvard Street, Brookline, Massachussetts.
[61c] Engelhard Industries Inc., 685 Ramsay Avenue, Hillside, New York.

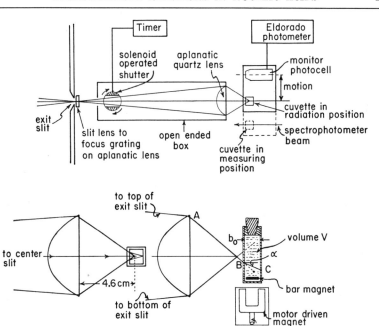

FIG. 5. Details of components at the exit slit of monochromator.

surface sees a *fixed* fraction of the radiation to be directed onto the cuvette regardless of its distribution over the exit slit.

In order to satisfy requirement (c), it is essential that the Eldorado photometer be linear over a factor of some $10^5$ in intensity. The Eldorado is equipped with a series of ranges (2 per factor of 10) and meets this linearity requirement. Furthermore, the instrument gives a rapid response and can be read with a precision of about 0.5% at full-scale deflection.

### 4. *Irradiation Cuvette*

For most purposes we have found a 3.5-ml quartz stoppered cuvette with a 1.0-cm light path most convenient for photochemical studies. The cuvette is held in a temperature-controlled block (Fig. 4) and stirred with a rotating magnet and magnet bar[61d] (Fig. 5). For low-temperature irradiations condensation of water vapor on the quartz surface can introduce problems. To overcome this, the cuvette, photocell and cuvette compartment of the Zeiss are enclosed in a metal box, and the air within the box is dried with $P_2O_5$. Any residual water vapor in the compartment can be condensed onto an aluminum cylinder filled with an alcohol-dry ice mixture situated in the box (Fig. 4).

[61d] Laboratory Plasticware Fabricators, Box 213, Kansas City, Missouri.

In our original design[60,61] the light from the exit slit was focused on the cell with either a spherical mirror or an elliptical "off axis" mirror. This has now been replaced by a quartz aplanatic lens[61e] with focal length 4.6 cm and aperture 7.0 cm (Fig. 5). This lens gives a much superior smaller image of the exit slit than the mirror. In addition it can be cleaned and does not deteriorate with time as does a front-surfaced mirror. Moreover, the use of a lens rather than a mirror gives one much more space in which to manipulate the monitor, cell, spectrophotometer, etc. For these reasons, it is much to be preferred. The radiation from the grating is focused on the aplanatic lens by the slit lens shown in Fig. 5. This radiation passes through a hollow open-ended box containing a shutter that is solenoid operated and controlled by an electric timer so that precise irradiations can be given.

## IV. Determination of Average Fluence (Exposure) for Photochemical Studies

### A. Relation between Average Fluence and Number of Incident Photons

The determination of cross section requires that every molecule in the irradiation vessel "see" the same known fluence, $\bar{L}$. Consider the irradiation vessel of Fig. 5 containing $V$ ml of solution and of thickness $b_0$. We will assume that the mixing is rapid enough that in each increment of fluence all molecules are equally exposed. Suppose further that the absorbance of the cell as measured in the parallel beam of the Zeiss at the wavelength of the radiation is $A_\lambda$. Suppose also that after a certain interval of time $\Delta t$ the number of photons incident on the solution in the cuvette is $\Delta N_i$ (methods for determining $\Delta N_i$ will be given in Section IV,B). Here we wish to relate $\Delta L$ to $\Delta N_i$.

The rays from the aplanatic lens converge to form an image of the exit slit near the center of the cell. These rays pass through the cell at all angles up to a maximum value $\alpha$ shown in Fig. 5. The ray making this maximum angle comes from the top (or bottom) of the exit slit and is $ABC$. Note that refraction bends the rays toward the normal on entering the cell. Thus rays pass through the cell with path lengths ranging from $b_0$ to $b_0$ sec $\alpha$. We require the average path length through the cell ($\bar{b}$). For the aplanatic lens $\bar{b}$ is 1.06 $b_0$ (methods for calculating $\bar{b}$ are given in Appendix A).

For rays traveling through the cell along a path $\bar{b}$, the absorbance is $A\bar{b}/b_0$, where $A$ is the absorbance of the liquid measured in the parallel

[61e] Leitz Co., Midland, Ontario, Canada. This lens has one spherical surface and a nonspherical surface. For a parallel beam it is corrected for spherical aberration and so gives an excellent image of the exit slit. The lens is very fast, having a diameter 50% larger than the focal length.

beam of the Zeiss spectrophotometer with a cell of path length $b_0$. The number transmitted $\Delta N_t$ may be related to the number incident $\Delta N_i$ using Eq. (7) to give

$$\Delta N_t = \Delta N_i \cdot 10^{-A\bar{b}/b_0} = \Delta N_i \exp(-2.30A\bar{b}/b_0)$$

The number of photons absorbed is

$$\Delta N_a = \Delta N_i - \Delta N_t = \Delta N_i \cdot f_{\mathrm{abs}} \tag{17}$$

where $f_{\mathrm{abs}}$ is the fraction of incident photons which are absorbed and is given by

$$f_{\mathrm{abs}} = \{1 - \exp(-2.30\, A\bar{b}/b_0)\} \tag{18}$$

Equation (17) gives the number of photons absorbed under the complex irradiation conditions at the focus of the aplanatic lens. Imagine now that the solution is vigorously stirred so that every molecule ($n/\mathrm{cm}^3$) is exposed to the same fluence $\Delta\bar{L}$. If the volume of the liquid in the cell is $V$, and if the absorption cross section is $\sigma_a$, then the total number of photons absorbed by the $nV$ molecules is

$$\Delta N_a = \Delta\bar{L} \cdot n \cdot V \cdot \sigma_a$$

Using the relation between $\sigma_a$ and $A$ of Eq. (14) this may be written

$$\Delta N_a = \Delta\bar{L} \cdot (2.30) \cdot \frac{A}{b_0} \cdot V \tag{19}$$

Since Eqs. (18) and (19) are both expressions for the number of photons absorbed, we may equate them and solve for $\Delta\bar{L}$ to give

$$\Delta\bar{L} = \left\{\frac{\Delta N_i}{(V/b_0)}\right\} \cdot \frac{f_{\mathrm{abs}}}{2.30\, A} \tag{20}$$

The quantity $(V/b_0)$ is the cross-sectional area of the liquid in the cell taken at right angles to the axis of the beam so the quantity in braces is the incident photon fluence $\Delta L_i$. We rewrite Eq. (20) as

$$\Delta\bar{L} = \Delta L_i \cdot F(A,\bar{b},b_0) \tag{20a}$$

where $F(A,\bar{b},b_0)$ is the factor by which the *incident* fluence must be multiplied to give the *average* fluence $\Delta\bar{L}$. The quantities $\Delta L_i$ and $F(A,\bar{b},b_0)$ are given by

$$\Delta L_i = \frac{\Delta N_i}{(V/b_0)} \tag{21}$$

$$F(A,\bar{b},b_0) = \frac{f_{\mathrm{abs}}}{2.30\, \bar{A}} = \frac{1 - \exp(-2.30\, \bar{A}\bar{b}/b_0)}{2.30\, \bar{A}} \tag{22}$$

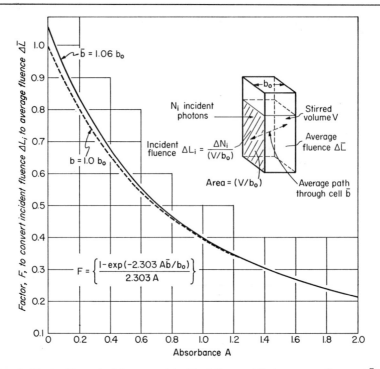

FIG. 6. Factor $F$ required to convert incident fluence $\Delta L_i$ to average fluence $\Delta \bar{L}$ as a function of absorbance. $F$ is evaluated from Eq. (20a). Graphs are shown for the average path through the cell $\bar{b}$ equal to $b_0$ and $1.06\,b_0$ where $b_0$ is the path length for a normal ray.

Since the absorbance of the solution changes with the increment in exposure, the absorbance to be used in Eq. (22) is the average absorbance $\bar{A}$ of the solution during the increment of exposure.

The quantity $F$ is plotted in Fig. 6 as a function of absorbance for $\bar{b} = b_0$ and $\bar{b} = 1.06\,b_0$ and is seen to vary by a factor of about 5.0 as the absorbance is changed from 0.0 to 2.0. Two limiting situations may be considered.

### 1. Solution Optically "Thin" $(A \simeq 0)$

If the solution has small absorbance, the exponential term of Eq. (22) may be expanded and $F$ reduces to $\bar{b}/b_0$ and

$$\Delta \bar{L} = \Delta L_i \cdot \left(\frac{\bar{b}}{b_0}\right) = \frac{\Delta N_i}{V} \cdot \bar{b} \qquad (\text{valid } A \simeq 0) \qquad (23a)$$

If $\bar{b} = b_0$ and $A \simeq 0$, the solution is being exposed to a parallel beam and the incident fluence $\Delta L_i$ and average fluence $\Delta \bar{L}$ are identical. This is the situation sometimes achieved when "optically thin" phage solutions are exposed in a petri dish at some distance from a germicidal lamp. More often, however, the solution is not optically thin and a correction should be made for absorption. For example, we see from Fig. 6 that if the absorbance is 0.4 (not an unlikely situation) the average fluence is 0.65 of the incident fluence, so that a 35% error may be made unless this factor is taken into account.

From the case where a converging beam is directed onto an optically thin solution as in the geometry of Fig. 5, the average fluence $\Delta L$ is *larger* than the incident fluence $\Delta L_i$ by the ratio $\bar{b}/b_0$ (see Eq. 23a).

## 2. Solution Optically "Thick" $A > 2.0$

When the solution is optically thick, the exponential term of Eq. (22) approaches 0 and

$$F \simeq \frac{1}{2.30\,A} \text{ and } \Delta\bar{L} = \Delta L_i \cdot \frac{1}{2.30\,A} \qquad \text{(valid } A > 2) \qquad (23b)$$

For such solutions $\Delta\bar{L}$ is inversely proportional to absorbance. For example, in irradiating polynucleotides to be used later in a biochemical assay, it is not unusual to irradiate solutions with an absorbance of 10.0. Under these circumstances the incident fluence $\Delta L_i$ must be divided by the factor 23 to give the average fluence through the irradiated mixture. In addition, we are likely to have another complicating factor since the absorbance is almost certain to change as the irradiation progresses. Methods for dealing with this situation will now be discussed.

## 3. Total Exposure after a Series of Increments of Exposure

In most photochemical studies the absorbance of the solution will change with exposure so the factor $F$ (Eq. 20a), which relates the incident fluence to the average fluence will also vary (see Fig. 6). To find the total fluence, $L$ "seen" by each molecule in the solution after a given sequence of short irradiations, we sum the increments of $\Delta\bar{L}$ (Eq. 20a) to give:

$$\bar{L} = \sum \Delta\bar{L} = \sum \frac{\Delta N_i}{V} \cdot b_0 \cdot \left\{ \frac{1 - \exp(-2.30\,\bar{A}\bar{b}/b_0)}{2.30\,\bar{A}} \right\} \qquad (24)$$

$\Delta\bar{L}$ depends inversely on the volume. In many irradiations the volume $V$ changes as aliquots are removed for assay purposes: the use of Eq. (24) will automatically correct for this effect.[61f]

---

[61f] However, one precaution should be observed: for the smallest volume, all rays must enter the cell below the surface of the liquid and no ray in passing through the cell emerges through the free surface of the liquid. In our irradiation geometry these conditions hold for $V > 1.5$ ml.

Equation (24) may be easily evaluated if the calculations are laid out in tabular form as shown in Table II. The term in braces can be evaluated with a slide rule or obtained from a graph such as that of Fig. 6. Dosimetry following this method will be used throughout the rest of the chapter.

## B. Absolute Calibration of Monitor Photocell Using Malachite Green Leucocyanide (MGL)

### 1. *Methods for Obtaining Absolute Calibration*

To determine $K(\lambda)$ given by Eq. (16) a determination of the absolute number $\Delta N_i$ of photons incident on the solution is required. This can be done in a number of ways. A calibrated thermopile can be used to absorb all the incident photons from which the number $\Delta N_i$ can be calculated. Although in principle this method is straightforward, there are a number of difficulties. First of all, it may be difficult to obtain a thermopile with a sensitive surface of the right shape and area to intercept all the photons at the focus of the aplanatic lens. In the second place, it is difficult to correct for the unknown number of photons that may be reflected from the surface of the thermopile. Finally, a thermopile with a precise absolute calibration is not easily obtained. Usually thermopiles are calibrated in the infrared region from a knowledge of the radiant energy from a standard lamp, and this calibration is then extrapolated to the UV region of the spectrum—at best a doubtful procedure.

A number of chemical methods are available. With these one absorbs a known fraction of the incident radiation in the solution and measures the amount of photoproduct produced. If the quantum yield for the production of the photoproduct is known, then the number of incident photons can be calculated. Uranyl acetate has been used for radiations in the range 365–406 m$\mu$, where the quantum yield is about 0.5.[62] The photoproduct is detected potentiometrically. We have found MGL to be a more useful dosimeter.[63] Its photoproduct is stable, has a large absorption at 622 m$\mu$, and can readily be detected by observing the change in absorbance at 622 m$\mu$. We have found that the quantum yield is $0.91 \pm 0.01$ over the range 225–289 m$\mu$.[63] The photoproduct has a peak absorptivity at 622 m$\mu$ with a value of $10.63 \times 10^4$. Our quantum yield and maximum absorptivity differ slightly from the earlier determinations of Harris and Kaminsky[64,65]

[62] C. A. Discher, P. F. Smith, I. Lippman, and R. Turse, *J. Phys. Chem.* **67**, 2501 (1963).

[63] G. J. Fisher, J. C. LeBlanc, and H. E. Johns, *Photochem. Photobiol.* **6**, 757 (1967).

[64] L. Harris, J Kaminsky, and R. G. Simard, *J. Am. Chem. Soc.* **57**, 1151 (1935).

[65] L. Harris and J. Kaminsky, *J. Am. Chem. Soc.* **57**, 1154 (1935).

and of Calvert and Rechen.[66] The latter obtained a quantum yield of 1.00 and a peak absorptivity of $9.49 \times 10^4$ at 620 m$\mu$.[66a]

## 2. Absorption Spectrum of MGL and Its Photoproduct

Figure 7 shows the absorption spectrum of purified MGL dye. The dye has a molar absorptivity of $5 \times 10^3$ or larger for wavelengths between 220 and 313 m$\mu$. At longer wavelengths the absorption becomes small so that dye would not be useful as a dosimeter above 320 m$\mu$. The absorption spectrum of the photoproduct is also shown in Fig. 7.[63] This was obtained by irradiating a dye solution of known molarity until all of it had been converted to the photoproduct. We have found that after the dye has stood in solution in the refrigerator for several months, its absorption spectrum changes slightly and less of it can be converted to the photoproduct. To avoid these changes we keep the dye frozen.

## 3. Preparation of Materials and Procedure

*Step 1.* Select two matched cuvettes made of good quality quartz with no absorption band.[70a]

*Step 2.* Redistill 95% ethanol to remove impurities that absorb in the UV. To the purified alcohol add enough HCl to make the solution $10^{-3}$ $N$. Call this stock solution S-1. The acid is added to stabilize the MGL photoproduct and prevent it from fading.[65] Measure the absorbance of S-1 against a water blank. Typical absorbances at 280, 260, 240, 230, and 220 m$\mu$ are <0.001, 0.004, 0.035, 0.09, and 0.17, respectively. Before purification of the ethanol, the corresponding values were 0.04, 0.06, 0.17, 0.27, and 0.50. This illustrates the need for redistillation of the ethanol.

[66] J. G. Calvert and H. J. L. Rechen, *J. Am. Chem. Soc.* **74**, 2101 (1952).

[66a] It should be noted that in using MGL to calibrate a photocell one would obtain very nearly the same result by using Calvert and Rechen's values for both $\phi$ and $\epsilon$ as the more recent values for both quantities. Our earlier papers,[7,24,28] whose dosimetry was based on Calvert and Rechen's values, are thus in agreement with our later papers.[67-70] In one paper,[29] exposures should be increased by 10% since we used an $\epsilon_{max}$ of 10.58 with $\phi = 1.00$.

[67] M. Pearson and H. E. Johns, *J. Mol. Biol.* **20**, 215 (1966).

[68] M. Pearson, D. W. Whillans, J. C. LeBlanc, and H. E. Johns, *J. Mol. Biol.* **20**, 245 (1966).

[69] M. Pearson and H. E. Johns, *J. Mol. Biol.* **19**, 303 (1966).

[70] H. E. Johns, M. L. Pearson, and I. H. Brown, *J. Mol. Biol.* **20**, 231 (1966).

[70a] Some quartz cells have an absorption band at 240 m$\mu$ and are not recommended. Cells without this absorption can be obtained from Zeiss, Germany; Thermal Syndicate Ltd. Wallsend, Northumberland, England.

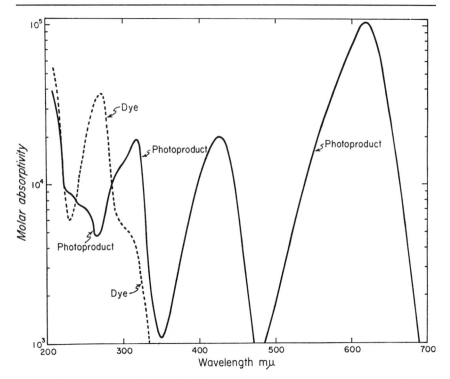

Fig. 7. Molar absorptivity of MGL dye and its photoproduct as a function of wavelength.

*Step 3.* Make a $10^{-3}$ *M* stock solution of MGL[70b] dye (mol. wt. = 355) in solution S-1 and call this solution S-2. Keep this stock solution in a light-tight container in the frozen state to avoid the changes discussed above. Under frozen conditions the material is stable for at least six months.

*Step 4.* Dilute S-2 with S-1 until the solution has an absorbance of about 1.5 at the wavelength $\lambda$ at which a calibration is sought (call this solution S-3). An estimate of the dilution required can be obtained from Fig. 7.

*Step 5.* Accurately measure a volume $V$ (2 ml is a useful volume) of solution S-3 into one of the matched cells and measure $A_\lambda$, and $A_{622}$ against a water blank.

*Step 6.* Clean the outer quartz window of the monitor (the quartz diffusers and 30R filter should remain clean since they are in a dust-tight container). Then position the monitor in the UV beam set for $\lambda$ and

---

[70b] MGL dye can be obtained from Fisher Scientific. It must be purified and converted to the reduced form before it may be used as a dosimeter.

obtain a monitor reading $M_1$. Now position the cuvette with the MGL dye in the UV beam; with the stirrer running, expose for $\Delta t$ minutes. The UV intensity should be adjusted so that an exposure of about 1 minute gives a change in $A_{622}$ of about 0.4. Note that the quartz cuvette should be positioned in the beam so that all photons enter the solution and none strike the side walls or the stirring bar. At the end of the exposure, make a second monitor reading $M_2$, stir the solution for at least 20 seconds, and then remeasure $A_{622}$ and $A_\lambda$ against the water blank. In measuring $A_{622}$, narrow slits should be used and the measurement should be made at the wavelength that gives the peak absorbance rather than at 622 m$\mu$, since the wavelengths dial of the spectrophotometer may be in error. These measurements, together with the data of Fig. 7, yield the calibration factor $K$.

## 4. MGL Calculations

During the irradiation, the solution S-3 contains three absorbing species—the dye, the alcohol and the photoproduct formed from the dye. Photons absorbed by the dye are detected while those absorbed by the alcohol and photoproduct are not. In this section we will assume that all photons which are absorbed, are absorbed by the dye. In Appendix B we will indicate how corrections may be made for the small fraction of photons that are "stolen" by the alcohol and the photoproduct.

Let $\Delta N_i$ represent the number of photons incident on the solution, i.e., those which have passed through the front layer of quartz and entered the solution.

The number of these absorbed by this solution during the exposure is

$$\Delta N_a = \Delta N_i \cdot f_{abs} \qquad (25)$$

where $f_{abs}$ is given by Eq. (18). Let the change in absorbance at 622 m$\mu$ produced by these photons be $\Delta A_{622}$.

Since all the absorbance at 622 m$\mu$ is due to the photoproduct (see Fig. 7), the number of molecules of photoproduct (pp) produced is

$$\Delta N_{pp} = 6.02 \times 10^{23} \cdot \frac{\Delta A_{622}}{\epsilon_{622}} \cdot V \qquad (26a)$$

If the quantum yield for the production of the photoproduct is $\phi$ (0.91) then $\Delta N_{pp}$ and $\Delta N_a$ are related by

$$\Delta N_{pp} = \phi \Delta N_a \qquad (26b)$$

Combining Eqs. (25), (26a), and (26b) yields

$$\Delta N_i = \frac{6.02 \times 10^{23} V \cdot \Delta A_{622}}{f_{abs} \cdot \phi \cdot \epsilon_{622}} \qquad (27)$$

The determination of $\Delta N_i$ thus requires only a knowledge of $V$, the volume irradiated, $\Delta A_{622}$, $f_{abs}$, and $\phi$. Combining Eq. (27) with Eq. (16) we obtain the calibration factor $K$:

$$K(\lambda) = \frac{\Delta N_i}{\bar{M}\Delta_t} = \frac{6.02 \times 10^{23} \cdot V \cdot \Delta A_{622}}{f_{abs} \cdot \phi \cdot \epsilon_{622}} \cdot \frac{1}{\bar{M}\Delta_t} \cdot \frac{photons}{monitor\ sec} \tag{28}$$

where $\bar{M}$ is the average of the monitor reading before and after the increment in exposure and $\Delta t$ is the length of the exposure. The calibration factor is expressed in photons per monitor second. The way these calculations may be laid out is illustrated in Appendix B. Table A-2 contains two calibrations at 230 m$\mu$ and three at 265 m$\mu$ and is included so that it may be used as a guide in performing an MGL calibration and also to indicate the internal consistency that may be obtained.

Column (11) of Table A-2 gives the value of $K$ calculated using Eq. (28). Column (18) of the same table gives the corresponding values for $K$ when corrections for the absorption of the alcohol and photoproducts are taken into account. At 230 m$\mu$ the alcohol absorbs about 5% of the photons [column (13)] and the photoproduct nearly 1% [column (16)], so that the corrected $K$ is about 6% larger than the uncorrected value. At wavelengths below 230 m$\mu$ the alcohol correction becomes important and may be very large if the alcohol is not predistilled. The photoproduct correction is largest for those wavelengths where the photoproduct has a large absorbance relative to the dye (see Fig. 7).

## 5. Calibration Factor vs. Wavelength

A complete calibration for our large monochromator is given in Fig. 8. We have found that this calibration remains constant over periods of months provided that the entrance window of the monitor photocell is kept clean. The factor $K$ increases with decrease in wavelength because the sensitive surface of the monitor photocell shows a decreasing sensitivity with decrease in wavelength below 400 m$\mu$. In addition, the diffusing screens, etc., shown in Fig. 4 tend to absorb more photons at short wavelengths than at the longer ones, thus increasing still more the variation of $K$ with wavelength. The variation of $K$ with wavelength creates no problems provided the factor remains constant with time. As a routine check we remeasure $K$ once every 2 weeks at one wavelength. This takes about 1 hour. The total calibration at all wavelengths takes about 2 days.

It is difficult to estimate the precision of $K$. By repeating each measurement about four times, results consistent to about 2% can be attained, but the absolute precision is certainly not this great since it depends on a knowledge that the quantum yield is indeed 0.91 and a precise value for the

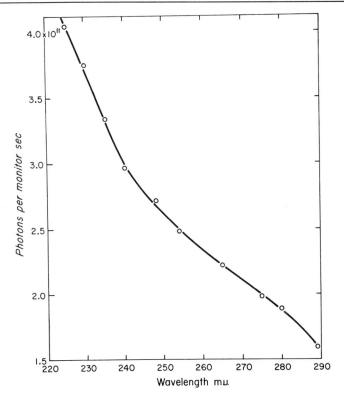

Fig. 8. Calibration factor in units of photons per monitor second as a function of wavelength for the arrangement shown in Figs. 4 and 5.

molar absorptivity at 622 m$\mu$. We estimate that the absolute accuracy is about $\pm 3\%$.

## V. Photochemical Reactions Involving Parent and One Photoproduct

In this and the following section we will discuss a number of photochemical reactions, using the basic ideas of dosimetry developed in the three preceding sections. Since the average fluence $\bar{L}$ (exposure) depends upon the absorbance (see Fig. 6 and Eq. 24), and since the absorbance changes with irradiation, it is essential to carry out the irradiation in a series of increments and determine $\Delta\bar{L}$ for each. This having been done, the total exposure $\bar{L}$ after a series of increments is $\Sigma\Delta\bar{L}$. Since this method is basic to the proper investigation of the kinetics of many photochemical reactions a detailed data sheet is given in Table II for a typical irradiation. If this method is followed carefully, complex reaction schemes can often be unraveled and cross sections determined precisely.

TABLE II

IRRADIATION 1,3-DIMETHYLURACIL IN WATER (pH 7) AT $\lambda = 265$ m$\mu^a$

| (1) Exposure increment | (2) Monitor $(M \times 10^{-4})$ | (3) $\Delta t$ (min) | (4) $M \times \Delta t \times 10^{-4}$ | (5) $V$ (ml) | (6) $\Delta L_i$ ($\mu E/cm^2$) | (7) $A_\lambda$ | (8) $\bar{A}_\lambda$ | (9) $f_{abs}$ | (10) $\overline{\Delta L}$ | (11) $L_i = \Sigma \Delta L_i$ | (12) $\bar{L} = \Sigma \overline{\Delta L}$ | (13) $A_{266}$ |
|---|---|---|---|---|---|---|---|---|---|---|---|---|
| 1 | 52.0 | 1.0 | 52.0 | 2.0 | 0.330 | 1.945 | 1.928 | 0.992 | 0.074 | 0.330 | 0.074 | 1.952 |
|   | 52.0 |     |      |     |       | 1.910 |       |       |       |       |       | 1.916 |
| 2 | 51.0 | 1.0 | 51.5 | 2.0 | 0.327 | 1.910 | 1.888 | 0.992 | 0.074 | 0.657 | 0.148 | 1.873 |
|   | 52.0 |     |      |     |       | 1.865 |       |       |       |       |       |       |
| 3 | 52.0 | 2.0 | 103  | 2.0 | 0.656 | 1.865 | 1.823 | 0.990 | 0.154 | 1.313 | 0.302 | 1.788 |
|   | 51.0 |     |      |     |       | 1.781 |       |       |       |       |       |       |
| 4 | 52.0 | 3.0 | 153  | 2.0 | 0.973 | 1.781 | 1.720 | 0.987 | 0.242 | 2.286 | 0.544 | 1.662 |
|   | 51.0 |     |      |     |       | 1.658 |       |       |       |       |       |       |
| 5 | 50.0 | 5.0 | 250  | 2.0 | 1.59  | 1.658 | 1.558 | 0.980 | 0.433 | 3.88  | 0.977 | 1.462 |
|   | 50.0 |     |      |     |       | 1.459 |       |       |       |       |       |       |

| | | | | | | | | | | | | |
|---|---|---|---|---|---|---|---|---|---|---|---|---|
| 6 | 50.0 / 49.0 | 10.0 | 495 | 2.0 | 3.15 | 1.459 / 1.082 | 1.270 | 0.960 | 1.030 | 7.03 | 2.007 | 1.085 |
| 7 | 49.0 / 49.0 | 10.0 | 490 | 2.0 | 3.12 | 1.082 / 0.743 | 0.912 | 0.900 | 1.335 | 10.14 | 3.34 | 0.744 |
| 8 | 49.0 / 48.0 | 10.0 | 485 | 2.0 | 3.08 | 0.743 / 0.461 | 0.602 | 0.782 | 1.730 | 13.22 | 5.07 | 0.462 |
| 9 | 48.0 / 48.0 | 8.0 | 384 | 2.0 | 2.44 | 0.461 / 0.320 | 0.390 | 0.627 | 1.700 | 15.54 | 6.77 | 0.285 |
| 10 | 48.0 / 48.0 | 7.0 | 336 | 2.0 | 2.14 | 0.320 / 0.178 | 0.249 | 0.467 | 1.735 | 17.68 | 8.51 | 0.179 |

$^a$ $K = 12.7 \times 10^{-7}$ $\mu E/M$ min;    $\Delta L_i = \dfrac{\Delta N_i}{(V/b_0)} = \dfrac{K \cdot \bar{M} \cdot \Delta t}{(V/b_0)}$    [Eqs. (21) and (28)]

$f_{abs} = [1 - \exp(-2.30\,\bar{A}_\lambda \bar{b}/b_0)]$ [Eq. (18)];    $\overline{\Delta L} = \dfrac{\Delta L_i \cdot f_{abs}}{2.30\,\bar{A}_\lambda}$ [Eq. (20)]

$V = 2.0$ ml;    $b_0 = 1.00$ cm;    $\bar{b} = 1.06$ cm.

A cross section determination requires a knowledge of the growth of a photoproduct with exposure. It also requires that we know the material *from which* the photoproduct is formed, and this implies a knowledge of the reaction scheme giving rise to it. Thus, for example, if irradiation converts A to B and then B to C, then a calculation of the cross section for the formation of C requires a knowledge of the concentration of B as a function of exposure. We will therefore be concerned with reaction schemes and with methods for separating and measuring the yield of photoproducts as a function of exposure.

From the reaction scheme it is usually possible to obtain a mathematical expression for the growth of a particular photoproduct in terms of the cross sections and the concentration of the precursor. The initial slope of such a growth curve gives an *approximate value* for the cross section. However, a much better way to obtain the cross section is to manipulate the data in such a way that a suitable plot gives a straight line whose slope can be determined accurately. This slope is then used to obtain the cross section. A number of examples of this rectification process will be given.

Reaction schemes involving only one photoproduct can often be analyzed by observing the changes in the absorbance with exposure. However, this approach can lead to serious errors in interpretation, as will be illustrated in this section, where we will deal in turn with the photochemistry of four systems of increasing complexity. These illustrate general principles that have much broader applications.

### 1. *One Photoproduct Formed Which Lacks Absorption at the Analyzing Wavelength*

A simple example of this type is 1,3-dimethyluracil (DMU), which Wang[71] has shown is converted by radiation to the water adduct with H added at the 5 position and OH at the 6 position to give 1,3-dimethyl-5-hydro-6-hydroxyuracil.

The photoproduct lacks the 5-6 double bond and so has negligible absorption at the $\lambda_{max}$ (266 m$\mu$) where the parent has maximum absorption. The simplest way to investigate this reaction is to irradiate at some wavelength $\lambda$ and follow the change in absorbance at $\lambda_{max}$. The way the data may be arranged is illustrated in Table II, where 8 sequential increments of exposure are recorded in such a way that the incident fluence $L_i$ and the average fluence $\bar{L}$ may be calculated. This table is shown in detail because it contains all the data necessary for accurate dosimetry. We have found it convenient to record *all* irradiations on data sheets like that given in Table II.

[71] S. Y. Wang, *J. Am. Chem. Soc.* **80**, 6196 (1958).

The data of Table II show the effects of UV irradiation at 265 m$\mu$ on 2.0 ml of DMU whose initial absorbance was 1.952. Before and after each increment of exposure, a monitor reading $M$ was obtained [column (2)]. When the average value of $M$ is multiplied by the time $\Delta t$, one obtains the exposure in units of monitor $\times$ minutes given in column (4). From these data and a knowledge of $K$ the *incident* exposure $\Delta L_i$ in $\mu$E/cm$^2$ is calculated and given in column (6). Since in this case $V$ is constant, $\Delta L_i$ would be directly proportional to the time of each increment of irradiation if the intensity from the monochromator remained constant. In the experiment shown in Table II the intensity, as measured by the monitor, dropped from 52.0 $\times$ 10$^4$ to 48.0 $\times$ 10$^4$ so that $\Delta L_i$ is not quite proportional to $\Delta t$.

Column (7) gives A$_\lambda$, the absorbance of the solution before and after each increment of exposure, and column (8) the average of these during the exposure. From $\bar{A}_\lambda$, we calculate $f_{abs}$ in column (9) and $\overline{\Delta L}$ in column (10). The sums of $\Delta L_i$ and $\overline{\Delta L}$ which are equal to $L_i$ and $\bar{L}$ are given in columns (11) and (12).

Column (13) gives the absorbance at $\lambda_{max}$ at the end of each increment of exposure. The absorbance given in column (13) has thus been determined as a function of the *incident* photon fluence $L_i$ (given in column 11) and the *average* photon fluence $\bar{L}$ (given in column 12). It should be noted that $L_i$ would increase linearly with time if the intensity remained constant, but that $\bar{L}$ increases slowly at first (when the absorbance is high) and then more rapidly as the absorbance decreases. For example, for the first increment of radiation $\Delta L_i$ is 5 times $\overline{\Delta L}$ while for the last one, $\Delta L_i$ is 1.23 times $\overline{\Delta L}$.

The data presented in the last three columns of Table II are plotted in Fig. 9 on a logarithmic scale. $A_\lambda$ versus $\bar{L}$ gives an excellent straight line while $A_\lambda$ versus $L_i$ is concave downward. This latter curve is quite impossible to analyze but has been included here to illustrate the difficulties in dealing with data that are based on incident fluence.

From the variation of absorbance with exposure given in Fig. 9, we may calculate the cross section for absorbance loss, $\sigma_{ab}$, as follows. This cross section could be defined by

$$dA = -\sigma_{ab} \cdot A \cdot d\bar{L} \text{ or } \frac{dA}{A} = -\sigma_{ab} \cdot d\bar{L} \qquad (29)$$

where $dA$ is the differential change in absorbance produced in a solution of absorbance $A$ by a differential of exposure $d\bar{L}$. This may be integrated to give

$$A = A_0 e^{-\sigma_{ab} \cdot \bar{L}} \text{ or } \log A_0 - \log A = \frac{1}{2.303} \sigma_{ab} \cdot \bar{L} \qquad (30)$$

where $A_0$ is the initial absorbance.

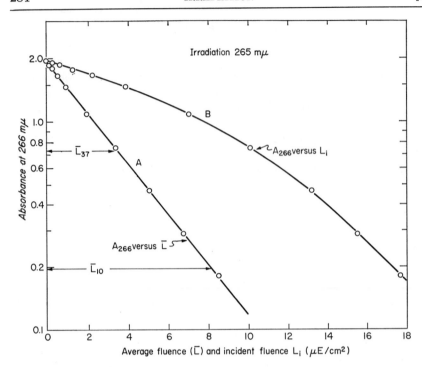

FIG. 9. Irradiation at $\lambda = 265\,m\mu$ of 1,3-dimethyluracil (pH 7.0, initial $A_{266} = 1.952$). Curve $A$: $A_{266}$ versus average fluence (exposure) $\bar{L}$. Curve $B$: $A_{266}$ versus incident fluence $L_i$.

Thus, if every interaction, or a constant fraction of the interactions, gave rise to the loss of absorbance, we would expect a straight line when the absorbance is plotted versus $\bar{L}$ on a log scale. Since we obtain such a straight line (Fig. 9), this is excellent evidence that a constant fraction of the interactions does give rise to absorbance loss. If we let $\bar{L}_{37}$ and $\bar{L}_{10}$ represent the exposures required to reduce the absorbance to $e^{-1}$ ($0.368 \simeq 37\%$) and 0.10, respectively, of the initial value, then from Eq. (30)

$$\sigma_{\mathrm{ab}} \cdot \bar{L}_{37} = 1.00 \text{ and } \sigma_{\mathrm{ab}} \cdot \bar{L}_{10} = 2.303 \qquad (31)$$

In radiobiology $\bar{L}_{37}$ or $\bar{L}_{10}$ are referred to as $D_{37}$ and $D_{10}$. $D_{37}$ is also referred to as the mean lethal dose. From Fig. 9 we can evaluate either $\bar{L}_{37}$ or $\bar{L}_{10}$ and substituting in Eq. (31) obtain:

$$\sigma_{\mathrm{ab}} = 0.284 \text{ cm}^2/\mu\mathrm{E}$$

In all the above discussion we have made no mention of the hydrate which we assume is being produced in the irradiation according to Eq. (29).

If we have proof that the *only* photoproduct involved in the reaction is the hydrate, then, we can with confidence state that the cross section for hydrate production, $\sigma_h$, is equal to 0.284 cm²/μE.

It must be emphasized that the straight line of Fig. 9 proves only that the reaction follows the differential Eq. (29) and that some reaction in the parent leads to its complete loss of absorbance, but it does not prove that only one photoproduct is involved. Thus, the 1,3-dimethyluracil could be converted into several types of hydrates with, for example, the OH added at the 5 position rather than at the 6, or the H and OH on the 6 position interchanged. It could even happen that a very different type of reaction was involved leading to the loss of the 5-6 double bond, as in the formation of a dimer. The data of Fig. 9 can in no way differentiate between these various possibilities. Questions of this kind can be answered by separating quantities of the photoproduct and determining whether a number of species are present. Wang has synthesized the hydrate of DMU and shown that it is identical to the UV photoproduct.[71] He has also shown[72] that with further irradiation the hydrate is converted to another non-absorbing species, a reaction that could not be detected by following the absorbance at 266 mμ.

### 2. *Photoproduct Has Less Absorption Than the Parent at Some Analyzing Wavelength*

In some cases, the photoproduct may have some absorption at the analyzing wavelength so that the irradiation will not reduce the absorption below some final value $A_f$. The absorbance of the solution may be written

$$A = \epsilon_p \cdot N_p + \epsilon_{pp} \cdot N_{pp} \qquad (32)$$

where $N_p$ and $N_{pp}$ are the concentrations and $\epsilon_p$ and $\epsilon_{pp}$ are the molar absorptivities of the parent and photoproduct, respectively. Since there are only two species present,

$$N_{pp} + N_p = N_0 \qquad (33)$$

where $N_0$ is the initial concentration of parent. The differential equation for the production of photoproduct, with cross section $\sigma_{pp}$ is

$$dN_{pp} = \sigma_{pp} \cdot N_p \cdot d\bar{L} = \sigma_{pp} \cdot (N_0 - N_{pp}) \cdot d\bar{L} \qquad (34)$$

which may be integrated to give

$$N_{pp} = N_0(1 - e^{-\sigma_{pp} \cdot \bar{L}}) \text{ and } N_p = N_0 - N_{pp} \qquad (35)$$

combining these two equations with Eq. (32), we obtain

[72] S. Y. Wang, *J. Am. Chem. Soc.* **80**, 6199 (1958).

$$(A - A_f) = (A_0 - A_f)e^{-\sigma_{pp}\cdot\bar{L}} \tag{36}$$

where $A_0 = \epsilon_p \cdot N_0$ is the initial absorption of the solution when only parent is present, and $A_f = \epsilon_{pp} \cdot N_0$ is the final absorption when all the parent has been converted to photoproduct. From Eq. (36) it is evident that a plot of $\log(A - A_f)$ against $\bar{L}$ will give a straight line from which $\sigma_{pp}$ may be calculated.

To illustrate a reaction of this type, data obtained from the irradiation of the 3'-isomer of cytidylic acid (Cp) are shown in Fig. 10 where the absorbance at 271 m$\mu$ and at 258 m$\mu$ is shown as a function of exposure to radiation at 265 m$\mu$. The main photoproduct of Cp is the hydrate Cp* which, under the conditions of this irradiation (pH 8.37, 0°C) is very stable so that it does not reverse during the irradiation.[37] The dosimetry for this irradiation was done in exactly the way illustrated in Table II and the exposure scale is the average photon fluence $\bar{L}$ "seen" by each molecule in the vessel.

The absorbance at 271 m$\mu$ ($A_{271}$) decreases linearly (on the log scale) initially but gradually approaches a final absorbance $A_f$ of about 0.065 at large doses. The final absorbance is due to the absorption of the hydrate at 271 m$\mu$, which is much smaller than the parent but *not* negligible. Subtracting $A_f = 0.065$ from the absorbance, we obtain a good straight line out to large doses with an $\bar{L}_{10} = 8.3$ $\mu$E/cm$^2$ which yields the required cross section of $2.303/8.3 = 0.278$ cm$^2$/$\mu$E according to Eq. (31). In this case $A_f \simeq 0$, so very nearly the same slope would have been obtained from the initial part of the curve of $A_{271}$ versus $\bar{L}$.

On the right side of Fig. 10, $A_{258}$ is plotted as a function of $\bar{L}$. At this wavelength, the absorbance of the hydrate is nearly as large as that of Cp so that the curve is not linear on a log scale, even at small doses, and quickly approaches an asymptotic absorbance of $A_f \simeq 0.60$. If we subtract this value from the measured absorbance, according to Eq. (36) we get a good straight line with an $\bar{L}_{10} = 8.4$ $\mu$E/cm$^2$ and a cross section of 0.274 cm$^2$/$\mu$E. This value for $\bar{L}_{10}$ and the cross section are in good agreement with the values obtained from studying the same reaction by following $A_{271}$ as a function of exposure. Both should give the same cross section but, of course, greater precision is possible when one measures at a wavelength (271 m$\mu$) at which the photoproduct has negligible absorption and the parent maximum absorption.

### 3. Photoproduct Produced and Reversed by the Radiation during the Irradiation

A simple example of this type of reaction is that involving TpT which to a first approximation can be represented by Eq. (1a).

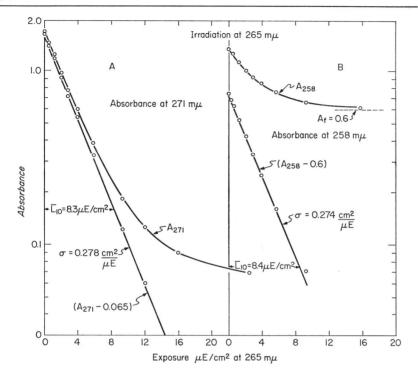

Fig. 10. Irradiation at $\lambda = 265$ m$\mu$ of the 3′ isomer of cytidylic acid (pH 8.37, 0.01 $M$ Tris buffer, 0°C; initial $A_{271} = 1.680$). (A) $A_{271}$ and $(A_{271} - 0.065)$ versus exposure in $\mu$E/cm$^2$, (B) $A_{258}$ and $(A_{258} - 0.6)$ versus exposure.

$$\text{TpT} \underset{r}{\overset{k}{\rightleftarrows}} \widehat{\text{TpT}} \tag{1a}$$

We will now show that this reaction will appear to be identical with that described in the last section, if one merely observes absorbance changes. Let P be the concentration of parent, D the concentration of dimer, and $P_0$ the initial concentration of parent. If we assume there are only two species present then

$$P + D = P_0 \tag{37}$$

and the absorbance of the solution may be written

$$A = \epsilon_p \cdot P + \epsilon_d \cdot D \tag{38}$$

where $\epsilon_p$ and $\epsilon_d$ are the molar absorptivity of parent and dimer. If we let $k$ represent the cross section for the conversion of P to D and $r$ the reverse cross section for the conversion of D to P, then the differential equation is

$$\frac{dD}{d\bar{L}} = kP - rD = k(P_0 - D) - rD \tag{39}$$

which may be solved to give

$$D = \frac{P_0 k}{(k + r)} (1 - e^{-(k+r)\bar{L}}) \text{ and } P = P_0 - D \tag{40}$$

Combining Eqs. (38) and (40) we obtain

$$A - A_f = (A_0 - A_f)e^{-(k+r)\bar{L}} \tag{41}$$

where $A_0$ is the initial absorbance $P_0\epsilon_p$, and $A_f$ is the final absorbance given by

$$A_f = \frac{A_0}{(k + r)} \left\{ r + \frac{\epsilon_d}{\epsilon_p} \cdot k \right\} \simeq \frac{A_0 r}{(k + r)} \tag{42}$$

Since the dimer has a small extinction relative to the parent, the above approximation is a good one.

It should be noted that Eq. (41) is exactly the same as Eq. (36) although the two cases represent very different physical situations. In this case, the absorption reaches a final constant value given by $A_f$ but the reaction is still proceeding with dimers being reversed at the same rate as they are being formed. For the situation described in the last section, no further reaction took place after the absorption reached its final value $A_f$.

Figure 11 shows data from the irradiation of TpT at 265 m$\mu$. The dosimetry was carried out in the way illustrated in Table II and, as before, the exposures are average photon fluence ($\bar{L}$). $A_{266}$ versus $\bar{L}$ gives a curve which approaches an asymptotic value of $A_f = 0.71$. The plot of $(A - A_f)$ gives an excellent straight line with an $\bar{L}_{10} = 2.55$ $\mu$E/cm$^2$, yielding a "cross section" of $2.303/2.55 = 0.90$ cm$^2$/$\mu$E. If we had no other information we would interpret this straight line as indicating that TpT was all converted to some photoproduct with a residual absorption of 0.71 and that the cross section for this reaction was 0.90 cm$^2$/$\mu$E. If on the other hand we knew that a dimer was involved and was reversing, then we would proceed as follows to obtain $k$ and $r$.

From Eqs. (41) and (42) it is evident that

$$k + r = 0.90 \tag{43}$$

$$\frac{k + r}{r} = \frac{A_0}{A_f} = \frac{1.817}{0.71} = 2.56 \tag{44}$$

which may be solved to give $k = 0.55$ cm$^2$/$\mu$E and $r = 0.35$ cm$^2$/$\mu$E.

It must be emphasized that $A_{266}$ in Fig. 11 varies in exactly the same way as $A_{271}$ or $A_{258}$ of Fig. 10, and from these kinetic data, one cannot

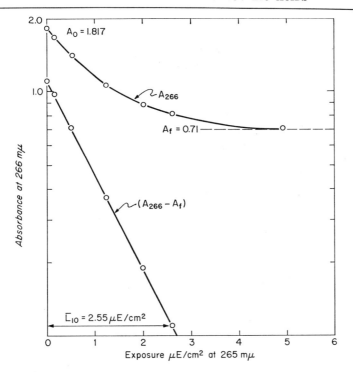

Fig. 11. Irradiation at $\lambda = 265$ m$\mu$ of TpT in water at pH 7. $A_{266}$ and $(A_{266} - A_f)$ is shown as a function of exposure.

distinguish between the simple reaction leading to a hydrate and the more complicated reaction leading to a reversible dimer. From Fig. 11, one could erroneously conclude that the cross section for the reaction was 0.90 cm$^2$/$\mu$E when in actual fact this cross section is the sum of the forward $k = 0.55$ and the reverse cross section $r = 0.35$. Again, it must be emphasized that even if we know a dimer is being formed and reversed, the analysis is quite incomplete for there might well be several types of dimers formed simultaneously and $k$ could well be the sum of a number of cross sections for the formation of several dimers. A proper analysis requires the separation of the photoproducts and a measure of each as a function of exposure. Such an analysis will be carried out in Section VI,(1), where we will show that two isomeric dimers $\widehat{TpT^1}$ and $\widehat{TpT^2}$ are formed from $TpT^{24}$ together with two other types of photoproducts referred to as $TpT^4$ and $TpT^3$. The cross section $k = 0.55$ obtained above, is close to the total cross section for the production of $(\widehat{TpT^1} + \widehat{TpT^2} + TpT^4)$ while the reverse cross section 0.35 is close to the average cross section for the reversal of $\widehat{TpT^1}$ and $\widehat{TpT^2}$.

## 4. *Photoproduct Thermally Reversed during the Irradiation*

Under some circumstances the photoproduct may undergo appreciable thermal reversion to the parent *during* the irradiation, and it is essential to correct for this reversal in analyzing the kinetic data. To illustrate this point C3'p was irradiated at pH 5.0 at 0°, where reversal is nearly negligible, and at 20°, where reversal is important.[37] The results are shown in Fig. 12 where $\log_{10}(A_{272})$ is plotted as a function of exposure. Curve D°, obtained at 0°, is very nearly a straight line, while curve D²⁰, obtained at 20°, is concave upward due to reversal during the irradiation. To correct these curves for reversal we derive the necessary equations as follows. The equation for the change in hydrate $\Delta H$ is

$$\Delta H = \sigma \cdot P \cdot \Delta \bar{L} - k_r' H \cdot \Delta t \tag{45}$$

where the first term gives the amount of parent P converted to hydrate by an exposure $\Delta \bar{L}$, and the second term gives the amount of hydrate which reverses in the time $\Delta t$, where $\Delta t$ is the elapsed time between successive

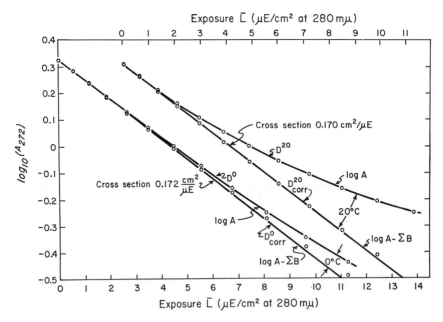

FIG. 12. Irradiation of the 3' isomer of cytidylic acid at 280 mμ (pH 5.0, 0.01 $M$ McIlvaine's buffer, at 0° and 20°C) D° = plot of $\log_{10}(A_{272})$ versus exposure. $D_{cor}°$ is a plot of $\{\log_{10}(A_{272}) - \Sigma B\}$ where $B$ is the correction term $k_r t / 2.303 \Sigma [(A_0 - A)/A] \cdot \Delta t$ for reversal of hydrate (see Eq. 50). D²⁰ and $D_{cor}^{20}$ are the corresponding curves at 20°C. The corrected curves give nearly the same cross section.

measurements of the absorbance of the solution. $\sigma$ is the cross section for production and $k_r{}^t$ the rate constant for reversal.[72a]

The hydrate reverses during the irradiation and during the manipulations between delivery of successive increments of exposure so that $\Delta t$ is longer than the time required to deliver $\Delta \bar{L}$. If we assume that only two species P and H are present, then

$$H = P_0 - P \tag{46}$$

where $P_0$ is the initial amount of parent. Substituting this value for H in Eq. (45) and rearranging we obtain

$$\frac{\Delta P}{P} - \frac{(P_0 - P)}{P} \cdot k_r{}^t \Delta t = -\sigma \Delta \bar{L} \tag{47}$$

This expression cannot be integrated analytically since it is a complicated function of $\bar{L}$. It will be remembered that $\bar{L}$ increases slowly with time initially, when the absorbance is high and then more rapidly as the absorbance decreases; in addition, the intensity from the monochromator varies with time and different times may elapse between successive increments of exposure. However, Eq. (47) can be evaluated by stepwise integration to give

$$\sum_{P_0}^{P} \frac{\Delta P}{P} - \sum_{0}^{t} \left( \overline{\frac{P_0 - P}{P}} \right) k_r{}^t \Delta t = -\sigma \sum \Delta \bar{L} \tag{48}$$

where $\overline{[(P_0 - P)/P]}$ stands for the average value of $(P_0 - P)/P$ during the interval $\Delta t$. Since the first term can be integrated, we replace the sum by the integral to give

$$\ln \frac{(P)}{(P_0)} - k_r{}^t \sum_{0}^{t} \left( \overline{\frac{P_0 - P}{P}} \right) \Delta t = -\sigma \sum \Delta \bar{L} \tag{49}$$

If we measure the absorbance of the solution at a wavelength where only the parent absorbs then (49) can be written in terms of absorbance thus

$$\log A - \frac{k_r{}^t}{2.303} \sum \left( \overline{\frac{A_0 - A}{A}} \right) \cdot \Delta t = \log A_0 - \frac{\sigma}{2.303} \cdot \sum \Delta \bar{L} \tag{50}$$

where $A_0$ is the initial absorbance and $A$ the absorbance measured after each increment of time $\Delta t$, and $\Delta \bar{L}$ is the increment in exposure delivered

---

[72a] Rate constants for reversal are usually represented by $k$ with the dimensions 1/time. Since, however, we have used $k$ with various subscripts for the cross section for dimer production we will represent the rate constant by $k_r{}^t$ suggesting a constant for reversal which occurs spontaneously with time.

within the interval of time $\Delta t$. The quantities $A_0$, $A$, $\Delta t$, and $\Delta \bar{L}$, can all be measured and if $k_r{}^t$, the rate constant for reversal is known, then the left-hand side of Eq. 50 can be evaluated. If this is plotted versus $\bar{L}$, a straight line of slope $\sigma/2.30$ will be obtained from which $\sigma$ can be calculated.

If the rate constant for reversal is small, then the second term of Eq. (50) is negligible and a plot of log $A$ versus $\bar{L}$ should give a straight line. This is very nearly the case for the data obtained at $0°$ (see curve $D°$ of Fig. 12). When reversal is not negligible, then one can still obtain a straight line if the correction term is subtracted from log $A$ (see curve $D_{cor}°$, Fig. 12). The details of these calculations are shown in Appendix C. Here we note that the corrected curves for $0°$ and $20°$ have almost the same slope, yielding a cross section of $0.171$ cm$^2$/$\mu$E. Examination of the $D^{20}$ curve at zero exposure will illustrate the difficulty of drawing the correct initial slope merely by inspection. The rectification procedure using Eq. (50) overcomes this difficulty and enables one to determine the cross section accurately.

The rate constant for reversal $k_2{}^t$ can often be determined by using a photochemical reaction. For example, if irradiation leads to a steady state when the rate of production of the hydrate is exactly balanced by the rate of reversal, then $\Delta H$ in Eq. (45) is zero, and this equation reduces to

$$\frac{k_r{}^t}{\sigma} = \left(\frac{P}{H}\right)_s \cdot \frac{\Delta \bar{L}}{\Delta t} \tag{50a}$$

where $(P/H)_s$ is the ratio of P to H under steady state conditions. This equation may be used to determine $k_r{}^t$ if $\sigma$ is known. Equation (50a) is particularly useful in determining $k_r{}^t$ when it is very large and the corresponding half-life of the hydrate $(0.693/k_r{}^t)$ is small. When the lifetime is very short (a few milliseconds) the problem can often be solved by using the rotating sector technique,[72b] or other type of intermittent illumination (see [9]), where photoproducts are produced in one interval of time and then allowed to reverse during the next interval of time as the sector cuts off the radiation.

## VI. Photochemical Reactions Involving Parent and Several Photoproducts

When more than one photoproduct is involved in a reaction, the measurement of absorbance changes usually does not give enough information to describe the reaction adequately. Absorbance changes can at best give only an average overall effect. In addition, when polynucleotides are involved, the production of photoproducts may alter the structure of the molecule and so alter its absorbance through hyperchromic effects. Thus, for example, irradiation of the complex poly (A + U) produces very little

[72b] G. M. Burnett and H. W. Melville, *in* "Technique of Organic Chemistry" (S. L. Friess, E. S. Lewis, and A. Weissberger, eds.), Vol. VIII, Part II, pp. 1107–1137. Wiley (Interscience), New York, 1963.

absorbance change,[67,73] which might lead one to think that the molecule is insensitive to radiation. When, however, the photoproducts in poly (A + U) are separated and their growth is studied, it is evident that both hydrates and dimers are produced. Evidently, the loss in absorbance resulting from the dimers and hydrates is almost balanced by the increase in absorbance resulting from the changes in hyperchromicity due to insertion of photoproducts into the poly U chain, which alters the structure of the complex. For these reasons photochemical studies based only on absorbance changes should be interpreted with caution.

To study systems involving several photoproducts, it is essential to separate the photoproducts and measure their growth as a function of exposure. The photoproducts may be separated from each other and from the parent using electrophoresis, paper chromatography, or columns as described in Section VIII. The amounts of each photoproduct can be measured if radioactive parent is used. We have found it useful to label nucleic acids and their components with [32]P. Here we will discuss a number of model systems.

## A. Systems Involving Several Photoproducts Where "Equilibrium" Is Approached

### 1. *Photoproducts of TpT and Reaction Scheme*

A good example of a reaction where equilibrium is approached is that involving thymidylyl-(3'-5')-thymidine (TpT), which we will refer to here as the parent P. On irradiation this gives two isomeric dimers,[73a] TpT[1] and TpT[2], the concentrations of which we will refer to as D[1] and D[2], as well as two other products TpT[3] and TpT[4].[24,78] The separation of these photo-

---

[73] G. De Boer, M. Pearson, and H. E. Johns, *J. Mol. Biol.* **27**, 131 (1967).

[73a] Four isomeric dimers for thymine were postulated by Wulff and Fraenkel[74] with the ring structures of Fig. 1 oriented in four different ways with respect to one another, and four dimers have been isolated by Weinblum and Johns.[75,76] In the dinucleotides (TpT, UpU, rUpU, CpC) one might expect that some of these dimers would be prevented from forming because of the 3' to 5' phosphodiester linkage. This seems to be the case, and to date two isomeric dimers have been produced in TpT,[24] 2 dimers dUpU,[28] 3 dimers in UpU,[29] and two dimers in CpC.[30] The structure of one of the dimers has now been proved by X-ray diffraction studies.[77]

[74] D. L. Wulff and G. Fraenkel, *Biochim. Biophys. Acta* **51**, 332 (1961).

[75] D. Weinblum and H. E. Johns, *Biochim. Biophys. Acta* **114**, 450 (1966).

[76] D. Weinblum, *Biochem. Biophys. Res. Commun.* **27**, 384 (1967).

[77] N. Camerman, S. C. Nyburg, and D. Weinblum, *Tetrahedron Letters* No. **42**, 4127 (1967).

[78] M. L. Pearson, F. P. Ottensmeyer, and H. E. Johns, *Photochem. Photobiol.* **4**, 739 (1965).

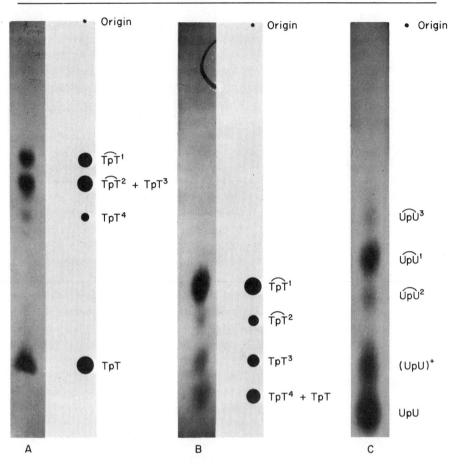

FIG. 13. (A) Separation of photoproducts of TpT by paper chromatography (see footnote 24) in solvent H. (B) Separation of photoproducts of TpT by paper chromatography (see footnote 24) in solvent G. (C) Separation of photoproducts of UpU by solvent H.

products in two solvents is shown in Fig. 13A and 13B. Once separated, the radioactive spots can be eluted and subjected to various tests. For example, when $D^1$ and $D^2$ were eluted and reirradiated, they were converted to TpT, indicating that they are dimers. $TpT^4$ on irradiation is converted not to TpT, but into a new product $TpT^3$. $TpT^3$ on separation was shown to have a peak absorption at 330 m$\mu$, and when it was irradiated at 330 m$\mu$ it could be reversed back to $TpT^4$, suggesting that it may be the dimer form of $TpT^4$.[24,78] By studies of this kind, the reaction scheme for TpT can be shown to be as follows:

$$
\begin{array}{ccc}
& k_1 \quad (\text{TpT}) \quad k_2 & \\
\text{D}^1 \rightleftharpoons & \text{P} & \rightleftharpoons \text{D}^2 \\
r_1 & \downarrow k_4 & r_2 \\
& \text{TpT}^4 & \\
& k_{43} \downarrow \uparrow r_{34} & \\
& \text{TpT}^3 &
\end{array}
\tag{51}
$$

In this scheme $k_1$, $k_2$, and $k_4$ are the cross sections for the production of $\text{D}^1$, $\text{D}^2$, and $\text{TpT}^4$; $r_1$ and $r_2$ are the reverse cross sections for $\text{D}^1$ and $\text{D}^2$; and $k_{43}$ and $r_{34}$ are the cross sections for the formation and reversal of $\text{TpT}^3$ from $\text{TpT}^4$.

## 2. Differential Equation Describing Reaction for TpT

Using the symbols of the reaction scheme of Eq. (51), the simultaneous differential equations describing the reaction are as follows:

$$
-\frac{d\text{P}}{d\bar{L}} = (k_1 + k_2 + k_4)\text{P} - (r_1\text{D}^1 + r_2\text{D}^2)
\tag{52a}
$$

$$
\frac{d\text{D}^1}{d\bar{L}} = k_1\text{P} - r_1\text{D}^1
\tag{52b}
$$

$$
\frac{d\text{D}^2}{d\bar{L}} = k_2\text{P} - r_2\text{D}^2
\tag{52c}
$$

$$
\frac{(\text{TpT}^4)}{d\bar{L}} = k_4\text{P} - k_{43}\text{TpT}^4 + r_{34}\text{TpT}^3
\tag{52d}
$$

$$
\frac{(\text{TpT}^3)}{d\bar{L}} = k_{43}(\text{TpT}^4) - r_{34}(\text{TpT}^3)
\tag{52e}
$$

The solution[23] of these five simultaneous equations is of the form

$$
\text{D}^1 = Ae^{\alpha\bar{L}} + Be^{\beta\bar{L}} + Ce^{\gamma\bar{L}} + De^{\delta\bar{L}}
\tag{53}
$$

where $A$, $B$, $C$, and $D$ are constants determined from the initial conditions and $\alpha$, $\beta$, $\gamma$, and $\delta$ are the roots of a fourth degree equation involving all the cross sections $k_1$, $k_2$, $k_4$, $k_{43}$, $r_1$, $r_2$, and $r_{34}$. The solutions for $\text{D}^1$, $\text{D}^2$, $\text{TpT}^3$, and $\text{TpT}^4$ are of the same form as Eq. (53) involving the same exponential terms $\alpha$, $\beta$, $\gamma$, and $\delta$ but different constants $A$ to $D$. Equation (53) involves *all* the cross sections, and it is practically impossible to determine the individual cross sections from it. Other methods for determining these will now be presented.

## 3. Forward Cross Sections $k_1$, $k_2$, $k_4$

To illustrate the determination of these cross sections, typical data for TpT are shown in Fig. 14, where the activities in $\text{D}^1$, $\text{D}^2$, and $\text{TpT}^4$ are

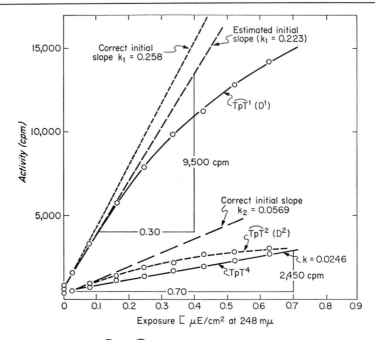

FIG. 14. Growth of $\widehat{\text{TpT}}^1$, $\widehat{\text{TpT}}^2$, and $\text{TpT}^4$ as a function of exposure at 248 m$\mu$.

plotted as a function of exposure $\bar{L}$. The dosimetry was exactly the same as that illustrated in Table II except that the volume $V$ of liquid in the cell continually decreased since samples were removed for chromatography after each increment in exposure. The initial slopes of these graphs give the cross sections as follows. Initially, $D^1 = D^2 = \text{TpT}^4 = \text{TpT}^3 = 0$ since all the activity is in the TpT spot. For $\bar{L} = 0$, Eqs. (52b), (52c), and (52d) reduce to the following:

$$\left(\frac{dD^1}{d\bar{L}}\right)_{\bar{L}=0} = k_1 P_0; \quad \left(\frac{dD^2}{d\bar{L}}\right)_{\bar{L}=0} = k_2 P_0; \quad \left(\frac{d(\text{TpT}^4)}{d\bar{L}}\right)_{\bar{L}=0} = k_4 P_0 \quad (54)$$

where $P_0$ is the initial activity in the TpT spot (142,000 cpm). Thus, a determination of the initial slopes of the curves of Fig. 14 and a knowledge of $P_0$ enables one to determine $k_1$, $k_2$, $k_4$. The growth of $\text{TpT}^4$ is linear with exposure over the range shown in Fig. 14 enabling one to evaluate

$$k_4 = \frac{2450 \text{ cpm}}{0.70 \ \mu\text{E}/\text{cm}^2} \cdot \frac{1}{142,000 \text{ cpm}} = 0.0246 \text{ cm}^2/\mu\text{E} \quad (55)$$

For the dimers a determination of the initial slopes is difficult because the curves approach saturation quickly. However, an estimate can be made in exactly the same way as above. From the estimated initial slope (dashed line) one obtains the cross section $k_1$ for dimer 1 as

$$k_1 \simeq \frac{9.500}{0.30} \cdot \frac{1}{142,000} \simeq 0.223 \text{ cm}^2/\mu\text{E} \tag{56}$$

Examination of Fig. 14 will show that the initial slope is determined by the first few points on the graph. If the points of Fig. 14 had shown substantial fluctuations, obtaining the initial slope by inspection would have been difficult and could have led to large errors in $k_1$. Even with the rather good data of Fig. 14, a more precise method of evaluating $k_1$ is desirable and may be obtained as follows.

The growth curve for TpT in Fig. 14 looks as though it could be fitted reasonably well by a single exponential approaching a final value D*, although we know from Eq. (53) that it is really the sum of four exponentials. This is a fairly good approximation because $k_4$, leading irreversibly to TpT⁴, is small compared with $k_1$ so the parent reaches a nearly constant value with increase in exposure. If it were a simple exponential, then a plot of $(D^* - D)$ versus $\bar{L}$ on a log scale should give a straight line. To test this, $D^* - D$ is plotted against exposure in Fig. 15 for three values of D* estimated from Fig. 14. If too large a value for D* is chosen the curve is concave upward, while too small a value gives a curve that is concave downward. The curve for $D^* = 20,000$ cpm gives a good straight line over the range of exposures studied. It must be emphasized that *no* value for D* can give a perfect fit over a wide exposure range because the curve cannot be fitted exactly by a *single* exponential. The slope of the straight lines of Fig. 15 can be used to give a better estimate of $k_1$ as follows.

The equations of the straight lines of Fig. 15 are of the form

$$D^* - D = (D^* - D_0)e^{-c\bar{L}} \tag{57}$$

where $C$ is a constant which can most easily be evaluated as $1/\bar{L}_{37}$ where $\bar{L}_{37}$ is the exposure required to reduce $(D^* - D_0)$ to 0.368 of its initial value. To relate $C$ to $k$ we differentiate, Eq. (57), to obtain

$$-\frac{dD}{d\bar{L}} = -C(D^* - D_0)e^{-c\bar{L}}$$

at $\bar{L} = 0$, this reduces to

$$\frac{(dD)}{(d\bar{L})_{L\to 0}} = C(D^* - D_0)$$

Now comparing with Eq. (54), it is clear that

$$C(D^* - D_0) = k_1 P_0$$

from which we obtain

$$k_1 = \frac{C(D^* - D_0)}{P_0} = \frac{(D^* - D_0)}{P_0} \cdot \frac{1}{\bar{L}_{37}} \tag{58}$$

Table III shows the calculations of $k_1$ using Eq. (58). While the value for $k_1$ depends upon the value for $D^*$ which is chosen, it is insensitive to the value for $D^*$. Thus, changing $D^*$ by 10% alters $k_1$ by only about 2.5%. The best value for $k_1$, 0.258 cm$^2$/$\mu$E is obtained with $D^* = 20,000$, since this gives the best straight line. It is interesting to compare this value for

TABLE III

ILLUSTRATING THE CALCULATION OF THE CROSS SECTION FOR

THE PRODUCTION OF $\overparen{TpT}^1$ FROM TpT

| $D^*$ (cpm) | $(D^* - D_0)$ (cpm) | $P_0$ (cpm) | $\bar{L}_{37}$ ($\mu$E/cm$^2$) (from Fig. 15) | $k_1$ (cm$^2$/$\mu$E) |
|---|---|---|---|---|
| 18,000 | 17,200 | 142,000 | 0.460 | 0.264 |
| 20,000 | 19,200 | 142,000 | 0.525 | 0.258 |
| 22,000 | 21,200 | 142,000 | 0.595 | 0.251 |

$k_1$ with the value 0.223 cm$^2$/$\mu$E obtained from the initial slope of the growth curve of Fig. 14. Our accurate value is about 12% higher, illustrating the difficulty of obtaining the correct initial slope from a curve such as that of Fig. 14. For comparison, the correct initial slope is shown as the dotted line. It would be difficult to draw this dotted line by inspection.

### 4. Determination of Cross Sections $r_1$, $r_2$, $r_{34}$, and $k_{43}$

To determine the cross section $r_1$ and $r_2$, one requires a separated sample of each of the dimers, which are then irradiated. At zero exposure, $P = 0$ and Eqs. (52b) and (52c) reduce to

$$\left(\frac{dD^1}{d\bar{L}}\right)_{L=0} = -r_1 D_0^1 \text{ and } \left(\frac{dD^2}{d\bar{L}}\right)_{\bar{L}=0} = -r_2 D_0^2 \tag{59}$$

where $D_0^1$ and $D_0^2$ are the initial activities in the dimer 1 and dimer 2 spots. From an estimate of the initial slopes, $r_1$ and $r_2$ may be determined approximately. By estimating the final value of each of the dimers, a more precise determination of $r_1$ and $r_2$ may be made by the method outlined in the last section.

The determination of $r_{34}$ and $k_{43}$ requires a separated sample of both TpT$^4$ and TpT$^3$. The cross sections $r_{34}$ and $k_{43}$ are then determined by

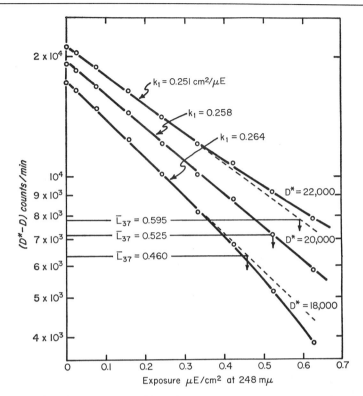

Fig. 15. Plot of $(D^* - D)$ on a log scale versus exposure at 248 m$\mu$. D is the activity in $D^1$ (Fig. 14) and $D^*$ is the estimated maximum activity. Graphs are shown for $D^* =$ 18,000, 20,000, and 22,000 cpm.

following the disappearance of $TpT^4$ and $TpT^3$ with exposure and estimating the initial slopes by the methods of the last section.

5. *Summary Concerning TpT*

For TpT the main reaction leads to the dimers $\overset{\frown}{TpT^1}$ and $\overset{\frown}{TpT^2}$ while the cross section for the formation of $TpT^4$ is smaller by a factor of about 10 at all wavelengths. This means that equilibrium is very nearly established between parent and dimer as illustrated in Fig. 14. Under these circumstances an estimate of this equilibrium amount of each of the dimers can be used to rectify the curves of Fig. 14 to give the straight lines of Fig. 15. These straight lines can be used to obtain a precise value for the cross section. In the next section we will deal with methods for handling data when the cross section leading irreversibly away from the parent is large. This is the case for UpU.

## B. Systems Involving Several Photoproducts Where "Equilibrium" Is Not Approached

An example of a system where equilibrium is not approached is UpU.

### 1. *Separation of Photoproducts and Reaction Scheme for UpU*

Figure 13C shows the separation of the photoproducts of UpU. These consist of three isomeric dimers, $\overset{\frown}{UpU^1}$, $\overset{\frown}{UpU^2}$, and $\overset{\frown}{UpU^3}$, which will be referred to as $D^1$, $D^2$, and $D^3$, as well as $(UpU)^*$. $(UpU)^*$ is a mixture of $UpU^*$, $U^*pU$, and $U^*pU^*$, where $U^*$ stands for the hydrate of uracil. Irradiation of either of the single hydrates gives rise to the double hydrate $U^*pU^*$. Since the formation of a double hydrate involves the loss of a single hydrate, the sum of $UpU^* + U^*pU + U^*pU^*$, which we refer to as H, is the *total* number of UpU molecules which have been hydrated at least *once*. The growth of this sum with exposure yields the cross section $h$ shown in the reaction scheme of Eq. (60).

$$
\begin{array}{c}
D^3 \\
k_1 \quad k_3 \uparrow\downarrow r_3 \quad k_2 \\
D^1 \leftrightarrows UpU(P) \rightleftarrows D^2 \\
r_1 \quad \downarrow h \quad r_2 \\
H
\end{array}
\qquad (60)
$$

Although this reaction scheme looks quite similar to that of TpT, the behavior of UpU is quite different because the cross section, $h$, leading to the hydrate is as large or larger than the dimer cross sections, so that the parent is irreversibly converted to the hydrate at a rapid rate. In contrast, in TpT, the cross section $k_4$ leading irreversibly to $TpT^4$ was small, so that once the dimers were formed the amount of parent remained nearly constant with further exposure.

The reaction of Eq. (60) is described in terms of the cross sections $k_1$, $k_2$, $k_3$, $h$, $r_1$, $r_2$, and $r_3$ by the differential equations

$$
-\frac{dP}{d\bar{L}} = (k_1 + k_2 + k_3 + h)P - (r_1 D^1 + r_2 D^2 + r_3 D^3) \qquad (61a)
$$

$$
\frac{dD^1}{d\bar{L}} = k_1 P - r_1 D^1 \qquad (61b)
$$

$$
\frac{dD^3}{d\bar{L}} = k_3 P - r_3 D^3 \qquad (61c)
$$

$$
\frac{dH}{d\bar{L}} = hP \qquad (61d)
$$

The solution of these equations is of the form of Eq. (53), but, as in the case of TpT, the complete solution is not of much help in allowing us to determine the cross sections. We will now discuss other methods for arriving at the various cross sections.

## 2. Cross Section for Hydrate Formation

In Fig. 16 is plotted the quantity $H/P_0$ versus exposure where $P_0$ is the initial amount of parent present. Since $P_0$ is constant, the curve $H/P_0$ is exactly the same shape as the curve of H versus exposure. The initial slope of this curve is

$$\frac{1}{P_0}\left(\frac{dH}{dL}\right)_{\bar{L}=0}$$

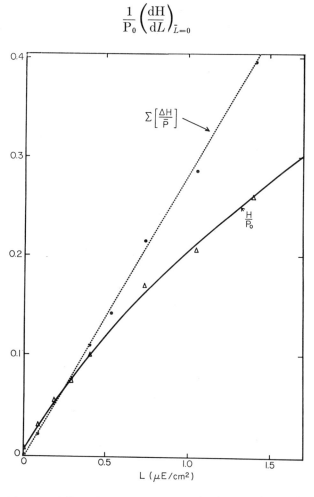

$$\Sigma\left[\frac{\Delta H}{\bar{P}}\right]$$

$$\frac{H}{P_0}$$

L ($\mu$E/cm²)

Fig. 16. Plot of $\Delta H/\bar{P}$ against $L$ at 248 m$\mu$. The slope of the straight line gives the cross section for hydrate formation 0.284 cm²/$\mu$E. A graph of $H/P_0$ is shown for comparison.

which from equation (61a) is the cross section $h$. An accurate determination of this initial slope is difficult, because the curve is not linear even for small exposures and the fact that experimental points all have errors associated with them. The nonlinearity is a consequence of the decrease in parent brought about by the irradiation. If only hydrate were being produced (or were the main product) a suitable log plot would give a straight line, but when the parent material is continually being converted to dimers as well, no log plot will give a straight line. The following numerical method, however, will yield a straight line from which $h$ can be determined.

For small increments in exposure Equation (61d) may be written

$$\frac{\Delta H}{\bar{P}} = h\Delta\bar{L}$$

where $\bar{P}$ is the average amount of UpU present during the increment in exposure $\Delta\bar{L}$. After $n$ increments in exposure we obtain

$$\sum_{n}^{n} \frac{\Delta H}{\bar{P}} = h \sum_{n}^{n} \Delta\bar{L} = h\bar{L} \tag{62}$$

so that a plot of $\Sigma\Delta H/\bar{P}$ against $\bar{L}$ should give a straight line with slope $h$. After each increment in dose a sample is chromatographed and the activity in the hydrate spot and in the UpU spot is measured. From these activities, $\Delta H$ and $\bar{P}$ may be calculated and $\Sigma\Delta H/\bar{P}$ determined. A convenient way to arrange these calculations can be found in the original publication.[29]

A plot of $\Sigma\Delta H/\bar{P}$ versus $\bar{L}$ gives a good straight line (Fig. 16) from which the hydrate cross section 0.284 cm²/$\mu$E is determined. It is obvious that this correct initial slope would be hard to determine from the growth of $H/P_0$. The numerical method has the advantage of using *all* the measured values for P and H while the plot of $H/P_0$ uses *all* the values of H but *only* the initial value of P.

### 3. Cross Section for Dimer Formation and Reversal

Figure 17A shows the growth of $D^1$ and the decay of the UpU as UpU is irradiated. It should be noted that the parent continues to disappear, even after $D^1$ becomes constant, owing mainly to the continual production of hydrate. Figure 17B shows the results of an experiment in which a pure sample of $D^1$ is irradiated. The amount of UpU increases to a maximum and then decreases slowly. The *two* sets of data shown in Fig. 17 may be used to obtain both the forward cross section $k_1$, and the reverse cross section $r_1$. These curves for the production and decay of the dimer show more rapid changes of slope than the curve for H (Fig. 16) so that an

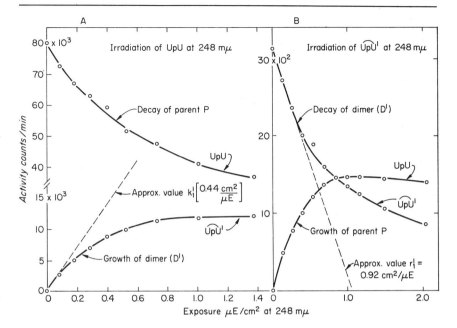

FIG. 17. (A) Irradiation at 248 mμ of UpU showing growth of U͡pU¹ and decay of parent. (B) Irradiation of U͡pU¹ at 248 mμ showing growth of UpU and decay of U͡pU¹.

accurate determination of the initial slope is even more difficult. Furthermore, because of the reaction of UpU to form several dimers and hydrates simultaneously, no logarithmic plot of dimer versus $L$ will yield a straight line. The numerical method introduced in the last section can be extended to cover this situation as follows.

For the growth curve (Fig. 17A), $D$[78a] is initially zero so that at zero dose, Eq. (61b) may be used to give an approximate value for $k_1$ which we will call $k'_1$ where

$$k'_1 = \left(\frac{dD/d\bar{L}}{P}\right)_{\bar{L}=0} = \frac{17{,}500 \text{ cpm}}{0.5 \ \mu E/cm^2} \cdot \frac{1}{80{,}000 \text{ cpm}} = 0.44 \text{ cm}^2/\mu E \quad (63)$$

Similarly, for the decay curve of Fig. 17B, P is initially zero so that at zero dose equation (61b) may be used to give an approximate value for $r_1$ which will be called $r'_1$.

$$r'_1 = -\left(\frac{dD/d\bar{L}}{D}\right)_{\bar{L}=0} = \frac{3130 \text{ cpm}}{1.09 \ \mu E/cm^2} \cdot \frac{1}{3130 \text{ cpm}} = 0.92 \text{ cm}^2/\mu E \quad (64)$$

[78a] For simplicity of notation we will refer to $D^1$ as D since we are at the moment interested in only one dimer.

We will now use the approximate value $k'_1$ to arrive at a better value for $r_1$. Dividing equation (61b) by D, rearranging and writing the differentials as increments, we obtain

$$\frac{\Delta D}{\overline{D}} - k'_1 \left(\overline{\frac{P}{D}}\right) \cdot \Delta \bar{L} \simeq -r_1 \Delta \bar{L} \tag{65}$$

where $\overline{D}$ and $(\overline{P/D})$ are the average values of these quantities over the increment $\Delta \bar{L}$ and $k_1$ has been replaced by its approximate value $k'_1$ determined by Eq. (63). This equation can be summed to dose $\bar{L}$ to yield

$$\sum_{D_0}^{D} \frac{\Delta D}{\overline{D}} - k'_1 \sum_{0}^{\bar{L}} \left(\overline{\frac{P}{D}}\right) \cdot \Delta L = -r_1 \bar{L} \tag{66}$$

where $D_0$ is the amount of dimer at $L = 0$ and D the amount at exposure $\bar{L}$. This can be simplied by noting that the first summation can be represented as an integral since,

$$\sum_{D_0}^{D} \frac{\Delta D}{\overline{D}} = \int_{D_0}^{D} \frac{dD}{D} = \ln D - \ln D_0 \tag{67}$$

Thus, we obtain for Eq. (66) its equivalent

$$\ln D - \ln D_0 - k'_1 \sum_{0}^{\bar{L}} \left(\overline{\frac{P}{D}}\right) \cdot \Delta \bar{L} = -r_1 \bar{L} \tag{68}$$

Now all the quantities on the left side of Eq. (68) can be measured as a function of $\bar{L}$. For details on how the calculations may be arranged see footnote (29). Figure 18B shows a plot of the left side of Eq. (68) against exposure $\bar{L}$, yielding an excellent straight line whose slope gives 1.10 cm$^2$/ $\mu$E for $r_1$, the reverse cross section. This is a much more precise value than can be obtained by drawing the dotted line of Fig. 17b, which gave 0.92 cm$^2$/$\mu$E.

We now manipulate Eq. (61b) to yield a better value for the forward cross section $k_1$; by dividing through by P and rearranging, we obtain

$$\frac{\Delta D}{\overline{P}} + r_1 \left(\overline{\frac{D}{P}}\right) \cdot \Delta \bar{L} = k_1 \Delta \bar{L} \tag{69}$$

where $\overline{P}$ and $(\overline{D/P})$ are averages over the increment $\Delta \bar{L}$. This can be summed to give

$$\sum \frac{\Delta D}{\overline{P}} + r_1 \sum \left(\overline{\frac{D}{P}}\right) \cdot \Delta \bar{L} = k_1 \bar{L} \tag{70}$$

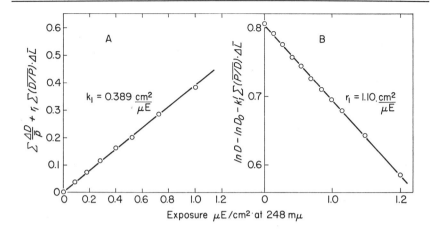

Exposure $\mu E/cm^2$ at 248 m$\mu$

FIG. 18. (A) Graph showing how the data for the growth of D in Fig. 17A can be rectified to give a straight line by plotting $\Sigma\ \Delta D/\bar{P} + r_1 \cdot \Sigma\ (\overline{D/P}) \cdot \Delta\bar{L}$ against exposure $\bar{L}$. (B) Graph showing how the data for the growth of P in Fig. 17B can be rectified to give a straight line by plotting $\ln\ D - \ln\ D_0 - k'_1 \Sigma\ (\overline{P/D}) \cdot \Delta\bar{L}$ against exposure.

Now all the quantities on the left side of Eq. (70) can be calculated since they depend upon our measured values of D and P (Fig. 17A) and our previously determined value of $r_1$. A plot of the left side of Eq. (70) is given in Fig. 18A yielding a good straight line, from which the cross section $k_1$ is determined as 0.389 cm²/$\mu$E. This value is about 12% larger than the approximate value obtained by attempting to draw the initial slope of Fig. 17A. This corrected value for $k$ could be used to further refine the value of $r$, but the effect of the correction would be small.

The numerical method of determining $k_1$ and $r_1$ yields much more precise results than can possibly be obtained by attempting to draw by eye the initial slopes of the growth and decay curves of Fig. 17. This is particularly true when experimental errors are present and at certain wavelengths where the amount of dimer which can be produced or reversed is small. This situation exists at short wavelengths for the formation of dimers and at long wavelengths for their reversal.

## VII. Photochemistry of Polynucleotides (Poly U)

The photochemistry of polynucleotides is much more complex than the simple model compounds dealt with earlier in this chapter. We will present a brief summary of our work on poly U.[67–70,73,79] When UpU is

[79] M. Pearson and H. E. Johns, *Proc. 7th Can. Cancer Res. Conf.*, *Honey Harbor*, *Ontario*, Vol. 7, p. 353. Macmillan (Pergamon), New York, 1967.

irradiated, three kinds of dimers and a hydrate are produced and one might therefore expect similar photoproducts to arise from the irradiation of poly U. Our experiments indicate that both dimers and hydrates are produced in poly U. However, in addition to the number and types of photoproducts produced one would also like to know the sequence of these products, for presumably the sequence would be important in determining the behavior of an irradiated polynucleotide in a biological assay. Such an assay might be for example the test of an irradiated poly-nucleotide when used as a messenger in a protein-synthesizing system. A complete discussion of the photochemistry of poly U should answer such questions as: Does a dimer in the chain increase the probability of putting either a dimer or hydrate next to it? Does a hydrate in the chain increase the chance of forming a dimer in the adjacent positions? Or, more generally, are photoproducts produced at random, or bunched in specific sequences? The question of bunching is important because of the information it yields concerning basic photochemistry and also because it will affect the length of sequences of unaffected bases, and this may be of biological importance.

Suppose a poly U chain consisting of 10 uracils represented by (71a) is irradiated until two dimers and two hydrates are produced. These photo-products could be arranged in many different sequences of which two possible sequences are shown in (71b) and (71c). Although (71b) and (71c)

$$1 \quad 2 \quad 3 \quad 4 \quad 5 \quad 6 \quad 7 \quad 8 \quad 9 \quad 10 \qquad (71a)$$
$$U \; p \; U \; p \; U \; p \; U \; p \; U \; p \; U \; p \; U \; p \; U \; p \; U \; p \; U$$

$$U \; p \; \overset{\frown}{U \; p \; U} \; p \; U \; p \; \overset{*}{U} \; p \; U \; p \; \overset{\frown}{U \; p \; U} \; p \; \overset{*}{U} \; p \; U \qquad (71b)$$

$$\overset{\frown}{U \; p \; U} \; p \; \overset{\frown}{U \; p \; U} \; p \; \overset{*}{U} \; p \; U \; p \; U \; p \; U \; p \; U \; p \; \overset{*}{U} \qquad (71c)$$

have the same amount of damage the two molecules are quite different from a photochemical point of view. None of the undamaged U's in (71b) have unaffected U's as nearest neighbors, so no more dimers can be produced in this molecule. On the other hand, molecule (71c) can still be dimerized between 6 and 7, 7 and 8, or 8 or 9, and we would say has three dimerizable sites, i.e., three combinations of pairs of unaffected U's. Inspection of (71a) indicates that the unirradiated polymer has nine dimerizable sites. In a chain with $N_0$ residues there are initially $(N_0 - 1)$ dimerizable sites or $(N_0 - 1)/N_0$ dimerizable sites per U. For large values of $N_0$ there is very nearly 1 dimerizable site per residue. Since every residue can be hydrated initially a chain of $N_0$ residues has $N_0$ hydratable sites.

Our results show that the presence of one photoproduct in the poly U chain increases the probability of placing another photoproduct next in the chain. Evidently then the concept of a cross section has only limited value since once a few photoproducts are produced in the chain each remaining dimerizable or hydratable site will have a different and unknown cross section for its alteration. Our results also indicate that energy transfer may well take place, i.e., a photon could be absorbed at one place in the chain; the excitation energy might then be transferred sequentially from one base to the next until the energy transfer is stopped by an existing photoproduct, at which point a second photoproduct might be produced. Presumably, each base in turn affected by the excitation energy would have a certain probability of being altered. From this discussion, it is evident that cross sections as we have defined them are not easily applicable to the photochemistry of polymers. Attempts to solve this problem using Monte Carlo techniques[70] have met with only limited success. In addition, our studies show that the structure of the poly U drastically alters its photochemistry.[67,73] If, for example, it is irradiated while complexed to poly A [the complex is referred to as poly (A + U)] it is very resistant to radiation, dimer production being reduced by a factor of 5 and hydrate production by a factor of 10. Our studies indicate that hydrates are probably not produced in poly (A + U) until a dimer is placed in the chain. This may lead to localized melting in such a way that a hydrate can be readily produced next to the dimer.[67] More experimental and theoretical work is required before the photochemistry of polymers such as poly U and poly (A + U) will be understood.

## VIII. Separation and Detection of Photoproducts

### 1. *Separation of Photoproducts*

Most of the work described in the preceding sections is based on the ability to separate the various photoproducts from the parent. For stable and long-lived photoproducts, chromatographic separation usually offers the most convenient method. For various chromatographic systems consult the following footnotes: 6, 19, 24, 28, 30, 69, 75, 80–84. We have found that solvent G (isopropanol, ammonia, water) and solvent H (isopropanol, ethanol, potassium tartrate) are particularly useful for dealing with the dinucleotides of thymine and uracil.

When milligram amounts of a photoproduct are required, column chromatography is useful.[37,69,74,81,84] For example, to produce large quan-

[80] R. B. Setlow, W. L. Carrier, and F. J. Bollum, *Biochim. Biophys. Acta* **91**, 446 (1964).
[81] H. Schuster, *Z. Naturforsch.* **19b**, 815 (1964).
[82] A. Wacker, H. Dellweg, and D. Weinblum, *Naturwissenschaften* **47**, 477 (1960).
[83] K. C. Smith, *Photochem. Photobiol.* **2**, 503 (1963).
[84] H. G. Zachau, *Z. Physiol. Chem.* **336**, 176 (1964).

tities of thymine dimers, it is convenient to irradiate milligram amounts of thymidine in frozen solution, separate the various dimers on a Dowex column, and hydrolyze them to remove the sugars. The fractions containing thymine can of course be detected by their absorbance at $\lambda_{max}$, but the dimers cannot be found in this way since they lack a convenient absorption band. However, dimers can be found by reirradiating a sample from each tube and detecting those that show an increase in absorbance resulting from the uncoupling of dimers.

The photoproducts of cytosine are usually unstable at room temperature. The hydrate Cp* can revert back to Cp or deaminate to Up* (see Fig. 1).[30,37,85-88] For the dinucleotide CpC, many possible photoproducts exist. We have shown, for example, that both the single CpC* or C*pC, and double hydrate C*pC* are found as well as the dimer CpC. The hydrates can reverse or deaminate. The dimer of CpC deaminates first to UpC then to UpU. All these photoproducts of Cp and CpC are unstable[20,26] having half-lives ranging from minutes to hours at 0° depending upon pH. Because of this instability, photoproducts of cytosine cannot be separated by either paper or column chromatography. We have found it convenient to separate such photoproducts using high speed electrophoresis. The separation is performed with the paper compressed (30 psi) between two insulated aluminum plates which are held near 0° by circulating antifreeze through the plates. By using about 4000 volts across a 20-cm width, adequate separations can be obtained in 20–30 minutes.[30,37]

Recently thin-layer chromatography (TLC), which may be used in many different ways, has largely replaced paper chromatography. Cellulose plates on plastic backing are particularly useful in separating nucleic acid photoproducts. The spots tend to remain small so that the concentration per unit area is high, making for ease in detection of the radioactive spot. The spot with its cellulose backing may be cut out and placed in scintillation fluid for counting. The plates give better resolution between close running photoproducts than the corresponding paper chromatographs. The short time to run a TLC plate is a real advantage. A few $R_f$ values for nucleic acid photoproducts on TLC are now available.[16]

## 2. *Detection of Photoproducts*

The detection of radioactively labeled photoproducts following paper chromatography or electrophoresis is most conveniently done by exposing the dried paper to Kodak KK X-ray film for about 24 hours. Following

[85] C. P. Hariharan and H. E. Johns, *Photochem. Photobiol.* **8,** 11 (1968).
[86] C. P. Hariharan and H. E. Johns, *Photochem. Photobiol.* **7,** 239 (1968).
[87] C. P. Hariharan and H. E. Johns, *Can. J. Biochem.* **46,** 911 (1968).
[88] C. P. Hariharan, G. Poole, and H. E. Johns, *J. Chromatog.* **32,** 356 (1968).

development the radioactive spots can be cut out and counted in a liquid scintillation counter. Greater precision is possible using this technique than by passing the paper through a strip recorder since the actual shape of the spots is determined, and the cut can be made close around the spots.

Under some circumstances one would like to carry out a number of tests on an unstable photoproduct after it has been separated, but before it has a chance to decay to a more stable form. This is particularly true of photoproducts of cytosine. In order to do this one requires a rapid way to locate the radioactive areas after electrophoresis. We have found that this can be done using a specialized scanner that can scan an area of $20 \times 40$ cm² in about 5 minutes to locate the radioactive regions. In our apparatus[88] the wet electropherogram is transferred to a metal plate at 0° and the scanning head consisting of 12 small Geiger counters moves rapidly over the area. For each disintegration in each counter a dot is placed on a sheet of telideltos paper on a recording table giving an exact reproduction of the original counts. With the rapid electrophoresis and scanner it is possible to separate, locate, and elute a species in about 40 minutes.

## IX. Summary

1. The concept of cross sections for photochemical reactions is discussed in detail in Section II.

2. The photochemical behavior of the nucleic acids components is wavelength dependent, dimers tending to be formed at one wavelength and reversed at another. We have, therefore, briefly described a suitable UV monochromator for such studies and indicated in detail how controlled irradiations can be performed (see Section III).

3. Quantitative photochemical studies require a knowledge of the number of photons incident on the irradiation cuvette. We have described in detail how a monitor photocell may be calibrated using MGL to give the number of incident photons for a given setting of the monochromator (see Section IV). In Table A-2 (Appendix B), the details of an MGL calibration are given.

4. In photochemical studies one should ensure, through adequate stirring, that each molecule in the irradiation vessel is given the same exposure. Knowledge of the exposure given the molecules at the surface of the vessel is of little value. Rather one requires the average fluence, or exposure "seen" by each molecule in the vessel. In Section IV,A this average exposure $\Delta \bar{L}$ is related to the incident exposure $\Delta L_i$. $\Delta \bar{L}$ can be calculated from $\Delta L_i$ [Equation (20)] if the absorbance of the liquid and the average path length of the photons in passing through the vessel are known. Methods for obtaining the average path through the cuvette are given in Appendix A.

5. The total exposure $\bar{L}$ after a series of increments in irradiation, $\Delta\bar{L}$, is obtained by adding the increments $\Delta\bar{L}$. Details on how these calculations can be arranged are given in Table II.

6. Photochemical studies based on absorbance changes can lead to erroneous conclusions unless there is proof that only one photoproduct is involved. This is discussed in Section V.

7. The photochemistry of dinucleotides, where many photoproducts are involved, is given in Section VI. Studies of this type require methods for separating, and measuring the amount of each photoproduct as a function of exposure. The interpretation of such data requires a knowledge of the reaction scheme if meaningful cross sections are to be obtained. The precise determination of a cross section is difficult when the parent can be converted simultaneously into a number of photoproducts. We have found it convenient to express the growth of such photoproducts by linear differential equations and then to manipulate the experimental data so that some quantity derived from the experimental data varies linearly with exposure. The slope of such a straight line can be determined accurately and used to yield the required cross section.

8. In Section VII we include a discussion on the photochemistry of poly U. The difficulties in the use of cross sections to describe the photochemistry of polymers are discussed.

9. Methods for separating and detecting photoproducts are given in Section VIII.

### Acknowledgments

The Author is indebted to Dr. Mark Pearson and to Professors G. F. Whitmore and A. M. Rauth for many helpful suggestions in preparing this chapter. He is indebted to the National Cancer Institute of Canada and the Medical Research Council of Canada for continuous support in his investigations of the photochemistry of nucleic acids.

### Appendix A. Calculation of Average Path Length in Cuvette

In Section IV,A we introduced the idea of an average path length $\bar{b}$ within the cell. We will now show how $\bar{b}$ may be evaluated. Figure A-1 shows the aperture of the aplanatic lens which forms an image of the exit slit at $I_1I_2$. A ray such as PQ makes an angle $\theta$ with the front face of the quartz cuvette where $\theta$ is given by

$$\cos\theta = \frac{z}{\sqrt{z^2 + (y+h)^2 + x^2}} \tag{A-1}$$

The angle $\alpha$ this ray makes with the normal to the surface of the cuvette within the liquid of refractive index $\mu$ is given by

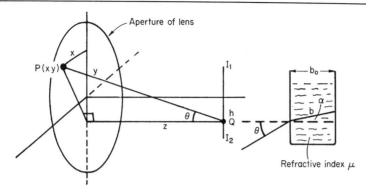

Fɪɢ. A-1. Schematic diagram to illustrate how the average path length through the cell may be determined. A typical ray from the point $P(xy)$ in the aperture of the lens makes an angle $\theta$ to the axis and is refracted through the cell at angle $\alpha$. The path length of this ray through the cell is $b$.

$$\sin \alpha = \frac{\sin \theta}{\mu} \tag{A-2}$$

The quantity required in Section IV,A was $\sec \alpha = b/b_0$ which can be obtained from Eqs. (A-1) and (A-2) as

$$\left(\frac{b}{b_0}\right) = \frac{1}{\sqrt{1 - \dfrac{1}{\mu^2}\left\{\dfrac{x^2 + (y + h)^2}{x^2 + (y + h)^2 + z^2}\right\}}} \tag{A-3}$$

Now each point $P(xy)$ on the surface of the lens will produce a ray to the point $Q$ in the image plane, so to obtain an average value of $b/b_0$ for point $Q$ we should average the value given by Eq. A-3 for all points $x$ and $y$ in the surface of the lens. We should then repeat this process for each point $Q$ on the image. Thus

$$\frac{\bar{\bar{b}}}{b_0} = \frac{1}{N}\sum_h \sum_y \sum_x \left(\frac{b}{b_0}\right) \tag{A-4}$$

where $\bar{\bar{b}}/b_0$ is the average value of $b/b_0$ after averaging $N$ rays. This averaging process is readily accomplished with a computer to give $\bar{\bar{b}}/b_0 = 1.061$ for our geometric arrangement.

Although this expression gives the correct average path through the cell, it is not exactly the quantity we require. We require the *average path* which will give the same absorption of radiation as the collection of actual paths. From Eq. (18) the fraction of photons which are absorbed in passing through the cell at angle $\alpha$, with path length $b$, is

$$1 - \exp(-2.303\ Ab/b_0) \qquad\qquad \text{(A-5)}$$

where $A$ is the absorbance. We wish to obtain the average value of the path length for $N$ photons and equate this to the amount that would have been absorbed if all the rays had passed through the cell with some average path length $\bar{b}$, thus

$$1 - \exp[-2.303\ (A)\bar{b}/b_0] = \frac{1}{N}\sum_{}^{N}\{1 - \exp(-2.303\ Ab/b_0)\} \quad \text{(A-6)}$$

where as before the summation is over all the pencils of rays. In this case $\bar{b}/b_0$ is a function of the absorbance $A$. The results of such a calculation on our geometrical arrangement for a series of absorbances are given in Table A-1.

TABLE A-1

CALCULATION OF AVERAGE PATH LENGTH $\bar{\bar{b}}/b_0$, AND PATH LENGTH, $\bar{b}/b_0$, AVERAGED TO GIVE THE SAME ENERGY ABSORPTION FOR RADIATION FOCUSED BY THE APLANATIC LENS OF FIG. 5

| Absorbance (A) | Number of rays averaged (N) | $\bar{\bar{b}}/b_0$ [Eq. (A-4)] | $\bar{b}/b_0$ [Eq. (A-6)] |
|---|---|---|---|
| 0.1 | 600 | 1.061 | 1.061 |
| 0.2 | 600 | 1.061 | 1.061 |
| 0.4 | 600 | 1.061 | 1.060 |
| 0.8 | 600 | 1.061 | 1.060 |
| 1.6 | 600 | 1.061 | 1.058 |

Table A-1 shows that $\bar{b}/b_0$ is a very slow function of the absorbance $A$ and gives nearly the same answer as $\bar{\bar{b}}/b_0$, so that it does not really matter whether one calculates $\bar{\bar{b}}/b_0$ or $\bar{b}/b_0$. The average value is 1.06.

## Appendix B.  Detailed Information on MGL Calibration

### 1. *Arrangement of Calculations*

A convenient way to arrange the calculations of the calibration factor $K$ using MGL is illustrated in Table A-2. Column (1) gives the wavelength, column (2) the monitor readings immediately before and just after the irradiation. Column (3) gives the duration of the irradiation, and column (4) the number of photons expressed in the arbitrary units of monitor seconds. Column 4 was found by taking the product of the average monitor reading and $\Delta t$. Column (5) gives the absorbance of the solution at the irradiation wavelength before and after the exposure. Column (6) gives

TABLE A-2

CALIBRATION OF MONITOR USING MGL (DATA SHOWN FOR TWO WAVELENGTHS)[a]

| (1) λ (mμ) | (2) Monitor (M) ×10⁻¹ | (3) Δt (sec) | (4) Photons (M·sec) ×10⁻⁴ | (5) $A_\lambda$ | (6) $A_{622}$ | (7) $\Delta A_{622}$ | (8) $\bar{A}_\lambda$ | (9) $f_{abs}$ Eq. (18) | (10) $\Delta N_i$ Eq. (27),[c] ×10¹⁵ | (11) $K$ Eq. (28), ×10⁻¹¹ | (12) $(A_{a1})_\lambda$[b] | (13) $f_{a1}$ Eq. (A-7) | (14) $\bar{A}_{622}$ | (15) $\frac{(\epsilon_{pp})_\lambda}{(\epsilon_{pp})_{622}}$ (Fig. 7) | (16) $f_{pp}$,[d] Eq. (A-8) | (17) $(1 - f_{a1} - f_{pp})$ | (18) Corrected $K$ Eq. (A-10), photons/ M·sec ×10⁻¹¹ |
|---|---|---|---|---|---|---|---|---|---|---|---|---|---|---|---|---|---|
| 230 |  | 60 |  | 1.733 | 0.008 |  |  |  |  |  |  |  |  |  |  |  |  |
|  | 17.0 |  |  |  |  |  |  |  |  |  |  |  |  |  |  |  |  |
|  | 16.5 |  | 1.00 | 1.743 | 0.314 | 0.306 | 1.738 | 0.985 | 3.51 | 3.51 | 0.085 | 0.049 | 0.161 | 0.084 | 0.008 | 0.943 | 3.72 |
| 230 |  | 60 |  | 1.725 | 0.004 |  |  |  |  |  |  |  |  |  |  |  |  |
|  | 17.5 |  |  |  |  |  |  |  |  |  |  |  |  |  |  |  |  |
|  | 17.0 |  | 1.035 | 1.738 | 0.319 | 0.315 | 1.731 | 0.985 | 3.62 | 3.49 | 0.085 | 0.049 | 0.161 | 0.084 | 0.008 | 0.943 | 3.71 |
| 265 |  | 60 |  | 1.513 | 0.001 |  |  |  |  |  |  |  |  |  |  |  |  |
|  | 30.0 |  |  |  |  |  |  |  |  |  |  |  |  |  |  |  |  |
|  | 34.0 |  | 1.92 | 1.411 | 0.370 | 0.369 | 1.462 | 0.970 | 4.31 | 2.24 | 0.005 | 0.003 | 0.186 | 0.044 | 0.006 | 0.991 | 2.26 |
| 265 |  | 60 |  | 1.517 | 0.000 |  |  |  |  |  |  |  |  |  |  |  |  |
|  | 34.0 |  |  |  |  |  |  |  |  |  |  |  |  |  |  |  |  |
|  | 33.5 |  | 2.02 | 1.414 | 0.374 | 0.374 | 1.461 | 0.970 | 4.36 | 2.16 | 0.005 | 0.003 | 0.187 | 0.044 | 0.006 | 0.991 | 2.18 |
| 265 |  | 60 |  | 1.515 | 0.000 |  |  |  |  |  |  |  |  |  |  |  |  |
|  | 33.0 |  |  |  |  |  |  |  |  |  |  |  |  |  |  |  |  |
|  | 33.0 |  | 1.98 | 1.411 | 0.366 | 0.366 | 1.458 | 0.970 | 4.27 | 2.16 | 0.005 | 0.003 | 0.183 | 0.044 | 0.006 | 0.991 | 2.18 |

[a] Calculated for $V = 1.82 \times 10^{-3}$ liter; $\epsilon_{622} = 10.63 \times 10^4$, $\phi = 0.91$; $\bar{b} = 1.06$.

[b] Determined by measuring absorbance of alcohol (solution S-1) against a water blank.

[c] $\Delta N_i = \frac{6.02 \times 10^{23} \cdot V \cdot (\Delta A_{622})}{f_{abs} \cdot \phi \cdot (\epsilon_{622})}$ [Eq. (27)]. Columns (10) and (11) values are uncorrected for absorption of alcohol and photoproduct.

[d] $f_{pp} = \frac{\bar{A}(pp)_{622}}{(\bar{A}_\lambda)} \cdot \frac{(\epsilon_{pp})_\lambda}{(\epsilon_{pp})_{622}}$ [Eq. (A-8)].

similar absorbances at 622 mμ. Column (7) records the change in absorbance at 622 mμ, and column (8) the average absorbance at λ over the irradiation interval. In column (9) the fraction absorbed is calculated using Eq. (18), and $\bar{b}/b_0 = 1.06$. This fraction is close to 1.00. The number of photons incident on the solution $\Delta N_i$ calculated from Eq. (27) is given in column (10), and the calibration factor $K$ using Eq. (28) in column (11). For many purposes, the $K$ value given in column (11) would be accurate enough.

## 2. Corrections for Photons Absorbed by the Alcohol and Photoproduct

The absorbance $A_\lambda$ of the solution results from the absorbance of dye, alcohol (al) and photoproduct (pp) so that it could be written as

$$A_\lambda = A_{dye} + A_{al} + A_{pp}$$

Dividing through by $A_\lambda$ and rearranging

$$f_{dye} = 1 - f_{al} - f_{pp} \text{ where } f_{dye} = \frac{A_{dye}}{\bar{A}_\lambda}, f_{al} = \frac{A_{al}}{\bar{A}_\lambda} \text{ and } f_{pp} = \frac{\bar{A}_{pp}}{\bar{A}_\lambda} \quad (A-7)$$

The $f$'s stand for the fraction of the absorbed photons which are absorbed by dye, alcohol, and photoproduct. Since these fractions are small and change only slightly with exposure we have replaced the absorbances by their average values during irradiation. The correction for alcohol absorption can be obtained from measured data similar to that given in Section IV,B and C, 3 (step 2). We have no direct way to measure $\bar{A}_{pp}$, the absorbance of the photoproduct at λ, but we can measure its absorbance at 622 mμ, since at this wavelength, only the photoproduct absorbs. Using the absorption spectrum of Fig. 7, we then calculate its absorption at λ to give

$$f_{pp} = \frac{(\bar{A}_{pp})_\lambda}{(\bar{A})_\lambda} = \frac{(\bar{A}_{pp})_{622}}{(\bar{A}_\lambda)} \cdot \frac{(\epsilon_{pp})_\lambda}{(\epsilon_{pp})_{622}} \quad (A-8)$$

All the quantities in Eq. (A-8) are measurable. The quantity $(\bar{A}_{pp})_{622} = \bar{A}_{622}$ is the average absorbance of the solution at 622 during the irradiation and the ratios of the molar absorptivity for the photoproduct can be obtained from Fig. 7. Combining (A-7) and (A-8) we have the following expression for the number of photons absorbed by the dye

$$(\Delta N_a)_{dye} = \Delta N_i \cdot f_{abs} \cdot (1 - f_{al} - f_{pp}) \quad (A-9)$$

The calibration factor corrected for these effects is then

$$K(\lambda) = \frac{6.02 \times 10^{23} \cdot V \cdot \Delta A_{622}}{f_{abs}(1 - f_{al} - f_{pp}) \cdot \phi \cdot \epsilon_{622}} \cdot \frac{1}{\bar{M} \cdot \Delta l} \quad (A-10)$$

The way these corrections may be made are illustrated in the right-hand part of Table A-2. Column (12) gives the absorbance of the acidified

TABLE A-3

CORRECTION FOR THERMAL REVERSION DURING IRRADIATION OF C3'p[a]

| (1) Time ($t$) (min) | (2) Absorbance ($A_{272}$) | (3) $\Delta \bar{L}$ ($\mu E/cm^2$) | (4) $\dfrac{A_0 - A}{A}$ | (5) $\dfrac{A_0 - A}{A}$ | (6) $\Delta t$ (min) | (7) $B^c$ | (8) $\Sigma B$ | (9) $\log_{10} A$ | (10) $\log A - \Sigma B$ | (11) $\bar{L} = \Sigma \Delta \bar{L}$ |
|---|---|---|---|---|---|---|---|---|---|---|
| 0 | 2.060[b] | | 0.000 | | | | 0.000 | 0.314 | 0.314 | 0.00 |
| | | 0.642 | | 0.061 | 2 | 0.001 | | | | |
| 2 | 1.838 | | 0.121 | | | | 0.001 | 0.264 | 0.263 | 0.64 |
| | | 0.705 | | 0.189 | 3 | 0.004 | | | | |
| 5 | 1.637 | | 0.258 | | | | 0.005 | 0.214 | 0.209 | 1.35 |
| | | 0.771 | | 0.338 | 3 | 0.007 | | | | |
| 8 | 1.453 | | 0.418 | | | | 0.012 | 0.162 | 0.150 | 2.12 |
| | | 0.847 | | 0.510 | 3 | 0.012 | | | | |
| 11 | 1.287 | | 0.601 | | | | 0.024 | 0.110 | 0.086 | 2.96 |
| | | 0.929 | | 0.707 | 3 | 0.016 | | | | |
| 14 | 1.137 | | 0.813 | | | | 0.040 | 0.056 | 0.016 | 3.89 |
| | | 1.022 | | 0.932 | 3 | 0.021 | | | | |
| 17 | 1.006 | | 1.050 | | | | 0.061 | 0.003 | -0.058 | 4.92 |
| | | 1.119 | | 1.189 | 3 | 0.026 | | | | |
| 20 | 0.885 | | 1.328 | | | | 0.087 | -0.053 | -0.140 | 6.03 |
| | | 1.206 | | 1.473 | 3 | 0.033 | | | | |
| 23 | 0.787 | | 1.618 | | | | 0.120 | -0.104 | -0.224 | 7.24 |
| | | 1.290 | | 1.786 | 3 | 0.039 | | | | |
| 26 | 0.698 | | 1.953 | | | | 0.159 | -0.156 | -0.315 | 8.53 |
| | | 1.365 | | 2.127 | 3 | 0.047 | | | | |
| 29 | 0.625 | | 2.300 | | | | 0.206 | -0.204 | -0.410 | 9.90 |
| | | 1.441 | | 2.460 | 3 | 0.055 | | | | |
| 32 | 0.569 | | 2.620 | | | | 0.261 | -0.245 | -0.506 | 11.34 |

[a] pH = 5.0; McIlvaine's buffer, 0.01 $M$; temperature, 20°C; $k_r{}^t = 1.7 \times 10^{-2}$ min$^{-1}$.

[b] $A_0$ = initial absorbance at 272 m$\mu$.

[c] $B = \dfrac{1}{2.303} \left( \dfrac{A_0 - A}{A} \right) \cdot k_r{}^t \cdot \Delta t; \qquad k_r{}^t = 1.7 \times 10^{-2}$ min$^{-1}$.

alcohol solution S-1 measured against a water blank. Column (13) gives the fraction of the absorbed photons that are absorbed by the alcohol and was calculated using Eq. (A-7). Column (14) gives the average absorbance measured at 622 mμ during the irradiation. Column (15) gives the ratio of the absorptivities of the photoproduct at λ and at 622 mμ taken from Fig. 7. Column (16) gives the fraction of the absorbed photons absorbed by the photoproduct, and column (17) the fraction absorbed by the dye. The corrected calibration factor is given in column (18).

### Appendix C. Corrections for Thermal Reversal of Cp Hydrate during Irradiation

The data required for the graphs of Fig. 12 are included in Table A-3, which shows the measured absorbances [column (2)] as a function of time [column (1)] and exposure [column (11)]. This table is arranged to enable us to evaluate the left-hand side of Eq. (50). The first column gives the time $t$ at which the absorbance $A$ given in the second column was measured. The third column gives the increment in exposure produced between absorbance measurements and was calculated by the method illustrated in Table II. The fourth column gives $(A_0 - A)/A$, and the fifth column the average value of this ratio during the interval of time $\Delta t$ given in column (6). The seventh column gives B, the product of columns (5), (6), and $k_r{}^t/2.303$ where $k_r{}^t$ is the rate constant for reversal. Thus, column (7) gives the value of the second term in Equation (50). This correction term is initially very small but increases with exposure as more hydrate is formed. Column (9) is log $A$ and column (10) is log $A$ less the correction term and gives the left-hand side of Eq. (50). According to Eq. (50), a plot of column (10) against (11) should give a straight line, as it does (see Fig. 12).

### [9] Measurement of Fast Reactions by Single and Repetitive Excitation with Pulses of Electromagnetic Radiation

*By* H. Rüppel and H. T. Witt

# I. Introduction

Fast *photochemical* reactions were investigated first by Norrish and Porter in 1949[1] by means of excitation with single flashes of light. In 1959 Czerlinski and Eigen succeeded in studying fast *chemical* reactions by means of excitation with rapid temperature jumps.[2] By use of flashes of light or temperature jumps with high energies the reaction signals normally have a magnitude of several times the statistical noise level in the measuring device.

*Sensitivity.* Numerous interesting reactions with reaction times in the range between $10^{-1}$ and $10^{-6}$ second are characterized by such a small turnover that they are far beyond the limit of measurement. To enable a measurement of these reactions, either the excitation energy or the sensitivity of the measuring device is to be increased considerably. As for high energies the possibilities for a further increase have been nearly exhausted.[2,3] On the other hand, high excitation energy may cause irreversible damage, especially in sensitive biochemical systems. Therefore, the main interest is in the experimental possibilities of increasing the sensitivity of the measuring devices, that is, by about one or two orders of magnitude compared with the earlier equipment.

*Time Resolution.* For most of the primary processes it is desirable to measure reaction courses with times shorter than $10^{-6}$ second—in fact, down to the order of $10^{-8}$ to $10^{-9}$ second. This time range corresponds to the lifetimes of excited singlet electronic states. To realize a measurement of such fast reactions, two essential points have to be considered: first, the duration of the excitation pulse must be kept extremely short; and, second, the sensitivity must be increased because the level of the statistical noise grows with the time resolution (see Fig. 1).

[1] R. G. Norrish and G. Porter, *Nature* **164**, 658 (1949).
[2] G. Czerlinski and M. Eigen, *Z. Elektrochem.* **63**, 321 (1959).
[3] S. Claesson and C. Lindquist, *Arkiv Kemi* **11**, 535 (1957).

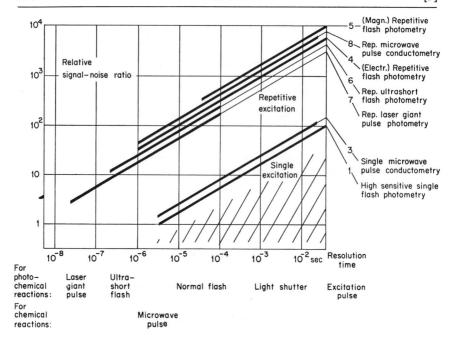

Fig. 1. Relative signal-to-noise ratio as a function of resolution time. For each type of equipment the approximate position in the relative signal-to-noise ratio range is depicted by a solid line. The hatched area indicates the range of earlier equipment. For this schematic diagram it is assumed that in each case the $S/N$ ratio grows strictly with the square root of the resolution time, which presupposes statistical noise as limit for measurement [see Eq. (12)]. Equipment employing repetitive excitation gives a 30–100 times higher $S/N$ ratio than that using single excitation. Flash conductometric method (equipment type 2) gives no direct time resolution and is therefore not depicted.

*Pulse Duration.* The time resolution is primarily given by the duration of the excitation pulse. For the excitation of the reactions the following sources are used (see Table I). For the excitation of *photochemical* reactions light shutters, flash tubes, and giant-pulse lasers are used. For the excitation

TABLE I
SOURCES FOR THE EXCITATION OF REACTIONS

| Source for excitation | Pulse duration, range (sec) |
|---|---|
| Photochemical reaction | |
| Light shutters | $10^{-1}$ to $10^{-3}$ |
| Flash tubes | $10^{-3}$ to $10^{-6}$ |
| Ultrashort flash tubes | $10^{-7}$ |
| Q-switched laser (giant pulse) | $10^{-8}$ |
| Chemical reaction | |
| Microwave pulse generator | $10^{-5}$ to $10^{-6}$ |

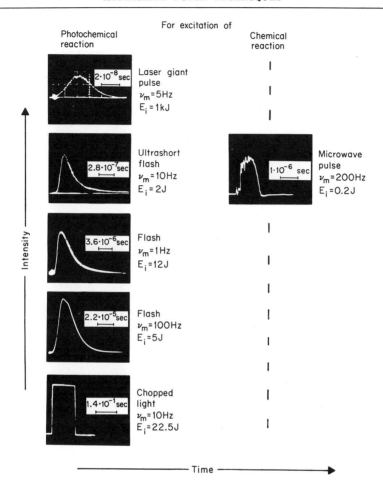

FIG. 2. Intensity time curves of different excitation pulses which are used with the equipment types 1–8 presented in Section II. $E_i$ = electric input energy, $\nu_m$ = maximum repetition rate. The time scales in the partial picture give the half-width time of excitation pulses. (See footnote 4, p. 320.)

of *chemical* reactions also, pulses of electromagnetic radiation are applied. The time courses of these different excitation pulses are shown in Fig. 2.

*Pulse Sequence.* Originally, fast chemical and photochemical reactions were initiated by single pulses only (equipment types 1–3). The sensitivity of measurements, however, can be increased considerably by repetitive excitations of the reactions. The signal-to-noise ratio can be increased by appropriate evaluation of the repetitive signals. The increase at $n$ repetitions is proportional to the square root of the number of excitations: $\sqrt{n}$.

The repetitive excitation technique—developed for the investigation of chemical reactions since 1962—is used in equipment types 4–8.

*Registration Technique.* As usual, the reaction courses are measured optically by absorption changes or electrically by conductivity changes.

In the case of *single* excitation the measuring sensitivity depends critically on the optimum adjustment of the different measuring parameters of the equipment. Optimum conditions are more nearly realized with equipment types 1–3 (see Fig. 1 and Sections II and III).

In the case of *repetitive* excitation a large number of identical but noisy signals are superposed in a special signal storage device to find out the average signal course with enhanced accuracy. For a time resolution not less than some $10^{-5}$ second, in most applications, the *magnetic* averager has best efficiency. For faster reaction signals ($10^{-5}$ to $10^{-6}$ second), the *electric* averager must be used. For extreme high time resolution ($10^{-7}$ to $10^{-8}$ second), sampling and averaging can be done with the aid of slightly modified *sampling oscilloscopes*. This is accomplished with equipment types 4–8 (see Fig. 1 and Sections II and III).

In Section II, eight types of equipment are briefly discussed by which a resolution time down to $10^{-8}$ second and a sensitivity increase up to one hundred times can be achieved. For each type of equipment, the measuring principle, specification details of performance, as well as an example of special application are given. Technical details are given in Section III.

A short survey of this equipment has already been published.[4] Numerous applications on biological reaction systems have already been reported.[5]

## II. Survey of Equipment

### A. Type 0. Flash Manometry

With this special photochemical measurement device, flash technique was used first in photochemistry to measure reaction times indirectly.[6] This technique is still indispensable if the intermediates of a photochemical reaction are not accompanied by direct measurable changes of physical properties (e.g., absorbancy, conductivity).

When introduced in 1932 by Emerson and Arnold[6] this method was capable of a resolution time of $10^{-2}$ second. The yield of the end product per flash is measured as a function of the dark time $t_d$ between the excitation

---

[4] H. T. Witt, *in* "Fast Reactions and Primary Processes in Chemical Kinetics (Nobel Symposium V)" (S. Claesson, ed.), p. 81. Almqvist & Wiksell, Stockholm; Wiley (Interscience), New York, 1967.

[5] (a) H. T. Witt, see footnote 4, p. 261; (b) H. T. Witt, B. Rumberg, and W. Junge, *Mosbach Colloq. 19* p. 262. Springer-Verlag, Berlin, 1968.

[6] R. Emerson and W. Arnold, *J. Gen. Physiol.* **15**, 391 (1932).

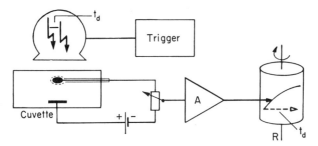

FIG. 3. Principle diagram: Flash manometric equipment. A Clark electrode is used for measurement of oxygen yield per flash (see Fig. 4). $A$, High-gain low-pass direct current amplifier; $R$, recorder; $t_d$, dark time between flashes.

flashes. In special cases kinetic data on intermediates can be evaluated from this dependency. Emerson measured the oxygen production in photosynthesis as a function of the dark time $t_d$. In this way, the rate limiting step of the overall reaction in photosynthesis was determined.

This method can be applied for different kinds of end products: Oxygen can be measured as a function of $t_d$ by means of manometers[6] or electrodes[7] (Fig. 3); protons by glass-electrodes; ATP, by tracer techniques,[8] etc. For this method the flashes initiating the photochemical reaction should have saturation intensity. Therefore, it is the magnitude of the power

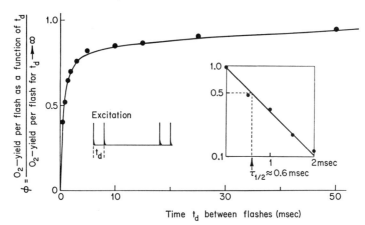

Time $t_d$ between flashes (msec)

FIG. 4. Example diagram: Relative oxygen yield per flash of spinach chloroplasts versus dark time $t_d$ between flashes. The excitations of the photosynthetic reaction was performed by repeated double-flash groups. Partial picture: relative oxygen yield per flash $\phi_r$ of rapid reaction only, $\log(1 - \phi_r)$ versus time.[9]

[7] K. Damaschke and L. Rothbühr, *Biochem. Z.* **327,** 39 (1955).

[8] M. Nishimura, *Biochim. Biophys. Acta* **57,** 88 (1962); H. Schröder, Thesis, Technical University of Berlin, 1969.

[9] J. Vater, G. Renger, H. H. Stiehl, and H. T. Witt, *Naturwissenschaften* **5,** 220 (1968).

supply which restricts the minimum dark time $t_d$; this in turn is responsible for the time resolution. By use of a cascade of flashes,[10] the time resolution is increased to $10^{-4}$ second in comparison with the original equipment of Emerson[6] (see principle diagram, Fig. 3).

The time resolution of $10^{-4}$ second was sufficient to determine the velocity of the photolysis of water *in vivo*.[9] This process is not accompanied by a measurable change of physical properties. The rate-limiting step of this reaction has a half-life $\tau_{1/2} = 6 \times 10^{-4}$ second at 20° (see Fig. 4).

### B. Type 1. High-Sensitive Single-Flash Photometry

*Method*

The principle of flash photometry was introduced by Norrish and Porter[1] (see Fig. 5). Monochromatic monitoring light irradiates a cuvette containing the reactive system and falls onto a photomultiplier ($PM$). The absorption changes which are caused by the reaction initiated by the flash ($Fl$) are transformed by the photomultiplier into corresponding voltage changes. These are amplified ($A$) and displayed on a cathode ray oscilloscope ($CRO$) as a function of time.

*Specification*

If the conditions of measurements (e.g., monitoring light intensity, sensitivity of photomultiplier) are elaborately adjusted for maximum

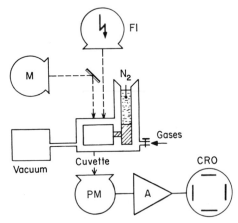

FIG. 5. Principle diagram: High-sensitive flash photometric equipment. $M$, Monochromatic monitoring light; $Fl$, excitation flash; $PM$, photomultiplier; $A$, amplifier; $CRO$, cathode ray oscilloscope; $N_2$, liquid nitrogen. (See footnote 11.)

[10] J. Vater, Thesis, Technical University of Berlin, 1969.

signal-to-noise ratio, the sensitivity can be increased up to more than 10 times in comparison with that of the earlier equipment.[11] A simplified diagram of the principle of the equipment is depicted in Fig. 5. The device permits measurements to be conducted also *in vacuo*, in different types of gases, and in the temperature range between $-180°$ and $+80°$.

TABLE II

LIMITING DATA FOR ABSORPTION CHANGES VERSUS RESOLUTION TIME

| Resolution time (sec) | Minimum absorption change $\Delta\alpha_{\min}{}^a$ |
|---|---|
| $10^{-6}$ | $3 \times 10^{-3}$ |
| $10^{-5}$ | $1 \times 10^{-3}$ |
| $10^{-4}$ | $3 \times 10^{-4}$ |
| $10^{-3}$ | $1 \times 10^{-4}$ |
| $10^{-2}$ | $3 \times 10^{-5}$ |

$^a$ See footnote 12.

Table II gives the variation of the realized limiting value $\Delta\alpha_{\min}$[12] for absorption changes versus the resolution time range. According to the relationship between noise level and time resolution (see Fig. 1), the sensitivity rises with slower processes.

*Example*

By means of this refined equipment, numerous different types of very small absorption changes of 0.1% have been detected which occur *in vivo* during photosynthesis in plants within the reaction time range between $10^{-1}$ and $10^{-6}$ second.[5] For this article, however, it seems appropriate to give an example of a more simple photochemical reaction.

Figure 6 shows the results obtained by measurement of a photochemical reaction between chlorophyll *a* and quinone in butanol.[13] The absorption changes show two relaxation times. The fast absorption change results in the dotted difference spectrum, the slow one in the difference spectrum represented by a solid line. The rapid change can be explained by the formation of the triplet state of chlorophyll *a* (Chl-*a*\*\*), the slower one by the formation of ionized chlorophyll *a* (Chl-*a*$^+$) and semiquinone (Q$^-$). A quantitative analysis results in the following reaction cycle:

[11] H. T. Witt, *Naturwissenschaften* **42**, 72 (1955); *Z. Elektrochem.* **59**, 981 (1955); H. T. Witt, R. Moraw, and A. Müller, *Z. Physik. Chem. (Frankfurt)* **20**, 193 (1959).

[12] $\Delta\alpha = -\ln(1 - \Delta I/I) \approx \Delta I/I$ with an error $<6\%$ ($<1\%$) if $\Delta I/I \leqslant 1/10$ ($\leqslant 1/100$); $I$ = intensity of monitoring light.

[13a] K. Seifert and H. T. Witt, *Naturwissenschaften* **5**, 222 (1968); [b] K. Seifert, Thesis, Technical University of Berlin, 1968.

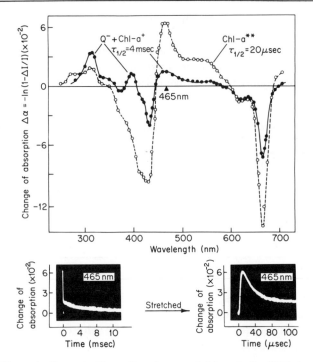

FIG. 6. Example diagram: Absorption changes of chlorophyll $a$ (*Chl-a*) and $p$-benzo-quinone $(Q)$ in butanol at $20°$ after single-flash excitation as a function of time versus wavelength. *Chl-a**\**, excited triplet state of chlorophyll $a$; *Chl-a+*, oxidized chlorophyll $a$; $Q$, $p$-benzosemiquinone. (See footnote 13.)

The excited chlorophyll $a$ reacts in its triplet state (Chl-$a$**) with quinone $(Q)$ to form $Q^-$ and Chl-$a^+$. Thereafter, a fast back reaction takes place. The formation of the semiquinone $Q^-$ can be recognized in the difference spectrum, *inter alia*, by the semiquinone bands in the 300 and 400 nm wavelength region (see Fig. 6).

## C. Type 2. Single-Flash Conductometry

*Method*

For the direct detection of electrically charged primary products (ions, electrons) of photochemical reactions, the method of flash conductometry was developed[14] (see Fig. 7). The conductivity cell for the photochemical active system is furnished with two electrodes, to which a voltage of some thousand volts (HT) is applied. As solvents high-insulating substances like $n$-hexane are used. The ions produced by flash excitation move during their

[14] H. Rüppel and H. T. Witt, *Z. Physik. Chem. (Frankfurt)* **15,** 321 (1958).

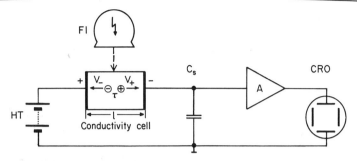

FIG. 7. Principle diagram: Flash conductometric equipment. *Fl*, Excitation flash; *τ*, ion life; $v_+$, $v_-$, ion velocities; *l*, length of conductivity cell; $C_s$, small charging capacitor; *A*, narrow-band pulse amplifier; *CRO*, cathode ray oscilloscope; *HT*, high tension. (See footnote 14.)

lifetime, $\tau$, the distance $s$ in the electric field $F$ and cause a current pulse $\Delta Q = (s/l)\Delta q$, which is gathered on the small input capacitance $C_s$ of the pulse amplifier $A$ ($\Delta q$ is the total amount of charges produced by the flash, and $l$ the length of the conductivity cell). The pulse amplifier forms the signal which is displayed on the oscilloscope, *CRO*. The maximum of this signal is proportional to the current pulse $\Delta Q$. As the charge separation is $s = (u^+ + u^-)F \cdot \tau$ ($u^+, u^-$ = mobilities of the ions) the current pulse $\Delta Q$ is proportional to the applied field strength $F$ (see Fig. 8).

With this measurement device, only the product (current times life time) (i.e., the current pulse $\Delta Q$) is recorded which is proportional to

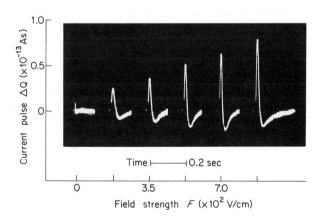

FIG. 8. Example diagram: Current pulses $\Delta Q$ versus field strength $F$. The current pulses were produced by flash photoionization of triphenylamine in *n*-hexane solution at 20°. Photoionization occurs in *n*-hexane solution by flash excitation with wavelength $\lesssim 320$ nm. The maxima of signal-time curves are a measure for the current pulse $\Delta Q$. (See footnote 14.)

$\Delta q \cdot \tau$. The high sensitivity (see below) is caused by the small transmission band of the pulse amplifier. If the time course of the ionization process $(\Delta q(t))$ is measured directly, this is accompanied by a decrease in sensitivity according to the necessary time resolution (see Section I).

## Specifications

The limiting sensitivity is given by a value of $\Delta Q_{\min} \approx 8 \times 10^{-16} A$ sec, which is just within the limit of measurement. From this amount the lower limit for the product $(\Delta c \cdot \tau) \approx 10^{-18} M$ sec is derived[14] ($\Delta c$ = transient ion concentration). Therefore, ions can be detected, e.g., with the extreme values of $\tau$ and $\Delta c$, which are given in Table III.

TABLE III

EXTREME VALUES OF ION LIFETIME AND CONCENTRATION

| Ion life time, $\tau$ (sec) | Ion concentration, $\Delta c$ (M) |
| --- | --- |
| $10^{-1}$ | $10^{-17}$ |
| $10^{-8}$ | $10^{-10}$ |
| $10^{-12}$ | $10^{-6}$ |

## Example

The reversible photoionization of aromatic amines in liquids at room temperature is followed by a very fast recombination reaction.

This process was investigated, for example, with the method described above. The current impulses $\Delta Q$ obtained from the photoionization of triphenylamine are depicted in Fig. 8 as a function of the field strength $F$. One remarkable result is the fact that ionization occurs at $\lambda \lesssim 320$ nm, i.e., with an ionization energy of only 4 eV,[15] whereas ionization is proved in the gaseous phase not below 7–10 eV.

## D. Type 3. Single Microwave-Pulse Conductometry

### Method

A chemical reaction can be initiated in an equilibrium or stationary state system by means of a rapid temperature jump. Eigen produced the temperature jumps by the ohmic heating of a condenser discharge through a solution with strong electrolytes.[2] Pulse heating with normal electronic flashes (resolution time $10^{-6}$ second) with aid of special light absorbers in the solution has already been realized in several laboratories (e.g., see

[15] U. Krog, H. Rüppel, and H. T. Witt, *J. Electrochem. Soc.* **107**, 966 (1960).

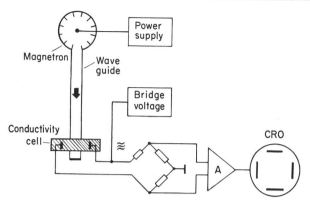

Fig. 9. Principle diagram: Single microwave pulse conductometric equipment. The microwave pulse is generated by a magnetron (9.4 GHz) and absorbed within a conductivity cell which is placed in through a rectangular waveguide. The conductivity cell is part of a radio frequency (100–300 kHz) measuring bridge circuit. A, Selective amplifier; CRO, cathode ray oscilloscope. (See footnote 18.)

footnotes 16 and 17). Ertl and Gerischer obtained a temperature jump by a microwave pulse that is absorbed without any additives, in polar solvents particularly in water, which has a broad absorption band near 10 GHz.[18]

In the equipment originally developed by Ertl and Gerischer, a pulse of microwaves with a frequency of 9.4 GHz is produced by a magnetron. This pulse is transmitted in a rectangular waveguide to an absorber tube within a conductivity cell which is a part of a high-frequency measuring bridge. The conductivity change initiated by the equilibration reaction is indicated as a bridge output signal, amplified (A) and displayed on an oscilloscope (CRO) (see Fig. 9).

*Specification*

The duration of the microwave pulses can be adjusted from 1 to 6 $\mu$sec. The temperature jumps in the conductivity cell have values of 0.2–0.8° per pulse. As bridge input voltage either a high frequency (100 kHz–1 MHz)[19,20] or a symmetrical rectangular pulse voltage with a microsecond rise time is used.[18] The sensitivity of the equipment was found to be $\Delta\kappa/\kappa \approx 3 \times 10^{-4}$ at a time resolution of $\approx 3 \times 10^{-5}$ second ($\approx 10$ kHz bandwidth).

[16] L. S. Nelson and J. L. Lundberg, *Nature* **179**, 367 (1957); *J. Phys. Chem.* **63**, 433 (1959).

[17] H. Strehlow and S. Kalarickal, *Z. Elektrochem.* **70**, 139 (1966).

[18] G. Ertl and H. Gerischer, *Ber. Bunsenges. Physik. Chem.* **65**, 629 (1961).

[19] P. Brumm, F. P. Kilian, and H. Rüppel, *Z. Naturforsch.* **20b**, 915 (1965).

[20] H. Hoffmann and K. Pauli, *Ber. Bunsenges. Physik. Chem.* **70**, 1052 (1966).

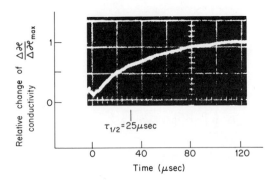

F<small>IG</small>. 10. Example diagram: Relative change of conductivity versus time. The conductivity change is caused in highly purified water by a rapid temperature jump of 0.7° which is produced by a microwave pulse (approximately 200 kW) with a duration of 6 $\mu$sec. The rise of conductivity with half-time of 25 $\mu$sec is originated by the shift to the new dissociation equilibrium of pure water. The portion of conductivity change which is caused by the considerable change of the ion mobility is compensated. (See footnote 18.)

### Example

Figure 10 shows the result of a measurement by Ertl and Gerischer, who heated extremely pure water with a temperature jump of 0.7° within 6 $\mu$sec.[18] The relaxation signal with a half-life $\tau_{1/2} = 25$ $\mu$sec is due to the neutralization reaction. The association rate constant $k_a$ can be calculated directly from the value of $\tau_{1/2}$: $k_a = 1.3 \times 10^{11}$ l(mole sec)$^{-1}$. This is in agreement with the earlier measurement by Eigen and De Maeyer using the field-jump method.[21]

### E. Type 4. Repetitive Flash Photometry: Electric Averager

### Method

In 1962 the repetitive averaging method was initially applied to photochemical reactions by excitation with repetitive flashes of light.[22,23] An electric averager type with one-address scanning was developed.[24,25] The performance of this time sampling and averaging system is illustrated

[21] M. Eigen and L. De Maeyer, Z. Elektrochem. **59,** 986 (1955).
[22] H. Rüppel and H. T. Witt, Presented at Symp. Rapid Reactions in Solution, Madison, Wisconsin, 1961 (unpublished).
[23] H. Rüppel, V. Bültemann, and H. T. Witt, Ber. Bunsenges. Physik. Chem. **66,** 760 (1962).
[24] H. Rüppel, Thesis, University of Marburg, Germany, 1962.
[25] H. Rüppel, V. Bültemann, and H. T. Witt, Ber. Bunsenges. Physik. Chem. **68,** 340 (1964).

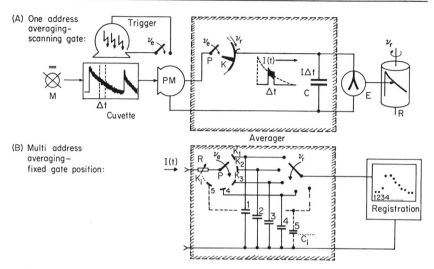

Fig. 11. Principle diagram: Repetitive flash photometric method. (A) One-address averaging, symbolized by rotating contactor with slowly moving contact point $K$. $M$, Monochromatic monitoring light; $PM$, photomultiplier; $E$, electrometer; $R$, recorder; $I(t)$, photo current; $\nu_e$, repetition rate of excitation trigger; $\nu_r$, revolution rate of recorder drum; $C$, integrating capacitance; $\Delta t$, switching time; $P$, $K$, contact points. (B) Multi-address averaging, symbolized by a rotating multipole contactor. A series of $m$ averaging units (symbolized by time constants $RC_i$) are connected stepwise with the input $I(t)$ by a rotating switch ($\nu_e$) with $m$ contact points $K_i$. Time sampling at $m$ fixed spots of the time axis (addresses) takes place during every excitation cycle. The averaged address values are "read" by a slow separate switching operation ($\nu_r$) and are indicated on an oscilloscope or recorder.

in Fig. 11A. The one-address time sampling is symbolized by a rotating contactor. (This principle was used as early as 1849 by Lenz[26] in order to increase the time resolution for electrical measurements.)

Let us assume that the recorder drum $R$ and, combined with it, the outer contact point $K$ of the contactor is set to a fixed-phase position. The trigger may stimulate $n$ excitations in the cuvette with the frequency $\nu_e$. The following $n$ reactions cause $n$ absorption changes which are transformed by the photomultiplier into $n$ current changes $I(t)$. Out of every signal $I(t)$ only one signal amplitude $I_1 = I(t_1)$ is recorded at a definite time $t_1$ which is fixed by the position of the contact point $K$. The contactor switches the photomultiplier $PM$ at the time $t_1$ for the time interval $\Delta t$ to the charging condenser where the current impulse $I_1 \Delta t = s_1$ is stored in the condenser $C$ and indicated on an electrometer $E$. The sum of $n$ current impulses at the reaction time $t_1$ results in the total signal amplitude $S_1 = ns_1$ at the time $t_1$.

[26] Reported by F. A. Laws, "Electrical Measurement," 2nd ed., p. 641. McGraw-Hill, New York, 1938.

The statistical noise level $N$, however, increases at $n$ repetitions only with $\sqrt{n}$. Therefore, a gain $\sqrt{n}$ in the signal-to-noise ratio is obtained for the total signal amplitude $S_1$ at $t_1$. The improved signal amplitude is recorded on the recorder $R$. Likewise, the total signal amplitude $S_2$ at the reaction time $t_2$ is evaluated at a corresponding phase position of the contactor point $K$. In this way, by a slow rotation of the outer contact $K$ with $\nu_r \ll \nu_e$, the entire time course of the absorption change is recorded with improved signal-to-noise ratio.

## Specification

Figure 12 demonstrates the efficiency of the first equipment. A signal (information) is mixed with noise of a ten times higher mean noise level. Therefore, the signal-to-noise ratio $(S/N)$ at the input of the averager (AV) is $S/N \approx 1:20$. The information is completely lost in noise. At the output, however, the signal is recorded with good accuracy: $S/N \approx 5:1$. This corresponds to an increase in sensitivity by a factor of about 100.

The time resolution of this technique is in principle limited only by the flash duration. In the special device originally developed in 1962[24,25] there was an additional limit by the width of the gating pulse (switching time) of approximately 1 $\mu$sec. During each excitation cycle, one signal and a corresponding zero-reference sample is gathered. The gain in $S/N$ ratio

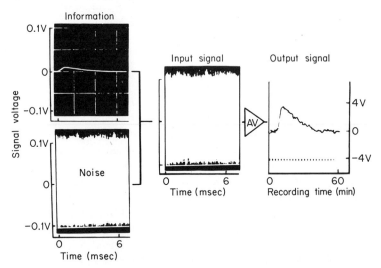

Fig. 12. Operation test diagram: Illustration of the gain in signal-to-noise ratio obtained with an electric one-address averager (equipment type 4). A small test signal (information) is optically mixed with white noise (left). The resulting $S/N$ ratio at the averager input is approximately 1:20. The $S/N$ ratio at the output after 126,000 single signals is approximately 5:1, gain in $S/N$ ratio $\approx 100$. Averager: impulse strobograph. (See footnotes 24, 25.)

improvement is in principle unlimited and is proportional to $\sqrt{n}$ ($n = \nu_e \cdot T_r$, $T_r$ = recording time).

### F. Type 5. Repetitive Flash Photometry: Magnetic Averager

*Method*

More than one time sample can be withdrawn from each signal if the scanning sampling gate (symbolized in Fig. 11A by a slowly rotating outer contact point $K$) is replaced by a series of gates at fixed phase positions (symbolized in Fig. 11B by contact points $K_i$ which are circularly arranged around the contactor $P$).

In 1964, two years after the first use of an electric one-address averager for fast photochemical reactions, this multiaddress sampling was realized by commercial instruments which use magnetic core memories for averaging. From each signal 100–1000 signal amplitude values are sampled at fixed phase positions (addresses), digitized, and registered in the magnetic core memory. Such digital magnetic averagers have been used since 1964 for the study of photochemical and chemical reactions with reaction times $\gtrsim 50$ $\mu$sec.[27–29]

A multiaddress electric averager using RC elements for analogous averaging according to Fig. 11B was realized in 1966 (see Section III).

*Specifications*

The data processing of the magnetic averager restricts the time resolution to about 50 $\mu$sec. The digital resolution of the magnetic averager defines the maximum gain $G$ in $S/N$ ratio improvement ($G_{max} \approx 50$–$100$ depending on the input $S/N$ ratio).

In the upper part of Fig. 13 the $S/N$ ratio amounts to $1:5$. The signal is lost in noise, after 2048 signals, however, the ratio is increased to $S/N = 12$, which corresponds to an enhancement of sensitivity by a factor of 60.

With repetitive equipment for flash photometric measurements in general (magnetic *and* electric averager), the approximate limiting data for absorption measurements given in Table IV can be achieved if between 2000 and 10,000 single signals are evaluated.

*Example*

The repetitive method has been applied mainly in photosynthesis, but also in other fields, e.g., for the investigations of fast chemical reactions

[27] H. Rüppel, V. Bültemann, and H. T. Witt, *Ber. Bunsenges. Physik. Chem.* **68,** 752 (1964).

[28] G. Döring, H. H. Stiehl, and H. T. Witt, *Z. Naturforsch.* **22b,** 639 (1967).

[29] H. H. Stiehl, Thesis, Technical University of Berlin, 1969.

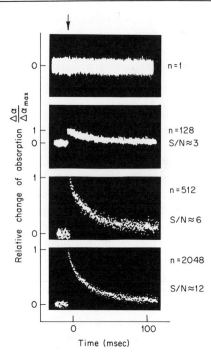

FIG. 13. Operation test diagram: Illustration of the gain in signal-to-noise ratio obtained with a magnetic multiaddress averager (equipment type 5). A small signal (absorption change) is buried in noise ($n = 1$, single signal, $S/N$ ratio at the averager input $\approx 1:5$). The $S/N$ ratio obtained at the averager output is $\approx 12$, gain in $S/N$ ratio $\approx 60$. Averager: Enhancetron, Nuclear Data. (See footnote 4.)

in monolayers. Various monolayers can be systematically built up on glasses.[30,31] Fast chemical reaction had previously not been detected in monolayers. Owing to its high sensitivity, however, such measurements are possible by the repetitive measuring technique. Layers of chlorophyll $a$

TABLE IV

LIMITING DATA FOR ABSORPTION CHANGES VERSUS RESOLUTION TIME

| Resolution time (sec) | Minimum absorption change, $\Delta\alpha_{min}{}^{a}$ |
| --- | --- |
| $10^{-6}$ | $10^{-4}$ |
| $10^{-4}$ | $10^{-5}$ |
| $10^{-2}$ | $10^{-6}$ |

$^a$ See footnote 12.

[30] K. B. Blodgett, J. Am. Chem. Soc. 57, 1007 (1935).

[31] K. H. Drexhage, M. M. Zwick, and H. Kuhn, Ber. Bunsenges. Physik. Chem. 67, 62 (1963).

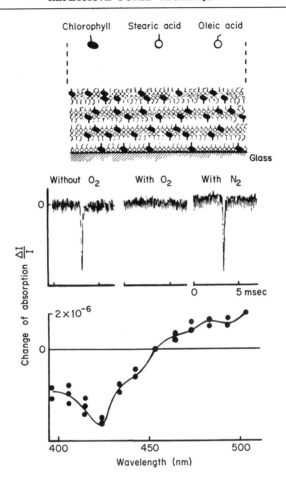

FIG. 14. Example diagram: Absorption changes in monolayers of chlorophyll $a$ versus time (middle). The wavelength dependence of these absorption changes is depicted (bottom). The structure of the three component monolayer on the glass carrier is shown above. The absorption changes with $5 \times 10^{-6}$ maximum were measured by repetitive excitation with approximately 10,000 flashes. Flash duration, 50 $\mu$sec. The absorption change appears *in vacuo* and in nitrogen atmosphere but diminishes in the presence of oxygen. (See footnote 32.)

and fatty acid salts were built up on glass slides (see Fig. 14, top). In these layers the reaction of the triplet state of chlorophyll $a$ with dia- and paramagnetic gases such as $N_2$ and $O_2$ have been measured;[32] about 10,000 single flashes have been evaluated to record the signals depicted in Fig. 14 (center) and also for each point of the difference spectrum (bottom).

[32] R. Reich, G. Döring, and H. T. Witt, *Z. Physik. Chem. (Frankfurt)* **53**, 387 (1967).

## G. Type 6. Repetitive Ultrashort-Flash Photometry

*Method*

For the study of photochemical reactions in the time range down to 0.1 $\mu$sec, flashes are to be used that have an emission time $\lesssim 0.1$ $\mu$sec. Such short flashes normally yield a low light energy output so that only very small chemical turnovers are produced. Moreover, the corresponding high time resolution (i.e., large bandwidth) causes a considerably high noise level. For both these reasons, flash photometric measurements could not be realized until now in this range of time resolution. With the aid of the repetitive technique, however, it becomes feasible to measure reaction courses directly in the 0.1 $\mu$sec range.[4,33]

With the so-called ultrashort-flash tube, a relatively high intensity is obtained in the visible spectral range in spite of the short flash duration (half-width 0.3 $\mu$sec, tenth-width 0.8 $\mu$sec). The equipment for the study of submicrosecond-photochemical reactions is illustrated in Fig. 15.

As the switching times of neither the magnetic nor of the electric averager are short enough for the necessary time resolution, a sampling oscilloscope is used which has a sample gate-width of some 0.1 nsec. The principle of time sampling is the same as with the electric averager illustrated in Fig. 11A. In order to collect a large number of samples in the signal course, the horizontal shift of the sample point (see in Fig. 11A: rotating of contact point $K$) is set extremely slow (e.g., $T_r = 20$ min recording time) by an external scanner (slow ramp). The distribution of sample values can be observed directly on the oscilloscope screen. These

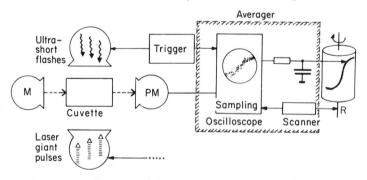

FIG. 15. Principle diagram: Repetitive ultrashort-flash and repetitive laser giant-pulse photometric equipment (types 6 and 7). $M$, Monochromatic monitoring light; $PM$, photomultiplier; $R$, recorder. Excitation wavelength of ruby laser, 694 nm. Averager: sampling oscilloscope with RC-integrating element. (See footnotes 4 and 33.)

[33a] Ch. Wolff, H.-E. Buchwald, H. Rüppel, and H. T. Witt, *Naturwissenschaften* **54**, 489 (1967); [b] Ch. Wolff and H. T. Witt, *Z. Naturforsch.* **24b**, 1031 (1969); [c] Ch. Wolff, Thesis, Technical University of Berlin, 1969.

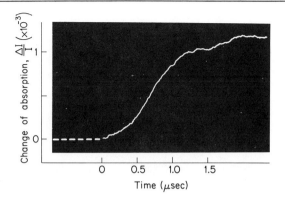

Fig. 16. Example diagram: Absorption change in cells of *Chlorella vulgaris* at 515 nm versus time. The rapid absorption change with $1.2 \times 10^{-3}$ maximum was measured by repetitive excitation with approximately 1000 ultrashort flashes. Signal-to-noise ratio $\approx 12$. The measured rise time of the absorption change is $\tau_R \approx 0.8$ $\mu$sec. This value corresponds to the tenth-width of the ultrashort-flash emission: $\tau_F \approx 0.8$ $\mu$sec. ($\tau_{1/2}$ = 0.3 $\mu$sec. (see Fig. 2). (See footnote 4.)

sample values are averaged by a simple RC-integrating element and registered on a recorder, $R$. A similar set-up for the direct measurement of fluorescence decay time was already developed in 1964 by Schäfer and Röllig.[34]

*Specification*

Absorption changes of less than $10^{-4}$ can be measured with a time resolution in the 0.1-$\mu$sec range by approximately 1000 excitations (see following example).

*Example*

Figure 16 shows the integrated sample values obtained from 1000 measuring signals. These signals are caused by absorption changes of a chlorophyll reaction at 515 nm in cells of *Chlorella vulgaris*.[4] The signal-to-noise ratio is about $S/N \approx 12$. The time course of the absorption change has a rise time of $\tau_R \sim 0.8$ $\mu$sec which corresponds to the flash duration (tenth-width).

## H. Type 7. Repetitive Laser Giant-Pulse Photometry

*Method*

Direct measurements of photochemical reactions with a time resolution of $\approx 10$ nsec could not be carried out before a short but high intense light pulse source was available by the Q-switched ruby laser (half-width

[34] F. P. Schäfer and K. Röllig, *Z. Physik. Chem. (Frankfurt)* **40**, 198 (1964).

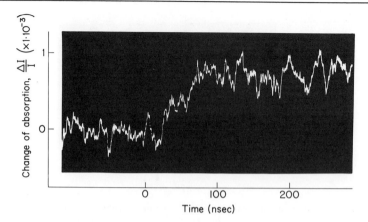

Fig. 17. Example diagram: Absorption change in a suspension of spinach chloroplasts at 515 nm versus time. The rapid absorption change with $\approx 1 \times 10^{-3}$ maximum was measured by repetitive excitation with approximately 3000 ruby laser giant pulses (694 nm). $S/N$ ratio $\approx 3$. The measured rise time of the absorption change is $\tau_R{}^m \approx 60$ nsec. The rise time of the absorption change derived from this value (see text) corresponds to the tenth-width of the Q-switched ruby laser emission: $\tau_F \approx 50$ nsec ($\tau_{1/2} \approx 20$ nsec; see Fig. 2). The result of both the measurements (Figs. 16 and 17) is that the real rise time of the chlorophyll reaction at 515 nm is $\tau_R < 50$ nsec. (See footnotes 5b, 35.)

20 nsec, tenth-width 40 nsec). The large bandwidth which is necessary for the transmission of these rapid signals causes such a high noise level that signals of $10^{-3}$ absorption changes, e.g., in photosynthesis are lost in noise. Therefore by single excitation with a Q-switched laser the sensitivity is too low for the realization of single measurements (see Fig. 1). For repetitive measurements with time sampling and averaging technique it is necessary to operate the Q-switched ruby laser at a relatively high repetition rate. This requirement has been a problem for laser technology. As such a laser system has been specially developed for a repetition rate of 5 Hz measurements could be realized in the time resolution range down to 10 nsec.[4,33a,35] The measurement of the rapid signal, time sampling, averaging, integration, and recording are performed in the same way as with the equipment described before (see type 6 and Fig. 15).

*Specification*

Absorption changes of $3 \times 10^{-4}$ can be measured in the time resolution range of 10 nsec by means of approximately 3000 excitations (see following example).

[35] Ch. Wolff, H.-E. Buchwald, H. Rüppel, K. Witt, and H. T. Witt, *Z. Naturforsch.* **24b**, 1038 (1969).

*Example*

Figure 17 shows the rise of an absorption change at 515 nm which is caused by a chlorophyll reaction in a suspension of spinach chloroplasts.[5a] The signal-to-noise ratio has a value of $S/N \approx 3$. The measured time course of the absorption change has a rise time of $\tau_R{}^m \approx 60$ nsec. Correcting this value for a slight bandwidth limitation the evaluated rise time of the absorption change is approximately 50 $\mu$sec which corresponds to the tenth-width of the tight pulse. Therefore the real rise time of the chlorophyll reaction is $\tau_R < 50$ nsec. (For a more elaborate evaluation of $\tau_R$ see footnote 35.)

*Conclusion*

In photochemistry numerous reactions with molecules in the excited triplet state could be studied having lifetimes of the order of microseconds.

With this repetitive technique it is now possible to follow directly reactions with molecules even in the excited singlet state with lifetimes of the order of 10 nanoseconds.

## I. Type 8. Repetitive Microwave-Pulse Conductometry

*Method*

The repetitive measuring technique can be used also for the investigation of chemical reactions by temperature jumps.[36] In this case, the reactive solution flows through the measuring cell. A flow system is necessary to avoid further heating of the solution by the repetitive temperature jumps. The microwave pulses can easily be generated in high repetition rates. Therefore, the method of Ertl and Gerischer[18] is especially suited for a repetitive temperature-jump apparatus (Fig. 18). The single microwave pulse equipment (equipment type 3, Fig. 9) is extended by a repetitive trigger generator, a flow system, an averager, and a recorder. The conductivity changes are measured repetitively by two ring electrodes which do not disturb the constant flow of the reactive solution. The bridge output signal is amplified by a selective band amplifier, averaged, and recorded.

*Specification*

The repetitive temperature jumps normally have the order of 0.4° in 100 $\mu$l of solution. The repetition rate is 0–20 Hz. With a time resolution of $10^{-5}$ second, conductivity changes $\Delta\kappa/\kappa \approx 10^{-5}$ can be just measured.

[36a] P. Brumm and H. Rüppel, *Z. Naturforsch.* **22b**, 980 (1967); [b] P. Brumm, F. P. Kilian, and H. Rüppel, *Ber. Bunsenges. Physik. Chem.* **72**, 1085 (1968).

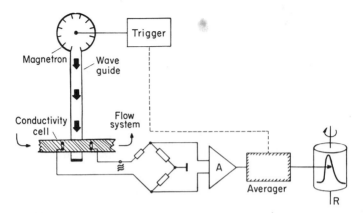

FIG. 18. Principle diagram. Repetitive microwave pulse conductometric equipment. *A*, Selective amplifier; *R*, recorder. The single microwave pulse equipment (type 3, see Fig. 9) is extended for repetitive excitation and registration by a trigger generator 0–20 Hz, flow system 0–50 cm/sec with ring electrodes, averager, and recorder. (See footnote 36.)

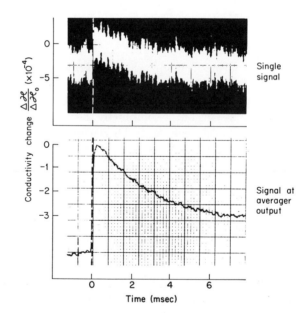

FIG. 19. Example diagram: Conductivity change versus time. The change of conductivity is caused by a water exchange reaction of the Ni-malate complex ($5 \times 10^{-3} M$) at 25°. *Top:* Signal after excitation by one microwave pulse with a temperature jump of 0.2°; $S/N$ ratio $\approx 1$. *Bottom:* Signal at averager output after excitation by 1200 microwave pulses: $S/N$ ratio $\approx 20:1$. (See footnote 36a.)

*Example*

The efficiency of the equipment was tested first using complexation reactions. The exchange of a water molecule out of the inner hydration shell around the central ion has been investigated. This water molecule is substituted by another ligand molecule. Figure 19 shows an example for such a reaction in the case of the system Ni-malate. Single temperature jumps give an $S/N$ ratio of only 1. By repetitive stimulation with 1200 microwave pulses the $S/N$ ratio is increased up to 25. From the averager output signal (Fig. 19, bottom) a relaxation time of $\tau = (2.9 \pm 0.1)$ msec can be definitely determined at 25°,[36a] which is impossible from the single signal (top).

*Conclusion*

The method of microwave pulse absorption for rapid heating may be extended to other polar solvents, e.g., alcohols.[37] As relatively homogeneous heating can be obtained in the microwave absorber tube (Fig. 18) without any disturbing secondary effects, this method seems to be very advantageous for measurements with a time resolution down to the sub-microsecond range. The method is especially promising for the investigation of biological systems, for instance, of the electron transport mechanism in the respiration chain of mitochondria.

For the measurement of faster chemical reactions, however, pulse heating with repetitive ultrashort flashes (equipment type 6) or repetitive laser giant pulses (equipment type 7) may be used to produce the necessary rapid temperature jumps. Thus, chemical reactions might be directly investigated even in the reaction time range down to $\approx 10$ nsec.

## III. Technical Details

### A. Type 1. High-Sensitive Single-Flash Photometry

*Monitoring Light Path*

The general principle of a flash spectrophotometric apparatus is depicted in Fig. 20A. The light of a constant monitoring light source $Q_1$ (see below) passes through a monochromator and is focused by the lens $L_2$ onto the cuvette $C$ which contains the photoactive sample. An approved type of measuring cuvette has the outer dimensions 20 × 20 mm with optically flat windows. For ultraviolet measurements the windows are of optical quartz. The irradiated area in the cuvette is approximately 15 × 15 mm square. By means of a Si-photoelement $Ph$ which is calibrated by a

[37] E. F. Caldin and J. E. Crooks, *J. Sci. Instr.* **44**, 449 (1967).

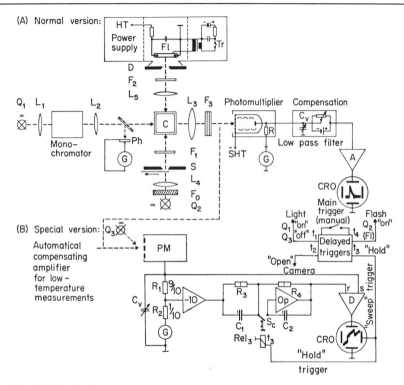

FIG. 20. Device diagram.

(A) Schematic drawing of single flash spectrophotometric (or fluorimetric) equipment; (See footnote 11.) $Q_1$, $Q_2$, monitoring light sources; $L_1$–$L_5$, lenses; $F_0$, heat filter (infrared filter or water cuvette); $F_1$–$F_3$, filters; $Ph$, photoelement calibrated by thermopile; $Tr$, flash trigger; $A$, amplifier; $R$, anode load; $G$, ammeter; $D$, diaphragm; $S$, light shutter; $HT$, high voltage; $SHT$, stabilized high voltage; $Fl$, flash lamp; $CRO$, cathode ray oscilloscope; $C_v$, variable capacitor for low-pass band limitation.

(B) Schematic drawing of the automatic compensating amplifier which is used for low temperature measurements instead of simple compensation (A). (See footnote 41.) $PM$, photomultiplier; $Rel$, relay; $Q_3$, auxiliary steady light; $Op$, chopper stabilized operational amplifier; $D$, difference amplifier; $S_c$, catch-hold switch; $C_v$, variable capacitor for low-pass band limitation.

thermopile, the radiation intensity of the monitoring light which enters the cuvette is controlled. Passing the cuvette the monitoring light is slightly scattered if a disperse suspension of biological cell material is investigated. In this case it is necessary to focus the light again by a lens $L_3$ onto the cathode of the photomultiplier $PM$. This may also be achieved by a light guide between the cuvette and the photomultiplier. For this purpose, the light cables of fiber optics (Mosaic Fabrications Inc., Fiskdale, Massachusetts; Schott & Gen., Mainz, Germany) may be used. In many cases, an external polished Plexiglas rod or, to include ultraviolet light also, an internal polished aluminum tube is sufficient. The diameter of the lens $L_3$ or the light guide should be large enough to assure an appropriate input aperture for the scattered monitoring light behind the cuvette.

On the other hand, the photomultiplier must be protected as far as possible from the scattering light of the excitation flash or the disturbing fluorescence light of the sample. Therefore, a narrow interference-band-filter or filter combination $F_3$ is brought into the light path just in front of the photomultiplier cathode. Its maximum transmission is matched to the monochromator setting. If the filter combination cannot fully cut off the scattered light (e.g., if the test solution gives fluorescent light just within passband of $F_3$), it is often useful to reduce the aperture in front of lens $L_3$. In this way the false signal may be suppressed more than the measuring signal is decreased by the corresponding loss in monitoring light intensity.

## Monitoring Light Source

The monitoring light source $Q_1$ is normally a direct-current energized filament lamp with a filament coil as small as possible. The lamp power is approximately 50–100 W; in special cases, 250 W. For high intensities or for measurements in the ultraviolet spectral range stabilized gas discharge lamps are used (e.g., Osram XBO 75, Hanovia Xenon-Mercury 200). In the case of filament lamps a small fluctuation $\delta U$ in the output voltage of the lamp power supply produces a fluctuation $\delta I$ in the light intensity $I$ in a ratio 10:1; $\delta I/I \approx 1/10 \cdot \delta U/U$. Therefore, transmission changes of $10^{-5}$ (see Table II) can be measured only if the voltage regulation is better than $10^{-4}$. With gas discharge lamps the fluctuations are transformed at least 1:1. In this case, the stabilization should be better than $10^{-5}$.

A relatively stable low-cost power supply is a well charged set of automobile batteries. But it is more convenient to use transistorized power supplies that have stabilities up to $10^{-6}$ (Knott, NSLV-BN 665, Munich, Germany; Hewlett-Packard-Harrison, Palo Alto, California).

Approved monochromators are the 125 mm grating units of Bausch & Lomb, Rochester, New York, type: "high intensity" which can be directly mounted into the light path on the optical bench (aperture 1:3.5; dispersion: visible 6 nm, UV 3 nm/mm slit width). Instead of monochromators, a set of interference filters (e.g., Schott & Gen., Mainz, Germany; Balzers AG, Liechtenstein; or Baird Atomic Inc., Cambridge, Massachusetts) may be used. These filters (50 mm square) have a half-width of approximately 10 nm and a maximum transmission of 30–40%. The expense for a good monochromator and a filter set with this wavelength resolution is nearly the same. For extreme high wavelength resolution very narrow interference filters with half-widths of 0.1 nm are available (Spectra Labs, Sylmar, California).

## Photoelectric Measurement

For photometric measurements including the visible as well as the ultraviolet spectral range, it is recommended to use a photomultiplier with a S20 spectral characteristic (trialkali photoelectric coating).[38] The photomultiplier types EMI 9558 Q (EMI Electronics Ltd., Hayes, United Kingdom) and RCA 7326 (Radio Corporation of America, Electronic Tube Division, Harrison, New Jersey) have been used with good success. The reason for the preferred use of photomultipliers instead of photoelectric cells is the high amplification up to the order of $A = 10^6$. An incoming light signal is raised by the internal amplification far above the noise level of normal amplifiers with this amplification. Such an amplifier is necessary if the signal is picked up by a photocell. In the photomultiplier the primary photo current $i_C$ at the photocathode is amplified to give an anode current $i_A = A \cdot i_C$. The amplification is an exponential function of the overall photomultiplier voltage $U$ (SHT) which can be approximated by $A = (U/U_0)^n$. The value of $n$ depends on the number and type of dynodes and has the order of 5–10; for both recommended photomultiplier types, $n \approx 8$. Therefore, a relative change $\delta U/U$ gives rise to a $\delta i_A/i_A = n\delta U/U$. This means that for kinetic measurements ($\tau \ll 1$ sec) with a limiting sensitivity $\Delta\alpha_{\min} = 3 \times 10^{-5}$ (see Table I) it is necessary to have a short time stability of the power supply (SHT) of at least $\approx 5 \times 10^{-6}$. Appropriate power supplies are available by several manufacturers (e.g., model 246 Keithley Instruments Inc., Cleveland, Ohio; Sorensen, South Norwalk, Connecticut; Nucletron NU 1250 and Knott NSLHV, both Munich, Germany).

The monitoring light with the intensity $I_0$ causes an anode current $i_A^0 = A \cdot i_C^0$ which is measured by an ammeter $G$. If the corresponding voltage drop $u_A^0 = i_A^0 \cdot R$ at the anode resistor $R$ becomes comparable with the potential $U_n$ of the last dynode, anode saturation occurs. If it is necessary to keep the intensity $I_0$ at a high constant level (for reasons see below), saturation can be avoided only by a decrease of the amplification $A$. Thus $u_A^0 = i_C^0 \cdot A \cdot R$ should be kept at least 10–20% below the critical value of $U_n$. The amplification is simply decreased by a reduction of the overall photomultiplier voltage (see above). At any change of the photomultiplier voltage, however, it is recommended to keep the voltage between cathode and first dynode (and focusing electrode) as well as that between last dynode and the anode bias constant (approximately 150 V

---

[38] Max. sensitivity: 0.07 A/W with 20% quantum efficiency at 420 nm; spectral range between half-value points: 330–600 nm.

and 100 V, respectively). This can easily be achieved if appropriate Zener diodes are placed in the corresponding positions of the photomultiplier voltage divider. In this case the overall voltage can be reduced to relatively small values of 10–20 V/stage without affecting the proper photomultiplier operation.

If the photomultiplier anode is direct-current coupled to the following measuring device, it is obvious that the cathode of the photomultiplier is to be connected to a negative high voltage and the anode resistor to be zero biased. In this case it is advisable to apply the same negative high voltage to an exterior shielding of the photomultiplier, e.g., the magnetic shield. Thus the cathode space is protected from the influence of a high electric field that may defocus primary photo electrons.

*Electric Registering System*

For a direct-current coupling it is necessary to compensate the voltage drop $u_A{}^0$ at the anode resistor, which is mostly high in comparison to the absorption change signal ($10^{-2}$ to $10^{-6}$; see Figs. 6 and 14). The compensation is performed in the compensation and filter unit. The absorption signal is picked up at the anode resistor $R$, passes the low-pass filter ($C_v$) for optimal band limiting (see below), is amplified in a preamplifier ($A$) and indicated on the oscilloscope $CRO$. A 300 kHz oscilloscope such as $hp$ 130 c (Hewlett-Packard, Palo Alto, California) is sufficient. Most of the oscilloscopes of this class have an input sensitivity of 1–5 mV/cm, which saves a separate amplifier $A$. For measurements of signals with a rise time of 10 $\mu$sec or less, it is necessary to have a low input time constant $RC_v$, i.e., a low input capacitance $C_v$ at a given anode resistor $R$ (for $R = 100$ kOhm, $C_v \lesssim 20$ pF, see below under "Optimum Bandwidth"). Therefore, short low-capacity connection cables to the amplifier or—if this is not possible— a cathode or emitter follower as impedance changer should be used. As impedance change is of interest for rapid signals only, the cathode follower can be alternating-current coupled to the anode resistor and mounted directly into the photomultiplier housing.

*Actinic Light Source*

The photoreaction is excited by the light of a short electronic flash $Fl$ or by steady light $Q_2$ which is switched on and off by a mechanical shutter or spinning wheel $S$ (1 sec–1 msec). The development of modern electrooptical switching and modulating systems (e.g., EOLM, Baird Atomic, Inc., Cambridge, Massachusetts) makes it possible that the inconvenient shutters may be replaced in the future by these light gate systems. The excitation light is normally filtered by glass filters ($F_1,F_2$), in special cases also by interference filters. The light of the flash tube $Fl$ or the filament lamp $Q_2$ is focused onto the cuvette $C$ by the lenses $L_5$ or $L_4$.

TABLE V

SPECIFICATIONS OF FLASH LAMPS USED IN EQUIPMENT TYPE 1, 2, 4-6

| Type | Manufacturer | Maximum energy $E_i$ (J) | Minimum[a] flash duration (half/tenth-width) (μsec) | Operation voltage $U_f$ (kV) | Gap width (cm) | Flashing mode | Used in equipment type |
|---|---|---|---|---|---|---|---|
| Photolysis quartz tube | Heimann PTW, Wiesbaden, Germany | 300 | 120/180 | 2.5 | 12 | Single | 2 |
| G 63 | Sylvania, Erlangen, Germany | 75 | 30/60 | 1 | 8 | Single | 1 |
| XIE 200[b] | Osram, Berlin, Germany | 6 | 22/50[e] | 2 | 15 | Single, repetitive | 1, 4, 5 |
| XE 10-4[c] | PEK Labs, Inc., Sunnyvale, California | 50 | 3.6/7.0[e] | 10 | 10 | Single | 1 |
| XE 9-3[d] | PEK Labs, Inc., Sunnyvale, California | 20 | 0.3/0.8[e] | 12 | 7.5 | Repetitive only | 6 |

[a] At reduced input energy.
[b] Used in AEG-Stroboscope LS5.
[c] Special developed discharge circuit (footnote 13b) using impulse capacitor Siemens B 25359 (1 μF).
[d] Special developed discharge circuit (footnote 33c), see type 6.
[e] Time course of light emission, see Fig. 2.

The type of flash lamp should be selected according to the special requirements of the photoreactive system being studied. A large number of different flash tubes are commercially available with flash energies of several microjoules up to the 10 kJ range and flash durations of nanoseconds to several milliseconds. The electrical input energy $E_i = \frac{1}{2}CU_f^2$ is varied by changing the capacity $C$ or the operating voltage $U_f$. The flash lamps are usually fired by a trigger-circuit $T_r$ via a small firing transformer (as in commercial units for flash photography). A short flash duration is required for a high time resolution in flash kinetic investigations. The flash duration is reduced if small capacitors $C$ are used and the circuit inductivity $L$ is kept as low as possible. In this case, however, the electric input energy can be kept sufficiently high only if a high operating voltage $U_f$ is applied. This is realized, first, by special flash tubes with high breakdown voltage (e.g., $FP$ 1, Xenon Corp., Watertown, Massachusetts, $U_f = 10$ kV, $E_i = 100$ J, $\tau_F = 5$ $\mu$sec); second, if several flash lamps with low breakdown voltage are connected in series (in a bifilar array to reduce circuit inductivity[39]); and third, if series firing via a high-voltage spark gap is applied (see also equipment type 6[33c]). The flash lamps that have been approved by operation in equipment type 1, 2, 4–6 are given in Table V.

*Two-Channel Measurements*

A second "measuring channel" with separate photomultiplier, amplifier, and, if necessary, oscilloscope (dual-beam scope) is used for a simultaneous measurement of absorption changes at different wavelengths or for a simultaneous study of both absorption and fluorescence changes (see below). Furthermore, this separate measuring channel is important if the scattered light or the disturbing fluorescence cannot be sufficiently suppressed by the filter combination $F_3$ or aperture constriction (see above in section on monitoring light path). In this case, the spurious light signal is separately measured in the second channel. After an appropriate tuning of both electrical channels it is possible to compensate the disturbing signal in a differential amplifier down to a rejection ratio of $10^{-2}$.

*Low Temperature Measurements*

*General Problem.* The main condition for the performance of spectrophotometric measurements is that the monitoring light does not essentially excite the photoreaction itself. If the reaction is reversible by a rapid back reaction the amount of stationary photoproduct can be kept sufficiently low at small light intensities. If the reaction is irreversible, however, the

---

[39] B. Rumberg, Thesis, Technical University, Berlin, 1966.

total amount of photoreactant reacts into the stable photoproduct within a definite irradiation time which depends on light intensity. Therefore, a kinetic measurement can only be conducted during this short time interval.

Under these conditions it is hardly possible to perform kinetic investigations by means of the normal flash photometric equipment (see Fig. 20A). The latter measuring conditions are, for example, given at liquid nitrogen temperatures at which all chemical reaction steps are stopped. Measurements in this temperature range are often of high interest in the detection of the primary photochemical reactions.[40] Photometric measurements of absorption changes at low temperatures[40] have been realized by the following method.[41]

*Automatic Baseline Compensation.* For measurements at low temperatures the prepared sample should be exposed to the monitoring light for as short a time as possible. Therefore the monitoring light is switched on immediately before the excitation flash is fired. After the monitoring light is switched on, however, there is a continuous voltage change at the photomultiplier output. This is caused, first, by the absorption change induced by the monitoring light and, second, by a formation process within the photocathode. The latter effect can be almost completely eliminated if the illumination of the photocathode is maintained at a constant level at all times. This is achieved with aid of a weak adjustable auxiliary light source $Q_3$, the light of which does not pass the sample cuvette. This auxiliary light is switched off whenever the monitoring light beam is on. The remaining linear change of the photomultiplier output voltage is due to the change of absorption of the sample which is induced by the monitoring light and causes a baseline drift on the screen of the *CRO*. At the high amplification range that is necessary for this measurement, this drift may be so rapid as to make it impossible to balance the compensation manually, i.e., to locate the baseline on the screen in a sufficiently short time. Not until then can the sweep be triggered and the flash be fired to display the signal. All these functions are realized automatically in a very short time by a modification of the original equipment.[41] The normal constant-value compensation (Fig. 20A) is replaced by a special compensation amplifier (Fig. 20B). The operation of the whole arrangement is as follows: At time $t_1$ the manually released delayed trigger generator gives the first trigger pulse which switches "on" the monitoring light from $Q_1$ and "off" the auxiliary light from $Q_3$ by electromagnetic shutters (not depicted in Fig. 20). The full photomultiplier-output voltage at $R = R_1 + R_2$ is fed into the signal input $s$ of a difference amplifier $D$. A tenth of it at $R_2 = \frac{1}{9} R_1$ is fed into an amplifier with an amplification

[40] H. T. Witt, A. Müller, and B. Rumberg, *Nature* **192**, 967 (1961).
[41] A. Müller, unpublished work.

$A = -10$. As long as the switch $S_c$ is closed ("catch" position) the following compensation amplifier acts as a normal amplifier with $A = -1$ (phase inverter, $R_3 = R_4$, $C_1 = C_2$). The output is applied to the reference input $r$ of $D$. Thus, at all times exactly equal voltages are applied to both inputs. At time $t_2$ a second trigger from the delayed trigger generator opens the shutter of the camera. At $t_3$ a third trigger opens the relay switch $S_c$ and starts the oscilloscope sweep. The compensation amplifier is now in the "holding" position and keeps the compensating voltage constant at that value which it had at time $t_3$. With very small delay a fourth trigger fires the excitation flash at time $t_4$ by opening the shutter $S$ in front of the steady light source $Q_2$ or else by triggering the flash lamp $Fl$ (see Fig. 20A and B). The flash-induced signal is now displayed on the $CRO$ screen superimposed to the drifting baseline. At the end of the sweep all the relays fall off, close the shutters and the switch $S_c$, and switch on the auxiliary light to clear for the next operating cycle. By adjusting the right values for the delay times, the difference amplification, and the sweep time it is possible to display the total absorption change on the oscilloscope screen within a short time after exposing the sample to the monitoring light.

*Fluorimetric Measurements*

Fluorescence investigations can be done in the same equipment (Fig. 20A) if the monitoring light source $Q_1$ is switched off and the photomultiplier is rearranged for low-intensity measurements. That is, maximum high tension is applied to the dynode system to get high amplification (see above).

*Adjustment of Optimum Conditions*

In principle, two independent data can be withdrawn from kinetic signal measurements: the maximum amplitude value $S$ and the reaction time $\tau$. The limiting values of both, however, are related: $\Delta S_{min} \sim 1/\sqrt{\Delta \tau_{min}}$ (see Section I and Table II). Thus, in any case, whether the accuracy $S/\Delta S_{min}$ or the time resolution $\tau/\Delta \tau_{min}$ of the measurement should be enhanced it is indispensable to improve the signal-to-noise ratio. Therefore, the first procedure is to adjust all measuring parameters to give optimum signal-to-noise ratio. The further improvement by the repetitive technique is the main feature in equipment types 4–8 (see below).

If $I_0$ is the intensity of the monitoring light at the wavelength $\lambda$ in front of the cuvette, $l$ the length of the cuvette, and $I$ the emergent intensity, the absorption $\alpha$ is given by the expression

$$\alpha = \ln I_0/I = \sum_{i=1}^{Z} \alpha_i = \sum_{i=1}^{Z} \epsilon_i c_i l \tag{1}$$

from which the optical density $D$ (i.e., the decadic absorption) is derived as $D = \alpha/\ln 10$; $\alpha_i$ is the partial absorption of the $i$th of $Z$ components $C_i$ with concentrations $c_i$ and the molar absorption coefficients $\epsilon_i$ at the wavelength $\lambda$. A reaction between $R$ components $C_i$ ($i = 1, \ldots, R, \ldots, Z$) gives rise to an absorption change $\Delta\alpha = \sum_i \Delta\alpha_i$. The $\Delta\alpha_i$ are linearly related, e.g., for a one-step reaction according to a stoichiometric equation $\sum_i \gamma_i C_i = 0$ ($\gamma_i$ = stoichiometric coefficient) by $\Delta\alpha_i/\Delta\alpha_M = (\epsilon_i\gamma_i)/(\epsilon_M\gamma_M)$ ($i = 1 \ldots R$, $M$ = reference index). In this case the absorption change can be expressed by $\Delta\alpha = (1/\epsilon_M\gamma_M) \sum_i \epsilon_i\gamma_i \cdot \Delta\alpha_M$.[42] Thus the reaction course may be characterized by the concentration change of the $M$th component, in general by

$$\Delta\alpha = \frac{d\alpha}{d\alpha_M} \cdot \epsilon_M \cdot \Delta c_M \cdot l \tag{2}$$

The absolute value of $\Delta c_M$ can be evaluated from $\Delta\alpha/l$ only if $\epsilon_M \cdot (d\alpha/d\alpha_M)$ (difference of molar absorption coefficients) is known. On the other hand, $\epsilon_M$ can be found only if $\Delta c_M$ and $d\alpha/d\alpha_M$ are known [the latter containing the $\epsilon_i$ of the residual $R$-1 reactants as well as all reaction coefficients (see above)].

The anode current $i_A$ of the photomultiplier produced by the monitoring light intensity $I$ is given by

$$i_A = A \cdot \sigma_\lambda \Phi I \tag{3}$$

$\sigma_\lambda$ is the spectral sensitivity of the photocathode at the wavelength $\lambda$ and $\Phi$ the irradiated area of the cuvette (see above "Photoelectric Measurements"). The measuring signal $S$ due to a light intensity change $\Delta I$ is

$$S \cong \Delta i_A = A \cdot \sigma_\lambda \Phi \, \Delta I \tag{4}$$

The statistical anode current fluctuation is mainly caused by the shot-effect of the photoelectrons. The mean square value is given by

$$\delta i_A{}^2 = A^2 \cdot 2q\sigma_\lambda \Phi I \, \Delta\nu \tag{5}$$

with $q$ = electron charge, $\Delta\nu$ = noise equivalent bandwidth.[43]

The noise level $N$ is usually defined as the width of the "noise band" which appears on the oscilloscope screen at a sufficiently high display frequency and is proportional to the root mean square value of the statistical fluctuation

$$N \cong p \sqrt{\overline{\delta i_A{}^2}} \tag{6}$$

[42] H. Rüppel, Habilitationsschrift Technical University, Berlin, 1968.
[43] M. Schwartz, "Information Transmission, Modulation, and Noise," Chapt. 5. McGraw-Hill, New York, 1959.

The proportionality factor $p$ has a value of 8 for a bandwidth of $\Delta \nu = 10$ kHz.[44] From Eqs. (4) and (6) follows

$$S/N = \frac{1}{p} \sqrt{\frac{\sigma_\lambda \Phi I}{2q\, \Delta \nu}} \cdot \Delta I/I \tag{7}$$

According to Eq. (1), for small intensity changes ($\Delta I/I \leq 1/10$) $\Delta I/I$ can be replaced by $\Delta \alpha$.[12] With Eq. (2) the $S/N$ ratio can be expressed as

$$S/N = \frac{1}{p} \sqrt{\frac{\sigma_\lambda \Phi I}{2q\, \Delta \nu}} \cdot \frac{d\alpha}{d\alpha_M} \epsilon_M \Delta c_M l \tag{8}$$

*Optimum Absorption of Monitoring Light.* In the case $\alpha \approx 0$ (zero initial absorption at wavelength $\lambda$, $I \approx I_0$) the $S/N$-ratio increases proportional to $l$ (for constant $\Delta c_M$). In this case a long cuvette type is recommended. In the case $\alpha \neq 0$, it is often possible to optimize $S/N$ by varying the initial absorption. In certain cases $\Delta c_M$ is proportional to $c_M$ and thus $\Delta \alpha_M / \alpha_M$ is constant. This relation is valid, for example, at a low absorption of actinic light (see below) or if only a definite fraction of excited molecules are able to react (e.g., in the primary process of photosynthesis). In this case Eq. (8) may be rewritten especially for $\alpha_M \neq 0$ as

$$S/N = \alpha_M e^{-\frac{\alpha_M}{2}} \cdot e^{-\frac{\alpha - \alpha_M}{2}} \sqrt{\frac{\sigma_\lambda \Phi I}{2q\, \Delta \nu}} \cdot \frac{d\alpha}{d\alpha_M} \cdot \frac{\Delta c_M}{c_M} \tag{9}$$

$S/N$ has an absolute maximum value at $\alpha = \alpha_M$, i.e., if the absorption is determined by component $M$ only. In this simple case the optimum $S/N$ ratio will be found at $\alpha_M = 2$, i.e., for an initial transmission of 14%.

In the general case $\alpha > \alpha_M$ only a lower, relative maximum is obtained for $\alpha_M = 2$. If the high initial absorption $\alpha$ is caused by nonreactive but high absorbing components (e.g., absorbing solvent) the case of optimum $S/N$ ratio ($\alpha = \alpha_M$) might be more nearly approached if a small cuvette length $l$ is used at a high concentration of $c_M$ such as to keep $\alpha_M = 2$. In this case a flat cuvette type is recommended which is placed at 45° to the monitoring light beam to give access for the actinic light.

*Optimum Absorption of Actinic Light.* Unlike monitoring light, no general rule can be given for the adjustment of actinic light absorption $\bar{\alpha} = \sum_i \bar{\epsilon}_i C_i \, d$ ($\bar{\epsilon}_i$ = molar absorption coefficient of component $C_i$ at wavelength $\lambda$ of actinic light, $d$ = thickness of cuvette). For high absorption $\bar{\alpha} \gg 1$ the excitation is restricted to a thin layer near the wall within the cuvette which is scarcely irradiated by the beam of monitoring light. Therefore it is appropriate to adjust $\bar{\alpha} \approx 1$ obtaining a relatively large measuring signal. On the other hand, for a homogeneous excitation it is

[44] V. P. Landon, *Proc. I. R. E.* **29**, 50 (1941).

necessary to adjust $\bar{a} \ll 1$, i.e., low absorption. This condition must be satisfied, for instance, if the kinetics of a second order reaction is directly studied or to assure the proportionality between $\Delta c_M$ and $c_M$ (see above).

*Optimum Bandwidth.* The main condition for the measurement of rapid photochemical reactions is to keep the tenth-width $\tau_F$ or the fall time of the excitation pulse short in comparison to the reaction time which is to be measured (see p. 318). Moreover, the time constant $\tau_S$ of the signal transmission system should be considerably smaller than the reaction time $\tau$. The measured reaction time $\tau^M$ is evaluated from the signal rise or fall time[45] is connected with the real reaction time $\tau$ by the equation:[46]

$$\tau^M = \sqrt{\tau^2 + \tau_S{}^2 + \tau_F{}^2} \tag{10}$$

According to Eq. (10) the real reaction time $\tau$ can be measured with a relative error

$$\Delta\tau/\tau = \frac{\tau^M - \tau}{\tau} \lesssim 0.05 \quad \text{if } (\tau^i)^2 \quad \tau_S{}^2 + \tau_F{}^2 \lesssim (\tau/\pi)^2$$

($\tau^i$ = effective input time constant).

On the other hand, for an RC filter, the bandwidth $\Delta\nu$ of a transmission system is connected with $\tau_S$ by the equation

$$\Delta\nu = \frac{1}{2\pi\tau_S} \tag{11a}$$

Finally, if $\tau_F \ll \tau_S$, the condition for an error smaller than 5% is $\tau_S \lesssim \tau/\pi$. This corresponds to a bandwidth $\Delta\nu \geq \frac{1}{2}\tau$. Therefore, the RC filter ($RC_v$ in Fig. 20A) should be tuned to a bandwidth as low as compatible with the accuracy requirements for reaction time measurement, e.g., for 5% deviation

$$\Delta\nu = \frac{1}{2\tau} \tag{11b}$$

The time resolution $\tau/\Delta\tau$ is thus limited by the bandwidth of the transmission system. This principle limitation is independent of the $S/N$ ratio. The effective accuracy of the $\tau$ determination, however, depends essentially on the accuracy of the signal registration, i.e., on the $S/N$ ratio. Therefore, in the case of poor $S/N$ ratios it is sometimes advantageous to decrease the bandwidth even further below the value given in Eq. 11b. In this case a distortion of the signal is tolerated for a better $S/N$ ratio.

Introducing condition (11b) the optimum $S/N$ ratio in the case $\alpha \neq 0$ is according to Eq. (9)

[45] For an RC-filter, the rise time from 10 to 90% is $\tau_R = 2.2$ RC (cf. also footnote 46).
[46] W. C. Elmore and M. Sands, "Electronics, Experimental Techniques," Chapt. 3. McGraw-Hill, New York, 1949.

$$(S/N)_{\text{opt}} = \frac{2}{e \cdot p} \sqrt{\frac{\sigma_\lambda \cdot \Phi \cdot I_0}{q} \frac{\Delta c_M}{c_M}} \cdot \sqrt{\tau_{\min}} \qquad (12)$$

if $\tau_{\min}$ is the minimum reaction time that is to be evaluated from the time course of the reaction signal.

For given values of $\Delta c_M/c_M$ and $\tau_{\min}$ the $S/N$ ratio can be enhanced by selecting a photomultiplier with high spectral cathode sensitivity $\sigma_\lambda$, and further by increasing light flux $\Phi \cdot I_0$ at wavelength $\lambda$. A high spectral sensitivity is especially recommended if the monitoring light intensity $I_0$ must be kept at a low level to avoid the influence of continuous excitation.

## B. Type 2. Single-Flash Conductometry

Figure 21 gives a block scheme of a flash conductometric apparatus which was especially developed for measurements in highly insulating organic solutions.[14] By a few variations in the cell assembly, this apparatus can be applied to the measurement of solid as well as of gaseous phase reactions. The negative electrode of the quartz cell is directly connected to the open grid of an electrometer tube. The positive electrode is connected to a high voltage capacitor $C_l$ which is charged and then disconnected from the power supply before the measurement. The windows of the conductivity cell are of highly transparent optical quartz. After excitation with the light of a quartz flash tube (max. 300 J), the transient photocurrent $i(t)$, which is produced in the photoreactive solution, is integrated on the input capacitance $C_s$ of the electrometer tube. The input time constant depends on the grid current according to the cell leakage. A typical value is about 1 second. The anode signal of the electrometer tube is a measure of the current

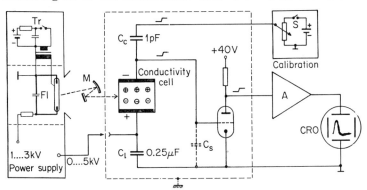

Fig. 21. Device diagram: Schematic block diagram of flash conductometric measuring equipment. (See footnote 14.) $Tr$, manual triggering; $C_c$, calibration capacitor; $C_l$, load capacitor; $S$, mercury switch; $Fl$, quartz tube flash lamp 300 J; $M$, mirror system; $C_s$, input stray capacitance $\approx$ 12 pF; $A$, pulse amplifier (RC = 5 msec); $CRO$, cathode ray oscilloscope.

impulse $\Delta Q = \int_0^\infty i(t) \, dt$ which is amplified by a pulse amplifier $A$ and indicated on an oscilloscope $CRO$. The maximum of the output pulse is proportional to $\Delta Q$ (see Fig. 8). The high sensitive pulse conductometer is calibrated by a small current impulse which is applied to the input capacitance $C_s$ by a mercury switch $S$ and a calibrated condenser $C_c$.

For optimum $S/N$ ratio, usually a selective RC-RC-pulse amplifier is employed.[47] However, in the first flash conductometric equipment[14] an RC-coupled pulse amplifier with an upper bandwidth of 40 kHz was used. The time constant for the lower band limit was $RC = 5$ msec. This pulse amplifier enabled the study of the time course of the charging process $q(t) = \int_0^t i(t) \, dt$ for $t < 5$ msec, which was of some interest in these investigations. The background noise, however, is by far higher than in the case of the selective RC-RC-amplifier.

### Time-Dependent Measurements

The effective time course of the ionization process is simply found in this technique by differentiating the output signal $\Delta Q(t)$. For time-dependent measurements it is more convenient in general, to use a high ohmic input resistor instead of the open grid.[48] The use of a wide-band amplifier results in an enhanced "noise level" which corresponds to a lower absolute sensitivity (see Fig. 1).

### The Current Impulse

For simplicity the following assumptions are made. After the flash excitation, a quantity $\Delta N = V \cdot \Delta c$ of reacting particles is split into one positive and one negative ion each ($V$ = volume of photoexcited area within the electrode distance $l$). These ions have mobilities $u^+$ and $u^-$, respectively. During the life time $\tau$, the ion concentration $\Delta c$ is constant all over the bulk of the electrode cell; this results in a conductivity change $\Delta\kappa = \mathfrak{F} \cdot (u^+ + u^-) \cdot \Delta c$ ($\mathfrak{F}$ = Faraday's constant). If an electric field $F$ is applied to the cell, a current change $\Delta i = \Delta\kappa \cdot S \cdot F$ is produced by which a charge

$$\Delta Q = \Delta i \cdot \tau = \mathfrak{F}(u^+ + u^-) \cdot F \cdot S \cdot \Delta c \cdot \tau \qquad (13)$$

is carried through the cross section $S$ during the life time $\tau$. The main features of the flash conductometric method can easily be discussed by this equation, which is valid for $\tau \ll l/(u^+ + u^-)F$ (ion drift time): (1) The current impulse $\Delta Q$ can be increased by the field strength $F$ (see Fig. 8); (2) $\Delta Q$ is proportional to the product $\Delta c \cdot \tau$.

[47] E. Baldinger and W. Haeberli, *Ergeb. Exakt. Naturw.* **27**, 248 (1953).
[48] H. S. Pilloff and A. C. Albrecht, *Nature* **212**, 499 (1966).

*Sensitivity of Flash Conductometric Measurements*

The lower limit of measurement is fixed by the statistical noise of the preamplifier tube. In the present case, the "noise band" corresponds to $\Delta Q_{noise} \approx 5000 \; q \gtrsim 10^{-20} \cdot \mathfrak{F}$ ($q$ = electron charge).[14] Substituting this value for $\Delta Q$ in Eq. (13) and also the known data for $F \approx 10^3 \cdot V/cm$, $(u^+ + u^-) \sim 10^{-4} \; cm^2/(Vsec)$, $V \approx S \cdot l = 100 \; cm^3$, one can estimate the lower limiting value of the product

$$(\Delta c \cdot \tau)_{min} \approx 10^{-18} M \; sec \tag{14}$$

(see also Table III).

By use of an RC-RC-selective amplifier, $\Delta Q_{noise}$ can be reduced to some hundred electrons; i.e., $(\Delta c\tau)_{min}$ is reduced to $< 10^{-19} M$ sec.

*Suppression of Interference*

For a critical consideration of the experimental technique it should be mentioned that the measuring apparatus is very subject to interference owing to the high sensitivity of the open grid circuit. Most of the disturbance is caused by the intense flash discharge which impairs the measurements by the electromagnetic field, by the sound, and mainly by any random light of the flash which gives rise to electrode effects. The interference can be suppressed to a large extent by electronic and magnetic shielding[14] coaxial construction of the cell and flash lamp circuit,[48] by diaphragms in front of the cell, and by light absorbers behind the cell, etc. According to the residual interference, the effective sensitivity may be slightly lower than the theoretical value given in Eq. (14).

Considering the manifold experimental difficulties, which are mainly caused by the single, high intense flash discharge, in a modern concept of this technique it is recommended to reduce the flash discharge energy considerably and to compensate the loss in signal amplitude by repetitive excitation and signal evaluation (compare equipment types 4–8).

Summarizing—the method of flash conductometry is convenient for high sensitive detection of photoionization processes, especially in insulating media where high electric fields can be applied.

## C. Type 3. Single Microwave-Pulse Conductometry

*Principle of Microwave Heating*

In the liquid phase, polar molecules such as water or alcohol have a high Debye absorption of electromagnetic energy in the centimeter wavelength range. This absorption is caused by dipole interaction in the liquid, which causes a relaxation effect of the dipole orientation in the electric

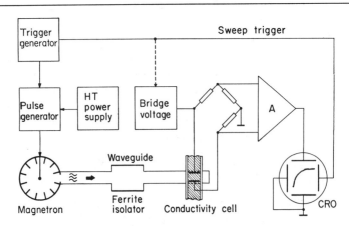

Fɪɢ. 22. Device diagram: Block scheme of temperature-jump apparatus using microwave pulse heating. The microwave pulse is produced by a magnetron and fed by a microwave guide into an absorber tube formed as conductivity cell. The conductivity cell is part of a radio-frequency bridge circuit. Normally a sine wave voltage (100 kHz) is used as bridge voltage. At extreme low conductivity (purest water, as in example of Fig. 10), a square-wave voltage is used which is triggered by the magnetron pulse trigger. *A*, Amplifier; *CRO*, cathode ray oscilloscope; *HT*, high voltage. (See footnote 18.)

alternating field. In water the relaxation time has the order of $10^{-11}$ second. The maximum of the broad absorption band is near 10 GHz. This is just the frequency which is commonly used by high-power radar magnetrons. For this frequency, the penetration depth $\delta = 1/(\partial\alpha/\partial z)$ in water is $\delta = 1.3$ mm (microwave propagation in $z$ direction).

A block scheme of the equipment for kinetic studies by microwave pulse excitation is shown in Fig. 22.

A trigger unit (manually released) fires a pulse generator consisting of a high voltage $LC$ pulse-timing circuit which is discharged via a thyratron into a pulse transformer. The high voltage output pulse is fed into the magnetron cathode, which generates microwaves during the pulse duration. The pulse duration is determined by the number of matched $LC$ elements. Maximum pulse length is limited by the magnetron itself. The microwaves are propagated to the absorber by a waveguide (X or H band).

The main problem is to get a sufficient energy absorption in the bulk of the measuring cuvette. This was achieved first by Gerischer,[18] who used a cylindrical cell with an outer diameter of 8 mm and an inner diameter of 3 mm. As tube material Jenaer Glas 16 III,[18] Nylon polyamide B,[19] or polyethylene[49] is used. This cylindrical cell is mounted in the center of the waveguide with the cylinder axis parallel to the $y$ direction (𝔉-vector). In

[49] M. Eigen, *Angew. Chem.* **80**, 892 (1968).

the $z$ direction (microwave propagation) the cell is aligned for optimum absorption near the short-circuit wave-guide terminal. After proper alignment, more than 95% of the microwave energy is absorbed in the cell. If microwave pulses are reflected, they may damage the magnetron. Therefore a ferrite isolator or a circulator is mounted between the magnetron and the absorber which allows microwave propagation only in one way ($z$ direction).

### The Conductometric Measurement

First investigations were performed by means of conductivity measurements.[18] For this purpose, two plain platinum electrodes are placed within the absorption tube perpendicular to the electric $\mathfrak{F}$ vector of the microwave. The small conductivity cell is one branch of a measuring bridge. The measuring bridge is fed either by a sine- or a square-wave voltage according to whether the conductivity is relatively high ($10^{-3}$ $M$ KCl solution) or extremely low (highly purified conductivity water). The bridge impedances (RC elements) are carefully balanced before the excitation. The trigger impulse starts the oscilloscope sweep just before the magnetron is fired. The rapid microwave pulse heating gives rise to a conductivity change $\Delta\kappa$. The corresponding bridge signal is amplified $A$ and indicated on the oscilloscope $CRO$.

The conductivity change $\Delta\kappa$ which is caused by the temperature jump in an electrolytic solution consists of two parts: the first one is due to the change of the ion concentration by the reaction [see Eq. (13)], the second one, however, to a change of the ion mobility. If both parts are of the same order, the time course of the ion reaction can be followed by the measurement with the simple bridge arrangement of Fig. 22 (compare the example of Ertl and Gerischer,[18] Fig. 10). For most of the ion reactions, however, the conductivity change due to the concentration change is very small compared with that of the mobility change. This ratio is further reduced if additional nonreactive ions are present in a multicomponent electrolytic solution. Thus, for a sensitive conductivity measurement it is necessary to compensate the relatively high conductivity change caused by the mobility increase. This is realized in the corresponding repetitive equipment type 8.

### Sensitivity of Conductometric Measurements

The limiting sensitivity of the measuring device alone is given by the noise in the bridge circuit, and also by tuning deviations caused by harmonics of the bridge frequency. A limiting value $(\Delta\kappa/\kappa)_{\min} \approx 2 \cdot 10^{-5}$ was measured as electronic circuit sensitivity. Additional noise, however, is produced within the bulk of the conductivity cell. Therefore, the effective sensitivity for measurements with single-pulse excitation is smaller by a factor of approximately 5 (bandwidth 10–200 kHz).[36]

### The Temperature Jump

The height of the temperature jump depends, first, on the maximum pulse energy of the magnetron pulse (power times pulse duration) and, second, on the effective volume of the absorber cell.

Typical values for temperature jumps are 0.2–1°. These are obtained with microwave pulse energies of 0.1–0.5 J according to a pulse length of 1–6 $\mu$sec in a water volume of 100 $\mu$l. The magnetrons used are: 4 J 50 with 400 kW,[18] Valvo 6972 with 200 kW[19] electric input power.

Extreme values are obtained by the commercial Raytheon "microwave modulator," which uses a 1 MW magnetron QKH 172 with a pulse duration of 3 $\mu$sec maximum, i.e., 3 J maximum pulse energy. This gives rise to about 6° in 100 $\mu$l of aqueous solution.[49]

### The Time Resolution

The time resolution is given by the pulse duration of the magnetron impulse. It has been proved that the temperature rise in a KCl solution takes place exactly within the microwave pulse duration.[19]

The minimum time resolution is therefore fixed by the smallest pulse that can be produced by the microwave generator. Limiting values are 0.3–1.0 $\mu$sec [0.3 $\mu$sec: microwave modulator, Raytheon Company, Waltham, Massachusetts; 1.0 $\mu$sec: microwave generator of Brumm[19] (see also Fig. 2)]. However, as pulse energy depends on the pulse length, these narrow pulses do have low pulse energies and produce accordingly low temperature jumps.

### Spectrophotometric Measurements

Conductivity measurements are very sensitive but relatively unspecific for kinetic measurements, especially in biological systems. Therefore, it is desirable to realize spectrophotometric measurements also for the microwave pulse equipment. In this case the light path has to be aligned through the length of the cylindrical absorber tube.[49]

### Comparison between Light and Microwave Heating

Temperature jumps up to about 10° can be produced also by light absorption in any solvent. A disadvantage of this method, however, is that a specific light absorber must be added to the solution.[16,17] As the light energy is transformed by secondary reactions into heat, the reaction products of the absorber system may disturb the measurement in the system.[17]

On the other hand, microwave absorption is restricted to polar solvents. Special cuvettes have to be used as microwave absorber. The time resolution is rarely kept lower than 0.5 $\mu$sec, and the temperature jump is only of the order of 1°. An essential advantage, however, is that the solvent itself is the energy absorber. Thus, no additional dyestuff must be added. As the radiation energy $h\nu$ is low there is no secondary reaction other than heating. These advantages may in many cases (especially for repetitive measurements, see equipment types 4–8) outweigh the inconveniences with the absorption cell, the low temperature jump, etc. Therefore, with respect to biological systems, equipment types are presented in this article which use heating by microwave absorption rather than by light absorption.

## D. Types 4 and 5. Repetitive Flash Photometry

### General Principle

According to the common rule of statistics the accuracy $S/\delta S$ of a test value $S$ is enhanced to a factor of $\sqrt{n}$ if $n$ independent tests of $S$ are taken. For direct kinetic measurements, however, not a single amplitude value $S_o$, but a complete time course $S(t)$, is to be measured from which one or more reaction times $\tau$ are evaluated with an accuracy $\tau/\Delta\tau$. Therefore, whole measuring signals are to be superposed to find out the average time course (see Fig. 23A). For slow processes ($\tau \gtrsim 1$ sec) this can be achieved by means

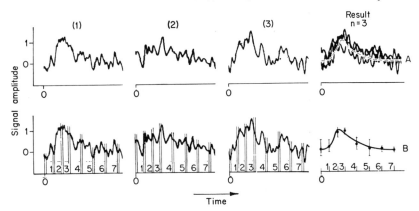

FIG. 23. Principle diagram: Illustration for the exhaustion (A) and the repetitive time sampling method (B) for $S/N$ ratio improvement. Three "noisy" signals $(1)$–$(3)$ which represent the same time course are evaluated by both methods. (The signals are original recorder plottings of the impulse strobograph, equipment type 4). (A) The signals $(1)$–$(3)$ are superposed photographically (result $n = 3$). The mean curve (dashed line) is found by visual averaging. (B) The same signals are sampled graphically at $m = 8$ different times (addresses). The samples are averaged separately for each address point. The mean values are plotted at the corresponding address points (with maximum variations). The solid line gives the mean curve of this plotting.

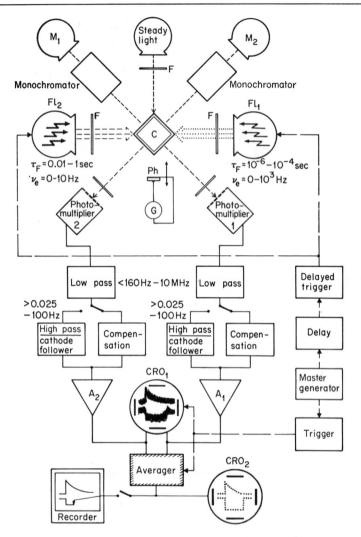

Fɪɢ. 24. Device diagram: General block diagram of repetitive flash photometric equipment. (See footnotes 28 and 29.)

*Excitation of photoreaction:* (1) Repetitive flash lamp $Fl_1$ [$0$–$10^3$ pulses/sec, e.g., specially developed (see footnotes 29 and 55) or AEG-stroboscope (footnote 55).] (2) Electromagnetic light shutter $Fl_2$ ($0$–$10$ pulses/sec with variable flash duration). $F$, Light filters (glass or interference; see Fig. 20). $\tau_F$, tenth-width of flash emission.

*Measurement of photoreaction:* Two separate monochromatic monitoring light beams $M_1$ and $M_2$ are used which irradiate the sample in the cuvette $C$ within the same area. In special cases additional steady light is used. The output of the photomultiplier (EMI 9558 Q) is shaped in both signal channels by a low pass. For slow signals the of a recorder; for rapid processes ($\tau \lesssim 1$ sec), by a photograph of the oscilloscope pattern[50] or by memory oscilloscopes (hp 141 A, Tek. 549). The utmost gain in signal-to-noise ratio, however, which is obtained by this simple averaging procedure (exhaustion technique[50,51]) is indeed not more than about 10. On the other hand, instead of superposing the whole signal course, it is sufficient to accumulate only a definite number of $m$ time samples[52] by which the time course of the signal is completely reproduced. The corresponding time sample values are superposed and averaged in special storage circuits ($RC$ elements, or—after digitizing—magnetic core memories). This time sampling and averaging technique[53] is illustrated by Fig. 23B. In contrast to the simple exhaustion technique, the time sampling method can be performed with high accuracy by modern impulse electronics. Instead of $m$ samples per signal (multiaddress sampling), it is also possible to take only one sample out of each signal (one-address sampling). In this case the sampling position is slowly delayed over the whole signal time $T_s$. Therefore, repetitive signal evaluation devices are not only characterized by the kind of averaging (electric or magnetic), but also by the sampling procedure: one-address or multiaddress signal analyzer.

The principles of both the averager types were illustrated in Fig. 11: (A) one-address averager (top); (B) multiaddress averager (bottom); the main features have already been discussed in Section II.

In the equipment types 5 and 8, multiaddress sampling with averaging by a magnetic core memory is mainly used, whereas in the equipment

direct-current level caused by the measuring light is compensated (see Fig. 20). For fast signals this level is eliminated by a high pass. The filter network is connected to the amplifier ($A \approx 10$) by a cathode follower. The amplifier outputs are controlled by a dual beam oscilloscope $CRO_1$ and fed to the averager; e.g., *Enhancetron*, CAT (two channels), *impulse strobograph* (one channel), etc. (see below and Table VI). The averager output is controlled by a second oscilloscope $CRO_2$ and read out on a recorder.—*Trigger assembly* (interrupted lines): The whole equipment is governed by a master generator (Tek. 162) which gives a direct trigger (Tek. 161) to the averager and $CRO_1$ and a delayed trigger (delay: Tek. 162; trigger: Tek. 161) to flash units $Fl_1$ and $Fl_2$.—*Intensity measurement:* directly within the cuvette by a small photoelement $Ph$ which is calibrated by a thermopile. $G$, $\mu$-ammeter—*Maximum bandwidth:* usually defined by the time resolution of the averager (see Table VI).

---

[50] Exhaustion technique, cited in A. V. Phelps and J. L. Pack, *Rev. Sci. Instr.* **26**, 45 (1955).

[51] H.-J. Einighammer, *Naturwissenschaften* **54**, 641 (1967).

[52] The minimum number $m$ is given by the sampling theorem of the transmission theory as $m = 2\Delta\nu \cdot T_s$ with $\Delta\nu$ = bandwidth and $T_s$ = signal time; see also M. Schwartz,[43] Chapter 4, on periodic sampling and pulse modulation.

[53] The time sampling and averaging technique was used first for instantaneous emission spectroscopy, by C. F. Hendee and W. B. Brown, *Phys. Rev.* **93**, 651 (1954).

types 4, 6, and 7 one-address sampling is applied which employs electric averaging.

A schematic block diagram of repetitive flash photometric equipment[28,29] is given in Fig. 24. The main features of the single flash equipment occur again in this set-up (compare Fig. 20). The essential differences are the

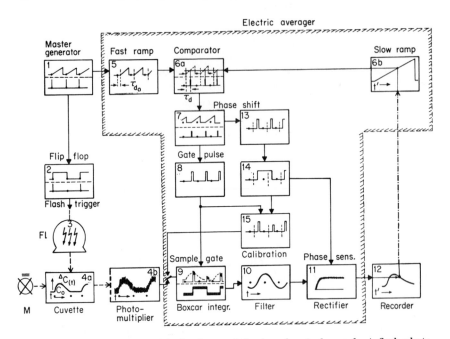

FIG. 25. Device diagram: Block scheme of the impulse strobograph. A flash photometric measuring device is added schematically (see Fig. 20) to illustrate the whole signal processing diagram. $t$, Signal time; $t'$, recording time. (1) Master generator, Tek. 162, repetitive rate $2\nu_e$. (2) Flash trigger unit with frequency reduction 2:1 (flip-flop circuit I). (3) Repetitive flash unit, repetitive rate, $\nu_e$. (4) $M$, monochromatic monitoring light; $a$, measuring cuvette, $b$, photomultiplier (see also Fig. 20). (5) Fast ramp with fixed delay $\tau_{do}$ = zero delay time, Tek. 162 (PM 5720). (6a) Comparator to produce variable delay $\tau_d$ with slow ramp of 6b: Tek. 161 (PM 5723). (7) Scanning gate pulse trigger, Tek. 162 (PM 5722). (8) Scanning gate pulse generator, Tek. 161 (PM 5728 a. pulse amplifier). (9) Sampling diode gate and boxcar integrator. (10) Tunable electronic double band filter amplifier, frequency $\nu_e$. (11) Phase-sensitive rectifier. (12) $x$-$t$-recorder. (6b) Slow ramp ring potentiometer coupled to recorder drum which controls externally the delay level of the comparator 6a. (13) Delayed trigger, Tek. 161. (14) Switching square-wave generator (flip-flop circuit II). (15) Calibration signal: ring modulator. For triggering and delaying operations the strobograph is normally furnished with Tektronix generators of the Tek. 160 series. For rapid operation ($\tau_d < 10\,\mu\text{sec}$, gate width $\tau = 0.05\,\mu\text{sec}$) Philips pulse generators are used (PM numbers in brackets) (4a) $c_0$, initial sample concentration, $\Delta c(t)$, concentration change. (See footnotes 24 and 25.)

two light beam systems $M_1$ and $M_2$ with separate measuring channels, the cathode follower for fast signal compensation, the trigger and delay units for the flash lamps $Fl_1$ and light shutter $Fl_2$, the averager, the readout oscilloscope $CRO_2$ and the recorder. There are several reasons for the requirement of two measuring channels for repetitive measurements. One reason is the disturbance by the scattered light of the flash, which has been discussed in detail for the single flash equipment (equipment type 1). According to the higher sensitivity, however, the repetitive flash equipment is much more subject to this interference. Another reason is the long recording time ($T_r \approx 30$ min) in comparison to the life of the photoactive samples. Therefore reference measurements have to be performed simultaneously, especially if a full difference spectrum of an absorption or fluorescence change is measured point by point. Both measuring beams pass the cuvette in the same area. The signals in both channels (absorption or fluorescence changes) are displayed directly on a monitor double-beam oscilloscope $CRO_1$, and fed into an averager. The averaged signals can be displayed on the oscilloscope $CRO_2$ or written out on an $X$-$Y$-recorder.[54]

With respect to the prolonged measuring time, precautions are necessary to stabilize the reaction system: thermostatting of the cuvette $C$, measures to prevent settling down of test suspension, chemical stabilization against activity loss, etc.[24] For further details about photometric measurement, compare equipment type 1 (Section III). According to the enhanced sensitivity (see Tables III and IV) it is necessary to stabilize the photomultiplier voltage and the light intensity to a correspondingly high extent. Moreover, in some cases it was necessary to improve the mechanical stability of the optical benches by mounting them shock free against ground vibrations.[29]

### Type 4. Electric Averaging

*Impulse Strobograph*

The principal set-up of a special arrangement for electric averaging is depicted in Fig. 25, which operates as a strictly periodic one-address signal analyzer.[24,25] For the elucidation of the performance by a complete signal processing plan, the excitation and detecting system of a flash photometric assembly (see Fig. 20) is also roughly outlined in Fig. 25. A typical version of complete repetitive flash photometric equipment is depicted in Fig. 24.

*Excitation System.* The whole equipment is governed by a master

---

[54] Further details about this equipment are given in the legend of Fig. 24.

generator (1) which delivers a strictly periodic sequence of trigger pulses (trigger frequency $2\nu_e$). The trigger pulse rate is reduced 2:1 (2) for the triggering of the excitation (3). As actinic light source usually a rod type flash tube[55] is fired with a repetition rate $\nu_e$.

*Detection System.* The photometric detection system is nearly the same as for single flashing (see equipment type 1; Fig. 20).

*Averaging System.* The trigger pulse series produced by the master generator of stage (1) is delayed in the stage (5) (fixed zero delay $\tau_{d0}$ for spreading of time at optional positions of time axis) and stage (7) (variable delay $\tau_d$ for signal scanning) and is used to trigger the narrow sample gate pulses (8), which open the measuring gate (9). By means of the 2:1 reduction in (2) there are two gate impulses per measuring signal. By the first gate impulse the signal is measured at the delaying time $\tau_d$, by the second impulse the corresponding zero-reference value. The series of measuring and reference impulses is fed into a boxcar integrator (9). The amplitude of the square wave is proportional to the difference between measuring and reference impulse. The square wave of the boxcar integrator is passed through an electronic tunable double-band filter (10) by which the basic wave (frequency $\nu_e$) is filtered out. The amplitude of the basic wave is proportional to the signal value at the delaying time $\tau_d$. This sine wave is rectified by a phase sensitive rectifier (11). The signal direct-current voltage is plotted on a recorder (12). The recorder drum is mechanically connected with the delay control potentiometer (6b) of the comparator (6a). The slow ramp which is thus produced at the output of (6b) defines the delay time $\tau_d$ for the sample gate position. This is achieved by an amplitude comparison between slow (6b) and fast ramp (5) in the comparator circuit (6a). The phase-sensitive rectifier (11) is operated by a square-wave switching voltage (14) which is exactly phase-matched, as derived from (6a) by means of a third delay unit [ramp generator (7), comparator (13)]. For calibration a periodic sequence of rectangular pulses is generated in (14) and (15) which has an adjustable pulse height difference.

### Time Resolution

The resolution time for the one-address averager with continuous sample scanning is determined only by the sample gate width, $\Delta t_{\min} \approx \tau_g$.[24] The gate width of the impulse strobograph is normally 1 $\mu$sec with Tektronix generators and $\approx 50$ ns with special Philips generators of the PM-series.

---

[55] Equipped with rod type quartz flash tube XIE 200 Osram, Berlin, Germany; $\tau_F \approx 50$ $\mu$sec (for further details see also Table V). For extremely short, intense flash sources see equipment type 6.

*Gain in S/N Ratio*

The numerical gain in $S/N$ ratio $G = \sqrt{n}$ for $n = \nu_e T_r$ repetitions,[56] cannot be obtained with a one-address analyzer because only one sample is withdrawn from each signal. According to the time sampling theorem[52] $m = 2\Delta\nu T_s$,[56] pieces of information (Fourier coefficients) are necessary for the complete reconstruction of a signal from the sample values. A gain in the $S/N$ ratio is thus given by the excess of available information $n/m$ only, i.e., by $G = (\nu_e T_r/m)^{1/2}$. If the output bandwidth $\Delta\nu^0$ is matched for optimum the output signal in the slowed down time scale $t^0$ on the recorder contains the same number $m = 2 \cdot \Delta\nu^0 T_s^0$ [56] pieces of information. As $T_s^0 = T_r$ and $\Delta\nu^0 = \frac{1}{4}\tau^0$,[57] the theoretical gain in $S/N$-ratio improvement is

$$G = \sqrt{2 \cdot \nu_e \cdot \tau^0} \tag{15}$$

By the phase-sensitive rectification which is used in the impulse strobograph, the output noise level is further decreased by a factor $1/\sqrt{2}$.[51] The gain is correspondingly further increased to

$$G = 2\sqrt{\nu_e \cdot \tau^0} \tag{16}$$

*Repetition Rate.* The equipment is arranged for repetition rates from 1 Hz to 2 kHz. The repetition rate of 1 Hz defines the slowest time deflection to about 0.5 second.

*Output Time Constant.* In principle, the same conditions are valid for the adjustment of $\tau^0$ as for the optimum bandwidth limitation of a single signal (see Section III, type 1, *"Optimum Bandwidth"*). The upper limit of $\tau^0$ is given by the inequality $\tau^0 < T_s^0 \leq T_r$. Within this range $\tau^0$ should be chosen as large as compatible with the accuracy requirements for the measurement of reaction times. A reasonable adjustment is given if $\tau^0$ is chosen in relation to the input time constant $\tau^i$ in the exact slow-down ratio of the bandwidth transformation by the time sampling procedure: $\tau^0 = (T_s^0/T_s)\tau^i$. Selecting further $\tau^i = \tau_{min}/\pi$ [for $\lesssim 5\%$ input signal distortion, see Eq. (11b)] and $T_s^0 = T_r$ (for optimum sampling operation) the output time constant is

$$\tau^0 = \frac{\tau_{min}}{\pi T_s} \cdot T_r \tag{17}$$

---

[56] $\nu_e$ = Repetition rate; $T_r = 1/\nu_r$ = signal recording time; $\Delta\nu^0$ = bandwidth of the detection system (rectangular); $T_s$ = signal time or signal sweep duration. The index (o) symbols are related to the slowed-down time scale ($t^0$) of the output signal.

[57] For noise transmission a simple RC-Low pass is equivalent to a rectangular low pass with $\Delta\nu^0 = \frac{1}{4}\tau^0$;[43] $\tau^0$ = RC = output time constant.

Typical values of $\tau^0/T_r$ are in the range 0.1–0.01. With condition Eq. (17) the output signal can be regarded as having twice passed a low pass of time constant $\tau^0$. Therefore, the resulting inaccuracy for the evaluation of reaction time is $<10\%$. In the case of poor output $S/N$ ratio, however, it is advantageous to increase $\tau^0$ tolerating an increased signal distortion [see Eq. (10)] for a better $S/N$ ratio.

*Sensitivity*

The limiting sensitivity is given by the electrical interference of the device itself, which has a value of 5–10 $\mu V$ nearly independent of the bandwidth. The maximum amplification is approximately $10^5$. Typical specifications are listed in Table VI.

*PAR—"Boxcar Integrator" CW 1*

A simple one-address electric averager is now also commercially available by Princeton Applied Research Corp., Princeton, New Jersey. The PAR "boxcar integrator" uses the same fundamental principle as the impulse strobograph but omits mainly reference impulse, filter, and phase-sensitive rectification. The overall amplification is 50. A variable time constant is used as integrating element at the boxcar integrator output. Other specifications are given in Table VI.

*PAR—"Waveform Eductor" TDH6*

This equipment too was developed by Princeton Applied Research Corp. and represents a 100-address electric averager according to the principle shown in Fig. 11B. Each time sampling value S is integrated during the gating time $\tau_g$ on separate integrating elements $RC_i$. The maximum value of the variable integrating elements $RC_i = \tau^0$ is 1 sec. The effective time constant for the output signal is $\tau_{\text{eff}} = 100\ \tau^0$. After $n = 2\tau^0/\tau_g = 2\tau_{\text{eff}}/T_s$ repetitions the charge on the capacitors $C_i$ has reached 86% of the corresponding input value ($T_s = 100 \cdot \tau_g$ = sweep time). Therefore, maximum gain in $S/N$ ratio is $G = (2\tau^0/\tau_g)^{1/2}$. During the time between the charging intervals the capacitors are disconnected from the signal source. In the case of rapid signals with low repetition rates (i.e., low duty cycle $\nu_e \cdot T_s \ll 1$) a long recording time $T_r$ up to ½ hour is necessary at $\tau_{\text{eff}} = 100$ sec to have a large number of signals averaged. During this long measuring time a loss of amplitude by condenser leakage cannot be avoided. Therefore, this equipment is recommended for repetitive signals with a duty cycle near 1. (Further specifications are given in Table VI.)

Type 5. Magnetic Averaging

For most of the biochemical reaction systems (photosynthesis, respiration chain, etc.) only low repetition rates can be applied although fast reactions are to be measured (see Fig. 17) under these conditions it is difficult in electric multiaddress averaging to store the averaged time samples over a sufficiently long time period on the charging capacitors (see above, PAR-TDH6). For this measuring problem the so-called "magnetic averager" is especially suited because the sampled signal information is digitized and can be stored for an unlimited time in a magnetic core memory. The mechanical size of such a core memory is relatively small and at least not essentially larger than an assembly of RC-elements, with a time constant of more than 200 seconds (see Fig. 11B), which is necessary for these measurements with low duty cycle. Magnetic averaging was used first for nuclear magnetic resonance investigations in 1962.[58]

The working principle of most of these commercial averagers can be summarized in five steps: (1) Samples of the signal are taken by a gate impulse at a preset number of phase positions which correspond to the "addresses" of the core memory. (2) Each of these sampled amplitude values is digitized by an analog-digital conversion device (a.d.c.). (3) The incoming digital amplitude information is arithmetically added to the last digital address value, which is read from the core memory. (4) The new address value is written into the core memory. (5) After the whole measurement, the averaged information for each address is read out of the core memory, decoded—i.e., digital-analog converted—and displayed on an oscilloscope or recorded on a $X$-$Y$-recorder.

*Analog-Digital Conversion*

For the analog-digital conversion (step 2) in most of the averagers a conventional amplitude controlled-frequency generator is used, e.g., CAT 1000, Technical Measurement Corp., or NS 544 Northern Scientific Instruments. In more recently developed averagers a "ramp type" a.d.c. is used. In this converter type a capacitor is charged by the sampled signal value and linearly discharged. During this amplitude-controlled discharge time a number of pulses from a constant high-frequency pulser is counted into the averager that is proportional to the sampled signal value (e.g., FT 1062, FABRI-TEK Instr. Inc., or $hp$ 5480A, Hewlett-Packard).

A quite different digitizing method is used in the averager Enhancetron (Nuclear Data, Inc.). The sampled amplitude values $S$ are compared with a reference pulse train of $m_0 = 256$ discrete alternating voltage steps

[58] M. P. Klein and G. W. Barton, Jr., *Rev. Sci. Instr.* **34**, 754 (1963).

which cover the input voltage range $V_0$ with intervals of $V_0/m_0$, but they do follow in a quasi statistical sequence ($m_0$ = digital resolution). At $m_0$ comparisons, the number of cases $m$ with signal values $S$ "larger than" the reference step $V_m$ (i.e., "yes") is a discrete measure for the relative height of the real signal amplitude value $S$ because $m/m_0 < S/V_0 \leq (m + 1)/m_0$. If in the "yes" case: $S > V_m$ one count is fed into the corresponding address in the memory, there is a digital representation of $S$ after $m_0$ comparisons.

Various magnetic averagers such as CAT or Enhancetron have the disadvantage that the sampled output signal rises with the number of excitations "totalyzers". Therefore the determination of the exact input signal value is sometimes tedious. This disadvantage is overcome by the NS 513 (Northern Scientific) or 5480 (Hewlett-Packard) which gives a successive improvement of $S/N$ ratio at constant mean value of the signal. The measuring pulse rate which is fed into each address is not proportional to the absolute sample value but proportional to the difference between the mean address value already stored and the signal value just sampled at the input. In this way a successive approximation to the exact mean value of the signal is obtained (real "averager").

*Time Resolution*

The problem of the magnetic averagers, however, is the finite time interval which is necessary for the total process of accumulating every sample value, i.e., digitizing, processing, and storing. This time interval defines the minimum time distance between two samples of the signal which determines the time resolution. A typical value for this memory cycle time with commercial magnetic averagers is about 30 μsec.[59] As a signal slope is clearly detected by three sample points, it is convenient to define the double cycle time as time resolution. The resolution times of some typical magnetic averagers are listed in Table VI. The time resolution of these relatively slow instruments may be enhanced by use of a time sampling unit which is attached ahead of the averagers. Suitable microsamplers are available from Technical Measurement Corp., and Northern Scientific which make measurements possible with time resolution down to 2 μsec.

*Gain in S/N Ratio*

In general, the gain in $S/N$ ratio $G = (S/N)^o/(S/N)^i$ ($o$ = output, $i$ = input) of a digital magnetic averager after $n$ repetitions cannot be

---

[59] A rapid analog-digital converter has recently been developed by Fabri-Tek Instruments Inc., which enables a sampling point distance of 1 μsec. The processing procedure is performed in the time interval between two following excitations (10 msec; see also Table VI).

expressed in a simple relation of $\sqrt{n}$ (in contrast to the analog electric averager). Certainly $G$ rises monotonically with $n$ but approaches a maximum value $G_{max}$[60] attaining 90% of maximum for $n = n_m$ (typical values: $n_m = 2000\text{--}10{,}000$). In particular, for small values of $n$ ($\lesssim 100$) $G$ mostly remains considerably lower than $\sqrt{n}$. This is caused by the effective internal interference level in relation to the input noise level. For small $(S/N)^i < 1$ $G$ rises linearly with $\sqrt{n}$ in a mean range $100 \lesssim n \ll n_m$:

$$G = \gamma\sqrt{n} \tag{18}$$

Usually $\gamma$ has values of 0.8–1.1. As the gain relations of the electric averagers given by Eqs. (15) and (16) are easily rewritten into the form of Eq. (18), $\gamma$ can be interpreted as a "gain efficiency" by which the different types of averagers may be compared (see Table VI). This comparison, however, can only be performed within the linear range of magnetic averagers. Moreover, as the gain efficiency of one-address electric averagers depends additionally on the duty cycle $\nu_e T_s$ ($\nu_e$ = repetition rate, $T_s$ = signal time) a numerical comparison must be based on equal operation data. The dependency of $G$ versus $n$ for magnetic averager described above is not valid for $(S/N)^i \gtrsim 1$. With increasing $(S/N)^i$ $G$ decreases for all values of $n$ and the linear range disappears. In contrast to the magnetic averager, however, the gain of the electric averager rises linearly with $\sqrt{n}$ at all conditions independently of the $(S/N)^i$-value (see graph in Table VI).

At equal duty cycles the gain efficiency $\gamma$ of one-address electric averagers is comparatively low (see Table VI) but may attain comparable values at extremely low duty cycles (recommended operation for one-address averagers).

*Sensitivity*

The input voltage range of the magnetic averagers is usually of the order of $\pm 1$ V. Therefore, in general a preamplifier must be used to give an adequate magnitude of input signal. It is suitable for best results at extremely poor $S/N$ ratios (see Fig. 13, $n = 1$) to set the amplification so that the noise band just fits the total input voltage range. Further specifications of the different averagers are given in Table VI.

### E. Type 6. Repetitive Ultrashort-Flash Photometry

Measurements of fast photoreactions with reaction times smaller than 10 $\mu$sec cannot be realized without modern impulse techniques. First of all,

---

[60] $G_{max}$ is given by the ratio $m_o/(S/N)^i$ if not restricted by too small a memory capacity, ($m_o$ = digital resolution) cf. footnote 42 and graph in Table VI.

TABLE VI
SPECIFICATIONS OF VARIOUS AVERAGERS[a]

| No. | Averager[b] Type | Resolution time ($\mu$sec) | Min. gate width ($\mu$sec) | Max. address/ channel number | Digital resolution[c] (bits) | Memory capacity (bits) | Repetition rate (pulses/sec) Max. | Repetition rate (pulses/sec) Min. | Gain $G = (S/N)^{out}/(S/N)^{in}$ Linear range | Gain $G = (S/N)^{out}/(S/N)^{in}$ Max. value | $\gamma = G/\sqrt{n}$ Linear range |
|---|---|---|---|---|---|---|---|---|---|---|---|
| 1 | NS 544 (m) | 100 | 25 | 1024/2 | 7 | 16 | 15 | $4 \times 10^{-3}$ | $\gamma \sqrt{n}$ | ? | ? |
| 2 | CAT 1000 (m) | 50 | 17 | 1024/4 | 4 | 20 | 30 | $6 \times 10^{-2}$ | $\gamma \sqrt{n}$ | 200 | 0.8–1.1 |
| 3 | Enhancetron | 50 | 27 | 1024/2 | 8 | 14 | 30 | $5 \times 10^{-4}$ | $\gamma \sqrt{n}$ | 150 | 0.8–0.9 |
| 4 | hp 5480/85A (m) | 20 | 1.2 | 1000/2 | 5 | 24 | 100 | $2 \times 10^{-3}$ | $\gamma \sqrt{n}$ | 100 | 1 |
| 5 | FT 1052 + 952 (m) | 2 | <1 | 1024/4 | 5 | 18 | 90 | $5 \times 10^{-4}$ | $\gamma \sqrt{n}$ | ? | ? |
| 6 | PAR-TDH6 "waveform eductor" (e) | 2 | 1 | 100/1 | — | — | $10^5$ | $10^{-1}$ | $\sqrt{n}$ | — | $1^f$ |
| 7 | PAR-CW 1 "boxcar integrator" (e) | 1 | 1 | 1/1 | — | — | $10^5$ | 1 | $\sqrt{\dfrac{2\tau^0}{T_r}} \cdot \sqrt{n}$ | Unlim. | $\approx 0.2$ |
| 8 | Impulse strobograph (e) | 1 $(5 \times 10^{-2})^d$ | 1 $(5 \times 10^{-2})^d$ | 1/1 | — | — | $2 \times 10^3$ | 1 | $\sqrt{2}\sqrt{\dfrac{2\tau^0}{T_r}} \, \sqrt{n}$ | Unlim. | $\approx 0.3$ |
| 9 | Averaging with sampling oscilloscope (hp 185 B) (e) | $10^{-3}$ | $10^{-3}$ | 1/1 | — | — | $10^8$ | 10 $(1)^e$ | $\sqrt{\dfrac{2\tau^0}{T_r}} \cdot \sqrt{n}$ | Unlim. | $\approx 0.2$ |

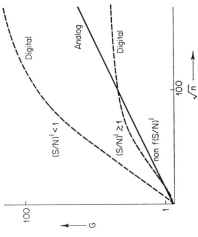

[a] $(m)$ = magnetic, digital; $(e)$ = electric, analog; $n = \nu_e \cdot T_r$ = number of repetitions, $\nu_e$ = repetition rate, $T_r$ = recording time; $T_s$ = signal time; $\tau^0$ = output time constant; $\tau_g$ = gate width. Maximum gain values for magnetic averagers are given for input $S/N \approx 1/10$. The numerical values of $\gamma$ are given for all averagers at equal duty cycles $(\nu_e \cdot T_s)$ and maximum time resolution, in case of electric averagers 7–9 especially for $\tau^0 = (1/60)\ T_r$.

[b] (1) Northern Scientific Inc.; (2) Technical Measurement Corp.; (3) Nuclear Data Inc.; (4) Hewlett Packard; (5) FABRI-TEK Instruments Inc.; (6) and (7) Princeton Applied Research Corp.; (8) Equipment type 4; (9) Equipment type 6 and 7.

[c] Analog digital converter: 1 and 3, weighed approximation type [by reference pulse train, full resolution at 128 (1) or 256 (3) input signals]; 2, freq. mod. type (digital resolution for standard modulator 250 kHz); 4 and 5, ramp type.

[d] With Philips generators PM.

[e] With selected gate diodes.

[f] For $n \ll \tau^0/\tau_g$.

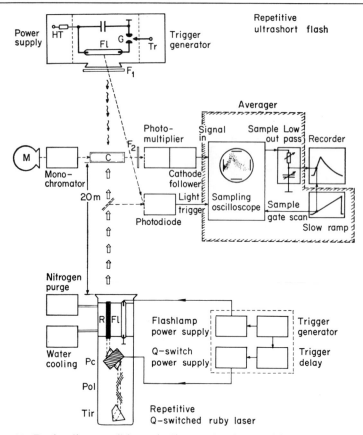

Fɪɢ. 26. Device diagram: Schematic diagram for the repetitive ultrashort flash and the repetitive Q-switched ruby laser spectrophotometric equipment (see footnotes 33b, 33c, and 35).

*Ultrashort flash:* Fl, Flash tube XE 9-3 (PEK); G, Spark gap GP 12 A (EGG); power supply: 12 kV, 100 mA. Tr, Trigger generator: normal flash lamp firing circuit (see Fig. 20) switched by a thyratron or controlled rectifier; firing pulse transformer: TR 60 (EGG). HT, high voltage. Repetition rate $\nu_e$: 0–10 pulses/sec (footnote 33c).

*Laser system:* Special development by Raytheon Company, Waltham, Massachusetts: $R$ = 6 inch ruby; Fl: Flash lamp FX 47 C-6.8 (EGG), $E_i \approx$ 900 J; Pc = pockels cell: KDP crystal. *Q-switch operation:* 15 kV, $\lambda/2$-retardation voltage, switched by kryotron circuit; Pol: Stacked plate polarizer; Tir: total internal reflection quartz prism; $\nu_e$: 0–5 pulses/sec. *Water cooling:* The flash lamp input power of approximately 5 kW at 5 pulses/sec is dissipated by a 4 gallons/min circulation flow of demineralized water driven by a circulating pump. Cooling water flows around ruby, flash lamp, and mirror cavity. Input temp: 6°C; maximum cooling power: 9 kW; cooling water reservoir: 200 gallons. *Nitrogen purge:* A flow of dry nitrogen protects the cooled optical parts from condensed water (footnote 35).

*Measuring system* (see also Fig. 20): M, monitoring light; C, cylindrical cuvette; $F_1, F_2$, filter. Sampling oscilloscope, hp 185 B; low pass: variable typ. val. $\tau_0$ = 20 sec.

a short but intense pulse of light is required for the photostimulation. Second, a narrow sampling gate pulse is necessary to resolve the repetitive rapid signals by the time sampling technique.

## Ultrashort Flashes

Extremely short electronic flashes or sparks[61,62] can be produced by special discharge circuits which are developed for low R-L-C-values. This circuitry normally exhibits only small geometrical dimensions. The discharge energies are extremely low. Besides, the plasma temperatures are relatively high and cause a light emission mainly in the far blue or ultraviolet spectral range. These flash sources are of interest for photochemical processes which are excited in the ultraviolet spectral range, but they do not emit sufficient light energy to excite photoreactions in the visible spectral range at all, e.g., in the primary process of photosynthesis. Furthermore, in spite of the electric discharge being kept extremely short, the light emission may be considerably longer because of the afterglow of the heated plasma. The afterglow can last a time up to the order of microseconds.

A relatively high power density in the visible spectral range and a short afterglow can be obtained with a discharge in a narrow channel. The small wall-to-wall space effects rapid deionization and thereby a low mean plasma temperature and a decrease of the afterglow.[63,64]

This principle is realized, for example, in the XE 9 series flash tubes of PEK Labs. Inc., Sunnyvale, California. The discharge is conducted in the small space between two concentric quartz tubes.

The discharge circuit of a XE 9-3 flash assembly is illustrated in the upper part of Fig. 26. A set of 10 low-inductance high-voltage ceramic capacitors is connected in parallel forming a 30 nF discharge capacity with very low inductivity. This capacity is charged with high voltage, $HT$, up to 12 kV. As this value is far above the holding voltage, the XE 9-3 is fired by a high voltage spark gap $G$ which itself is triggered by a normal flash tube trigger pulse, $Tr$, from a usual firing pulse transformer. The whole flash discharge circuit is elaborately mounted so as to keep inductivity as low as possible. With this circuit a flash duration of 0.3 $\mu$sec half-

---

Slow ramp: generated by a motor-driven potentiometer. Minimum rise time of optical detector system: 30 nsec corresponding to 10 MHz bandwidth. (PEK: PEK Laboratories, Inc., Sunnyvale, California; EGG: Edgerton, Germershausen & Grier, Boston, Massachusetts.)

[61] Q. A. Kerns, F. A. Kirsten, and G. C. Cox, *Rev. Sci. Instr.* **30**, 31 (1959).

[62] H. Fischer, *J. Opt. Soc. Am.* **47**, 981 (1957).

[63] G. Glaser and D. Sauter, *Z. Physik* **143**, 44 (1955).

[64] I. S. Marshak, *Proc. 3rd Intern. Congr. High Speed Photography, London 1956*, pp. 30–41. Butterworths, London and Washington, D.C., 1956.

width and 0.8 $\mu$sec tenth-width (between 10%-light power points) is obtained at an electric input energy of $E_i = 2$ J (see Fig. 2). The flash tube $Fl$ is mounted parallel to the cuvette $C$ in a focus line of an elliptic mirror.

### Averaging with Sampling Oscilloscopes

Commercial sampling oscilloscopes are developed for the repetitive measurement of extremely rapid signals. A bandwidth of about 1 GHz is achieved by a sampling gate width of less than 1 nsec. These sampling oscilloscopes can be used for one-address averaging of very rapid repetitive signals. The only condition is that the horizontal phase shift of the sample point can be controlled externally by a slow ramp. The sample-gate scan time $T_r$ (slow ramp duration) must be large in comparison to the time between two following excitations ($\nu_e \gg \nu_r = 1/T_r$). In this case, a sufficient number of samples can be taken for the averaging procedure (compare Fig. 25). A slow ramp is easily produced by a rotary potentiometer fed by direct-current voltage and driven by a servomotor via a variable reduction gear. The slow ramp controls the horizontal phase shift of the sampling point as well as the $X$-deflection of a $X$-$Y$-recorder (see Fig. 26). The vertical output of the sampling oscilloscope (stretched sample pulses) is fed into the $Y$-input of the recorder via a low-pass filter (RC-element) with a time constant $\tau^0$. The selection of optimum value for $\tau^0$ is discussed in Section III, type 1 [see Eq. (17)]. The sampling oscilloscope is triggered by a trigger signal derived directly from the light impulse by a separate photodiode or photomultiplier.

If repetition rates lower than 50 pulses/sec are used for excitation, normally most of the sampling oscilloscope exhibit a perceptible decrease of the sample amplitude from one to the next sample. At the $hp$ 185 B sampling oscilloscope of Hewlett Packard, the holding time of the sampling diode gate could be increased considerably. Thus, repetition rates down to 1 pulse/sec can be used for stimulation without remarkable decrease in overall sensitivity (see below) or deterioration of the signal shape (see also Table VI).

### Optimum Bandwidth

The maximum bandwidth of the measuring channel is determined by the photomultiplier (EMI 9558 AQ), which has a transit time of about 8 nsec. This value corresponds to a bandwidth of 40 MHz. In order to avoid excess input noise, the input bandwidth is kept as low as is just necessary for an undistorted signal transmission. For instance, at the measurement given in Fig. 17 the transmission time constant was $\tau_s = 15$ nsec which corresponds to a bandwidth of 10 MHz. The bandwidth can be simply adjusted by varying the photomultiplier anode resistor. But as this measure is always combined with a change in amplification, it is more convenient to use a variable high frequency low pass behind the cathode follower (see below "sensitivity").

### Gain in S/N Ratio

The operation of the slow-scanning sampling oscilloscope corresponds to that of the other analog one-address analyzers which employ the stroboscopic principle. Therefore the gain in $S/N$ ratio is given by Eq. (15)

$$G = \sqrt{2\nu_e \tau^0} \tag{15}$$

With this assembly a gain in $S/N$ ratio $G \approx 20$–$30$ has been obtained.[33c]

### Sensitivity

The input sensitivity is fixed by the sampling oscilloscope, which has a maximum sensitivity of 1 mV/cm. The internal interference level is of the order of $\lesssim 0.5$ mV. Independent of the white shot noise, the real input signal should be considerably larger than this interference level. For small signals this can be achieved to a small extent by a higher amplification in the photomultiplier, i.e., by higher voltage per stage or—if compatible with bandwidth requirements—by a higher anode resistor. Since recently wideband amplifiers, especially for PM-connection are commercially available (e.g., Model 105, 15 Hz–180 MHz, Keithley Instr. Inc., Cleveland, Ohio), it is suitable to use the following amplifier bench: a 1:1 impedance changer (emitter follower) connected to the PM anode resistor followed by one or two 10:1 low-impedance amplifiers. Two of these units can be connected via the variable low-pass filter to adjust optimum bandwidth (see above). According to the high preamplification the amplitude losses which are necessarily combined with a band limitation can be fully compensated by excess amplification.

All other details of this equipment are similar to those of type 1.

### F. Type 7. Repetitive Laser Giant-Pulse Photometry

The measuring and averaging part of this equipment is the same as in equipment type 6. Therefore both types of equipment have the same values of input sensitivity and gain in $S/N$ ratio. Only the ultrashort flash source is replaced by a repetitive giant-pulse ruby laser. The maximum bandwidth of 40 MHz (rise time 8 nsec) is fully sufficient to resolve signals that rise within the giant-pulse duration.

### Laser Giant Pulses

For rapid photochemical reactions a minimum light energy is necessary during the short excitation time to give a measurable turnover. In special cases, for example, in photosynthesis, it is important to have at least full saturation energy for the photochemical primary process. On the other hand, there is a rule that maximum radiant power density of every flash tube or spark gap discharge is constant.[63, 65] That means, the light energy in the pulse becomes necessarily too low for kinetic measurements if the flash duration is decreased below a definite value. This limit seems to be attained, for example, for the primary process of photosynthesis with the ultrashort flash tube with 0.3 $\mu$sec half-width time. For shorter excitation times the radiant power density is to be increased considerably. This is impossible with normal discharge circuits. At a giant-pulse laser, however, the light energy of a high energy flash discharge with relatively long emission time is stored in the ruby crystal. The ruby is placed in a light resonator system which has a closed light "Q"-switch between the reflectors. If maximum population of excited chromium atoms in the ruby is gained shortly after flash maximum, the "Q"-switch of the resonator is opened within a few nanoseconds. Now, the stored energy is immediately released by stimulated emission within the extremely short time of 10–20 nsec. By this Q-switched laser technique, extraordinarily high power densities of many GW/cm² have been obtained.

Obviously, all requirements for high light-excitation energies can be satisfied by far with such laser giant pulses. The problem for photochemical excitation is that these light pulses can be produced only at the ruby wavelength of 694 nm or—with reduced intensity—via a frequency doubler crystal at 530 nm if a YAG[66]-crystal laser with original emission at 1060 nm is applied.[67] After the first results with liquid dyestuff solution, however, it seems very probable that, in the future, lasers could be operated at optional wavelengths with additional fine tuning.[68]

The essential parts of a Q-switched ruby laser system for high repetitive giant-pulse operation with the required specifications are depicted in the schematic diagram Fig. 26 (bottom). The physical problem is that the requirements of shortest emission time and that of high life and reliability contradict each other. A short emission time implies high amplification, which itself is obtained with (a) low threshold energy and (b) high pumping energy. Low threshold is achieved at low losses in the resonator system and by a relatively long ruby rod. High amplification, however, yields high power densities, more by far than is necessary for rapid excitation of photochemical reactions. The high power density ($\gtrsim 20$ MW/cm²) causes damage to the surfaces or phase boundaries. The exceptional technological

[65] G. Glaser, *Optik* **61** (1950) especially paragraph 12, p. 79.

[66] YAG = yttrium-aluminum garnet.

[67] Frequency doubling takes place if the intense laser light passes special crystals such as potassium dihydrogenphosphate (KDP). An efficiency for the energy transformation to second harmonic up to 100% has been reported: J. E. Gensic, H. J. Levinstein, S. Singh, R. G. Smith, and L. G. van Uitert, *Appl. Phys. Letters* **12**, 306 (1968).

[68] W. Schmidt and F. P. Schäfer, *Z. Naturforsch.* **22a,** 1563 (1967).

problem is to operate a giant-pulse laser repetitively with a frequency of 5–10 pulses/sec in an uninterrupted series of 5000–10,000 shots. Within this series there should not occur any remarkable energy drop and no change in pulse duration. Therefore, all additional optical surfaces must be of high-quality quartz.

The optical resonator system is given by the ruby $R$ front planar at one side and a total internal reflection prism $Tir$ at the other. A pockels cell, $Pc$, serves as Q-switch which is closed by a voltage for $\lambda/2$-retardation in combination with a stacked plate polarizer $POL$. The firing pulse for the pockels cell operation is derived from the main trigger via a delay circuit which is adjusted for optimum giant-pulse operation (see above). The firing pulse triggers a rapid discharge circuit which drops the pockels cell voltage down to zero and opens thus the Q-switch within a few nanoseconds. The flash tube (EGG[69] FX 47C-6.8) is fired with 5 pulses/sec maximum. The electric input energy is about 1 kJ; the light output energy, however, is only about 0.1 J. Therefore, at 5 pulses/sec, nearly 5 kW of heat energy is dissipated during flashing. The optimum laser temperature is about 5–10°. Therefore, an efficient water cooling system is necessary. A water flow of 4 gallons/min in a circular system cools ruby, flash-lamp, and the mirror cavity directly. The flow system is thermostatted by a 9 kW cooling machine (Escher Wyss, Lindau, Germany) via a special heat exchanger. The flash lamp with its high voltage contacts is fully immersed in the circulating cooling water. Therefore it is recommended that demineralized water be used for the cooling flow system, to minimize electrocorrosion of the flash lamp seals.

The mirror cavity (elliptical cylinder: one focus line, ruby $R$; the other, flash tube $Fl$) and the whole resonator system is purged by dry nitrogen in order to protect the parts of the cooled optical system from condensed water.

The high peak discharge currents of the pumping flash lamp ($\approx 1$ kA) as well as the high recharge currents of the power supply ($\approx 200$ A solid state controlled rectifier circuits) give an enormous electric interference to the measuring part of the equipment and especially to the gate input of the sampling oscilloscope. Only by rigorous measures was it possible to avoid this disturbance: (1) The power which is necessary for the measuring system (Fig. 26, center), especially for the sampling oscilloscope, is taken from a fully separate power line, e.g., 2.5 kW alternating current power generator driven by a direct-current motor. (2) The laser system with power supply and cooling system is operated more than 20 meters apart from the measuring assembly. The laser giant pulses are guided to the measuring cuvette by a mirror system. To reduce the beam divergence, a

[69] EGG: Edgerton, Germershausen & Grier, Boston, Massachusetts.

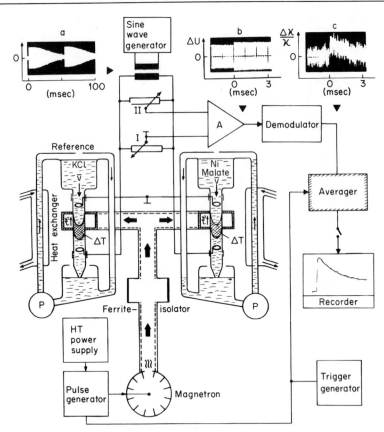

Fig. 27. Device diagram: Schematic diagram for repetitive conductivity measurements after temperature-jump excitation in a flow system. The temperature jumps are produced by short microwave pulses that are absorbed in aqueous solutions. For microwave generator and waveguide system see type 3. Frequency 9.4 GHz. The sample solution circulates in a flow system which consists of upper reservoir, absorber tube through the waveguide with electrodes for conductivity measurements, receiver, pump $P$, heat exchanger, overflow. $\bar{v}$ = mean flow velocity (0–50 cm/sec), $\varepsilon$ = electric vector of microwave field, $\Delta T$ = temperature jump, HT-high voltage. Two identical flow systems are used. Reference system: KCl solution of the same zero conductivity as the measuring system (e.g., Ni malate). Both conductivity cells are connected to the radio-frequency bridge which is balanced by a potentiometer (point $II$). In this way the effect of ion mobility change (partial picture a) is mainly compensated. The bridge is fed by 100 or 300 kHz sine wave voltage of approximately 3 $V_{pp}$. The bridge difference signal is amplified by a selective amplifier $A$ (partial picture b), demodulated (partial picture c), and fed into the averager which is triggered by the magnetron pulse trigger. The averaged signal is written out on a $X$-$Y$-recorder. (See footnote 36; *also* P. Brumm, Thesis, Technical University, Berlin, Germany, 1968.)

1:5 Galilean telescope is used which broadens the beam 5-fold but reduces the beam divergence 1:5 to 1–2 mrad. As the telescope is mounted in the beam near the ruby output, the first negative lens should be of quartz.[35]

The irradiation density at the cuvette can be reduced (a) by proper adjustment of a second stacked polarizer just in front of the ruby output; (b) by a negative lens which increases the beam divergence before the cuvette; and (c) by neutral glasses just in front of the cuvette.

General specifications of the repetitive Q-switched ruby laser system are given in Section II, type 7.

Finally, for a critical consideration of this equipment it should be noted that the main experimental difficulty with the repetitive ruby laser is the necessarily high power density in the resonator (see above) which impairs the operation stability and gives rise to a relatively high failure rate of the optical parts. Therefore, a more reliable laser system should operate within the required specifications above all with lower power densities. First tests with a repetitive YAG laser system with KDP doubler (Laser Associates, Iver, Bucks, United Kingdom)[67] gave a promising result in a stable repetitive "Q"-switched operation at 10 pulses/sec of at least 10 mJ light pulse energy at 530 nm in 15 nsec half-width time.

## G. Type 8. Repetitive Microwave-Pulse Conductometry

The repetitive equipment is extended with respect to the single-pulse equipment type 3 only by the flow system and the averager (cf. Fig. 9 and Fig. 18; see Fig. 27). The microwave generator is additionally equipped by a repetitive trigger source (Tek. 162) and has a slightly larger power supply (200 W output at maximum 200 pulses/sec). In the flow cell, ring electrodes are used instead of the plate electrodes in normal cells for resting solutions (see below "flow system").

The bridge output consists of a r.f.-voltage modulated by the measuring signal, which has to be demodulated before being fed into the averager (see detail pictures b and c in Fig. 27).

### Flow System

In a resting system it is impossible to investigate reactions by repetitive excitation which involve irreversible effects. In these cases it is necessary, for repetitive measurements, to exchange the solution after each excitation. A simple exchange method is a constant flow of the test solution. For temperature-jump excitation it is useful to have a closed circulation flow system. This flow system is thermostatted by a heat exchanger which dissipates the input heat power of the repetitive temperature jumps. With regard to the conductivity measurements, only a laminar flow was used in order to avoid the additional noise of a turbulent flow.[36b]

The flow system is schematically depicted in Fig. 27, which consists of a reservoir, the "falling" tube with electrode cell and absorber part, the receiver, pump, and heat exchanger. The dimensions of the absorber part are given by the requirements of optimum microwave absorption (see Section II, and equipment type 3). The electrode rings are fixed tightly to the wall to avoid turbulence areas behind the electrode wire. The boundary layers between the platinum electrode and the electrolyte are influenced by the flow velocity. Therefore such electrolyte cells are very sensitive to velocity changes and it is necessary to stabilize the flow velocity at least to less than $10^{-2}$. In the system depicted in Fig. 27, the flow through the measuring cell is caused by gravity fall in the "falling" tube between the levels of the upper reservoir and the lower receiver. Its velocity is controlled by the size of the jet at the end of the "falling" tube.[36] The effect of the unavoidable pump shocks is further decreased by a restriction of the inflow.

*Heat Flow*

The heat transport in the flow system[36b] depends on the heat distribution[70] in the cell, the parabolic velocity distribution of the flow, and the geometric data. Let the heated area have the length $l$, the distance $d$ between the electrodes, and the maximum velocity $2\bar{v}$ in the tube center ($\bar{v}$ = mean velocity). Thus the heated volume stays within the electrode gap during the time $t_M = (d-l)/(4\bar{v})$ (if the electrode gap is arranged symmetrically to the absorber part as depicted in Fig. 27). Within this time interval $t_M$ the mean temperature rise in the electrode cell is constant (see Fig. 27,a) so that kinetic measurements can be performed. After this time interval the mean temperature decreases because the heated volume leaves the electrode cell. The temperature decrease approaches a function proportional to $1/\bar{v}t$. If the mean temperature rise in the cell has decreased to 10–20% of the original value, a new temperature jump can be applied to the absorber. Therefore, the highest repetition rate is nearly proportional to the flow velocity. Maximum repetition rate is given by the maximum velocity of 50 cm/sec and is about 20 pulses/sec (see Fig. 27,a).

*Difference Measurements by Double Flow and Absorber System*

A temperature jump applied to an electrolytic solution produces a conductivity change which depends on a change of both the ion mobility and the ion concentration (see equipment 3, "Conductometric Measurements"). In order to measure the ion reaction only the conductivity increase

---

[70] A thorough investigation of the heat transport under the conditions of parabolic velocity distribution gives information about the primary heat distribution in the absorber bulk as well as in axial and in radial directions. The results in this case are a deviation from homogeneous distribution: axial <10%, radial <25%. (cf. footnote 36b).

caused by the mobility change is compensated with aid of a second identical flow system. The reference system is filled with an indifferent aqueous solution, e.g., KCl. Both absorbers are placed symmetrically into the branches of a T-shaped waveguide. The cells are connected in series to form one branch of the measuring bridge (see Fig. 27). By means of the potentiometers *I* and *II* as well as by two trim capacitors, the bridge is balanced for maximum compensation. For a complete compensation, the temperature profiles in both cells (see Fig. 27,a) must be identical. This is obtained only by exactly equal flow velocities. Therefore, two identical pumps are mounted on the same motor well. The heat exchanger is formed in a double helix to give equal heat flow.

*Signal Processing*

The bandwidth of the selective amplifier should be kept as small as possible for undistorted transmission of the signal. The demodulator is of a simple peak rectifier type. The averager, e.g. (Enhancetron), can be used for signals with a rise time not below 50 $\mu$sec. The averaged signal is read out on an $X$-$Y$-recorder. The output signal of the averager (see Figs. 27 and 19) is smoothed by an additional time constant. The main disturbance with these measurements is given by slow variations of the bridge output voltage, probably caused by small statistical fluctuations of velocity and temperature in the cells.

Some special data particularly of the flow system are tabulated below (see Sec. II.).

| | | | |
|---|---|---|---|
| | | Jet diameter | 0.1–0.2 cm |
| Heated volume | 100 $\mu$l | Max. flow velocity | 50   cm/sec |
| Total volume | 100 ml | | |
| Level difference | 13 cm | Microwave pulse duration | |
| between reservoir | | Min. | 1.0 $\mu$sec |
| and receiver | | Max. | 5.5 $\mu$sec |

## Note Added in Proof

Since this paper was written the equipment described above has been used extensively in numerous biological reaction systems. The following types of molecular events can now be analyzed: 1. metastable states[33b]; 2. light reactions[28,71]; 3. electron transfers[5a]; 4. electrical fields across membranes[5b,72]; 5. ion transfers[5b]; and 6. proton transfers[73,74] (in addition to those previously mentioned).

[71] B. Rumberg and H. T. Witt, *Z. Naturforsch.* **19b,** 693 (1964); B. Rumberg, *ibid.*, p. 707; G. Döring, G. Renger, J. Vater, and H. T. Witt, *Z. Naturforsch.* **22b,** in press (1969).

[72] H. M. Emrich, W. Junge, and H. T. Witt, *Naturwissenschaften* **9** (1969).

[73] W. Schliephake, W. Junge, and H. T. Witt, *Z. Naturforsch.* **23b,** 1571 (1968).

[74] H. H. Grünhagen and H. T. Witt, *Z. Naturforsch.*, in press.

# [10] Fluorescence Methods in Kinetic Studies

## By Gregorio Weber

## I. Introduction

The absorption of light by organic molecules results in the appearance of an excited state characterized by structural and functional properties often very different from those of the ground state. Although it is possible to reach upon absorption any of several excited states of the singlet manifold, redistribution of the energy among the degrees of freedom of the excited molecule, as well as energy exchange with the surrounding molecules, results in appreciable population of only the lowest excited singlet or triplet. The excited singlet has a lifetime of the order of nanoseconds whereas the triplet state persists for a much longer period: microseconds to milliseconds in solution and up to seconds in rigid media. The excited singlet may be characterized by the fluorescence emission spectrum, the absolute quantum yield, the lifetime of the excited state, and the polarization of the emitted radiation. Lifetime and polarization are of interest in the use of fluorescence as an indication of molecular kinetic events.

Since the singlet excited state lives only some tens or at most hundreds of nanoseconds, the rate of fluorescence emission is of the order of its reciprocal: $10^8$ to $10^7$ sec$^{-1}$. It follows that it can be perceptibly affected only by those processes with rates of this order or faster. The translational diffusion processes of small molecules may be characterized by the displacement undergone in a given time. According to Einstein's (1906) relation the average displacement $\langle \Delta X \rangle$ in $\delta t$ time is $(2D\delta t)^{1/2}$ where $D$ is the linear diffusion coefficient. For $D = 10^{-6}$ cm$^2$/sec and $\delta t \simeq 10^{-8}$ sec $\Delta X \simeq 10^{-7}$ cm. We thus see that the interactions with the surroundings by the excited molecule will be limited to those structures within 10 Å or so. This distance is of the order that is of particular interest in the study of biological phenomena where knowledge of the precise local molecular events and their time sequence is desirable. The polarization of the emitted fluorescence reflects the intensity of the Brownian rotational motion that takes place

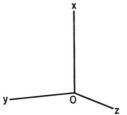

FIG. 1. System for depolarization by transport of oscillators from one orthogonal direction to another.

between absorption and emission. This may be characterized by the rotational relaxation time $\rho$ of the emitting molecule, and it is clear that observable effects will occur only if the rate of rotation $R = 1/2\rho$ is of the order of $\lambda$, the rate of emission.

## II. Rate Equations for Rotational Depolarization

It has been customary to deduce the kinetics of fluorescence phenomena by describing the time-dependent rotations or translations first and then combining these with the rate of emission to yield the appropriate expressions.[1-4] It is possible, however, to use the familiar rate equations to express the differential relations and obtain from this either the time course or the steady-state equations of the system. As an example that applies to polarization—as well as to other effects—we can consider the depolarization as a transport of oscillators from one orthogonal direction to another. Consider the system of Fig. 1, where the exciting light arrives along $yO$ to the solution placed at the origin of coordinates. The exciting light is polarized along $Ox$, and observations of the emission are made from $z$. The light emitted from O in all directions may be considered the result from emission by three orthogonal dipoles, vibrating along $Ox$, $Oy$, and $Oz$ of which only the first two ($Ox = F_{\parallel}$ and $Oy = F_{\perp}$) are seen from $z$. We can set the ordinary differential rates.

$$\frac{dF_x}{dt} = -(\lambda + 4K)F_x + 2F_yK + 2F_zK + F_{x0}$$

$$\frac{dF_y}{dt} = 2F_xK - (\lambda + 4K)F_y + 2F_zK + F_{y0}$$

$$\frac{dF_z}{dt} = 2F_xK + 2F_yK - (\lambda + 4K)F_z + F_{z0} \qquad (1)$$

[1] F. Perrin, *Ann. Phys. (Paris)* **12,** 169 (1929).
[2] G. Weber, *Advan. Protein Chem.* **8,** 415 (1953).
[3] R. Memming, *Z. Physik. Chem. (Frankfurt)* **28,** 168 (1961).
[4] G. Weber and S. Anderson, *Biochemistry,* in press (1969).

$F_{x0}$, $F_{y0}$, and $F_{z0}$ are the rates of excitation to the respective vectors, $K$ the rate constant for transport among the vectors, and $\lambda$ the rate of emission. Since the components $F_y$ and $F_z$ are equal by symmetry, the system (1) contains only two independent equations.

$$\frac{dF_x}{dt} = \frac{dF_\parallel}{dt} = -(\lambda + 4K)F_\parallel + 4KF_\perp + F_{\parallel 0}$$

$$\frac{dF_y}{dt} = \frac{dF_z}{dt} = \frac{dF_\perp}{dt} = 2KF_\parallel - (\lambda + 2K)F_\perp + F_{\perp 0} \tag{2}$$

The steady-state relations may be derived by setting $dF_\parallel/dt = dF_\perp/dt = 0$ solving the resulting simultaneous equations gives:

$$\frac{F_\parallel - F_\perp}{F_\parallel + 2F_\perp} = \left(\frac{F_{\parallel 0} - F_{\perp 0}}{F_{\parallel 0} + 2F_{\perp 0}}\right)\frac{\lambda}{\lambda + 6K} \tag{3}$$

which may also be written as

$$\frac{1}{p} - \frac{1}{3} = \left(\frac{1}{p_0} - \frac{1}{3}\right)\left(1 + \frac{6K}{\lambda}\right)$$

$$p = \frac{F_\parallel - F_\perp}{F_\parallel + F_\perp}; \qquad p_0 = \frac{F_{\parallel 0} - F_{\perp 0}}{F_{\parallel 0} + F_{\perp 0}} \tag{4}$$

Equation (4) is the Perrin equation. The rate constant $K$ is—from his derivation—twice the reciprocal of the average relaxation time of rotation of the fluorescent molecules and therefore equals the rotary diffusion constant.[1,5]

In applications of polarization of fluorescence using the steady-state approximation the rotational relaxation $p$ (or $R$) is obtained. It is possible to obtain $K$ directly by kinetic measurements in real time. The result is also contained in Eq. (2). In this more general case $F_{\parallel 0}$ and $F_{\perp 0}$ are functions of the time. The system of differential equations, (2), is then

$$(D + \lambda + 4K)F_\parallel - 4KF_\perp = F_{\parallel 0}(t)$$
$$- 2KF_\parallel + (D + \lambda + 2K)F_\perp = F_{\perp 0}(t) \tag{5}$$

with $D = d/dt$.

Setting $F_{\parallel 0}(t) = F_{\perp 0}(t) = 0$ we obtain a system of linear homogeneous equations appropriate to describe the free decay of the system. The solution is

$$F_\parallel(t) = \frac{F_0}{3}\exp(-\lambda t) + \frac{2}{3}(F_{\parallel 0} - F_{\perp 0})\exp[-(\lambda + 6K)t]$$

$$F_\perp(t) = \frac{F_0}{3}\exp(-\lambda t) - \frac{1}{3}(F_{\parallel 0} - F_{\perp 0})\exp[-(\lambda + 6K)t]$$

$$F_0 = F_{\parallel 0} + 2F_{\perp 0} \tag{6}$$

[5] G. Weber, In "Fluorescence and Phosphorescence Analysis" (D. M. Hercules, ed.) Wiley, New York, 1965.

Eqs. (6) show that the polarized components do not have a simple exponential decay, but a complex one that is dependent upon the rate of rotation of the fluorescent molecules; Jablonski[6] pointed out this fact for the first time and obtained Eqs. (6) in a slightly different form. These equations show that from the polarized free decays, that is the decays after excitation has ceased, one may isolate the two decay constants $\lambda$ and $\lambda + 6K$, thus obtaining the rate of rotation by observations in real time. Jablonski applied also Eqs. (6) to the measurements by phase fluorometry.[6,6a] This is not justified. The correct solution[6b] for excitation with light modulated at frequency $\nu = 2\pi\omega$

$$F_{\parallel 0}(t) = a_\parallel + b_\parallel \sin \omega t$$
$$F_{\perp 0}(t) = a_\perp + b_\perp \sin \omega t \tag{7}$$

where $b$, $a$, are determined by the degree of modulation of the excitation. The phase delay ($\tan \delta$) and degree of modulation ($B/A$) are then given by:

$$\tan \delta = \frac{\omega}{\lambda} \; \frac{(\lambda + 6K)^2 + \omega^2 - 2K(2 - x)(2\lambda + 6K)}{(\lambda + 6K)^2 + \omega^2 - 2k(2 - x)\left(\lambda + 6K - \dfrac{\omega^2}{\lambda}\right)}$$

$$\frac{B}{A} = \frac{(b/a)\lambda}{\sqrt{\lambda^2 + \omega^2}}\left(\frac{[\lambda + 6K - 2K(2 - x)]^2 + \omega^2}{[\lambda + 6K - 2k(2 - x)]^2} \; \frac{(\lambda + 6K)^2}{(\lambda + 6K)^2 + \omega^2}\right)^{1/2} \tag{8}$$

$$\tan \delta = \tan \delta_\parallel; \qquad B/A = (B/A)_\parallel \text{ if } x = x_\parallel = 2\left(\frac{1 - p_0}{1 + p_0}\right)$$

$$\tan \delta = \tan \delta_\perp; \qquad B/A = (B/A)_\perp \text{ if } x = x_\perp = \frac{2}{1 - p_0} \tag{9}$$

Equation (4) has been applied in a considerable number of cases to studies of the rotational relaxation of proteins in solution.[2,6c] Recently Wahl[7,7a] and Stryer[8] have applied to protein conjugates Eq. (6) and Pasby and Weber[9] Eqs. (8) and (9) to the same problem. For the purpose of studying the molecular rotations, proteins are made fluorescent by conjugating them with fluorescent compounds. The conjugating bonds are covalent bonds, e.g., carboxamide, sulfonamide, so that the conjugates are stable to almost the same extent as the amino acid sequence in the protein. The lifetime of the fluorescent labels determines the range of rates of rotations that may be studied. Rotational motions that are fast in comparison with the rate of emission result in an initial depolarization that is difficult or impossible to separate from the loss of polarization due to the existence of a finite angle

[6] A. Jablonski, Z. Physik. **95**, 538 (1935).

[6a] R. K. Bauer, Z. Naturforsch. **18a**, 718 (1963).

[6b] R. Spencer and G. Weber, Ann. N. Y. Acad. Sci. **158**, 361 (1969).

[6c] R. F. Steiner and H. Edelbach, Chem. Rev. p. 457 (1962).

[7] P. Wahl, Compt. Rend. Acad. Sci. **263**, 1525 (1966).

[7a] M. Frey and P. Wahl, Compt. Rend. Acad. Sci. **262**, 2653 (1966).

[8] L. Stryer, Science **162**, 526 (1968).

[9] T. Pasby and G. Weber, to be published.

between the oscillators of absorption and emission. Rotational motions that are slow by comparison with the rate of emission will not appreciably affect the depolarization. As an example, consider three possible kinds of rotations: (1) that of the labeling groups and/or the amino acid residue to which it is attached (local rotation), (2) the rotation of a large fragment of the protein molecule; (3) the rotation of the whole molecule.

The first may take place during a fraction of a nanosecond whereas the second and third will take many nanoseconds, depending on the volume and shape of the kinetic unit in question.

If the fluorescent group has a lifetime of less than 4 nsec, the influence of 1 will be the most marked with hardly any influence from the others in the case of an average-sized globular protein. This is the case with the tryptophan fluorescence, intrinsic to the protein. Because of the short lifetimes involved (2–5 nsec), it can give only an indication of the local rotation of the tryptophan residues inside the protein matrix.[10]

If the label has a lifetime of 10–15 nsec, it is possible to distinguish, as Wahl[11] and Gottlieb and Wahl[12] have shown, the local label rotations from the rotations of the protein as a whole.

It is clear that if the protein is very large the rotational relaxation corresponding to the whole of the protein may be missed, but it will appear clearly in depolarization studies if the label is replaced by another with a longer lifetime. Using dimethyl amino naphthalene sulfomyl chloride (DNS) as label, Weltman and Edelman[12a] and Wahl and Weber[13] showed the overall as well as the local label rotations in $\gamma$-globulin. With the use of a longer-lived label (pyrene butyric anhydride, 100 nsec), Knopp and Weber[13a] were able to determine not only the rotational relaxation of $\gamma$-globulin $\sim$200 nsec but that of 19 S $\gamma$-globulin $\sim$900 nsec. The latter would not have been possible by the use of DNS because of the too short lifetime.

## III. Instrumentation

The methods for the direct determination of fluorescence lifetimes may be discussed by considering each single fluorescent species as a detector in which the response to an infinitely narrow pulse of light decays in time $t$ according to the exponential law $I(t) = I(0) \times \exp(-Kt)$. From the relations of the exciting light with the light emitted by such exponential detector the rate constant $\lambda$ or its reciprocal $\tau$, the lifetime of the excited state may be determined. Figure 2 shows the Fourier spectrum characteristic of an

[10] S. Anderson and G. Weber, *Arch. Biochem. Biophys.* **116,** 207 (1966).
[11] P. Wahl, Thesis, University of Strasbourg, 1962.
[12] Y. Gottlieb and P. Wahl, *J. Chim. Phys.* **60,** 849 (1963).
[12a] J. A. Weltman and G. M. Edelman, *Biochemistry* **6,** 1437 (1967).
[13] P. Wahl and G. Weber, *J. Mol. Biol.* **30,** 371 (1967).
[13a] J. Knopp and G. Weber, *J. Biol. Chem.* **242,** 1353 (1967).

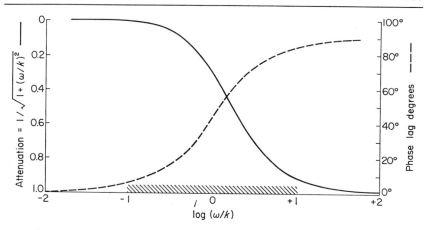

FIG. 2. Frequency response and phase lag of exponential detector $\omega = 2\pi f$.

exponential detector (e.g., McLachlan[13b]). The detector responds without attenuation or change of phase to all frequencies from zero to those approaching $1/K$, while frequencies of this order or exceeding $1/K$ are progressively more attenuated. The phase of the emitted light lags behind that of the excitation by 45° when $\omega = K$, and tends to 90° lag toward complete attenuation. It follows that for the determination of λ the fluorescence must be excited by light of one or more frequencies in the neighborhood of $k$ (hatched area of Fig. 2). The former is practically realized by excitation with sinusoidally modulated light, and the latter by excitation with light pulses the frequency spectrum of which is made up of a narrow band of frequencies; λ is of the order of $10^8$ sec$^{-1}$ (or $\tau$ of order $10^{-8}$ sec). Therefore the frequencies to be employed in the excitation must be of the order of $10^8/2\pi$ or in the region of 1–50 MHz. Light modulated at a single frequency in this range may be produced by various electro-optical methods,[14-17] and techniques for the production of approximately Gaussian pulses of half width 0.3 to 3 nsec[18,19] have been described. Each type of method has its own advantages and disadvantages. The pulse method is particularly useful when several modes of decay are possible, whereas the methods employing a single frequency are undoubtedly the more accurate because of the extreme frequency selection attainable by electronic techniques. They are

[13b] N. W. McLachlan, "Complex Variable Theory and Transform Calculus," 2nd ed., p. 184, Cambridge University Press, London, 1963.

[14] E. Gaviola, Z. Physik. **42**, 852 (1927).

[15] O. Maerks, Z. Physik. **109**, 685 (1938).

[16] R. Bauer and M. Rozwadowski, Bull. Acad. Polon. Sci. Classe III **7**, 365 (1959).

[17] A. Müller, R. Lumry, and H. Kokubu, Rev. Sci. Instr. **36**, 1214 (1965).

[18] J. T. D'Alessio, P. K. Ludwig, and M. Burton, Rev. Sci. Instr. **35**, 1015 (1964).

[19] L. Hundley, T. Coburn, E. Garwin, and L. Stryer, Rev. Sci. Instr. **38** (1967).

also better adapted to the measurement of lifetimes in the subnanosecond region.

The theory of the fluorometer, so named by Gaviola, who constructed the first apparatus of this kind in 1926, has been given in detail by Dushinsky.[20] If a fluorescent species is illuminated with light modulated with frequency $f$, describable by the expression

$$E(t) = A + B \cos 2\pi f t$$

where $A \geqslant B$, the modulation of the exciting light is given by $B/A$. The fluorescent light due to emission by a population with exponential decay $\exp(-t/\tau)$ is then described by the expression,

$$F(t) = A + B \cos \delta \cos(2\pi f t - \delta) \tag{10}$$

where

$$\tan \delta = 2\pi f \tau \tag{11}$$

Relative modulation $D = \dfrac{\text{modulation of fluorescence}}{\text{modulation of excitation}}$

$$= \cos \delta = (1 + 4\pi^2 f^2 \tau^2)^{-1/2} \tag{12}$$

Thus it is possible to determine the lifetime either by a measurement of the phase lag $\delta$ or of the relative modulation of the emitted light (Fig. 3). The above equations were derived on the assumption of a single homogeneous emitting population with a unique decay constant, and it is often cited as one of the shortcomings of the single frequency method, that the exist-

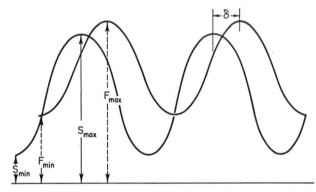

FIG. 3. Alternative methods of measurement of $\tau$: (1) phase difference with scatterer; (2) relative modulation. Modulation frequency $= f$; phase difference $= \delta$; $\tan \delta = 2\pi f \tau$

$$D_F = F_{\max} - F_{\min}/F_{\max} + F_{\min}$$
$$D_S = S_{\max} - S_{\min}/S_{\max} + S_{\min}$$

relative modulation $= D =$

$$D_F/D_S = 1/\sqrt{1 + 4\pi^2 f^2 \tau^2} = \cos \delta$$

[20] F. Dushinsky, Z. Physik. **81**, 7 (1933).

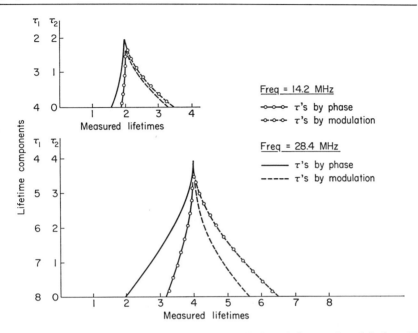

Fig. 4. Measurements of lifetime by two methods: (1) degree of modulation; (2) phase difference with scatterer.

ence of a unique decay is presupposed but not demonstrated by the measurements. This is indeed the case if only one of the two quantities mentioned above is measured. If both are measured, the existence of an exponential decay may be demonstrated. In a heterogeneous emitting population the lifetime measured by the degree of modulation will almost always be longer than the weighted average of the component lifetimes while the lifetime determined from the phase lag will always be shorter than the weighted average. This fact is graphically demonstrated in Fig. 4. It is only when there is a single decay that both methods give the same results, so that obtaining, within experimental error, the same value of $\tau$ by the two methods is a sufficient proof of the simple exponential character of the decay, and indirectly of the homogeneity of the emitting population.

## The Cross-Correlation Method[6b]

The measurements of phase differences is simplified, and the reliability of the measurements increased, if the photocurrent produced by the fluorescence is mixed with a voltage of fixed frequency near but not equal to the modulating light frequency and the phase is measured in the amplified difference signal.[21,22] In these heterodyning procedures the direct current

[21] E. A. Bailey and G. K. Rollefson, *J. Chem. Phys.* **21**, 1315 (1953).
[22] A. Schmillen, *Z. Physik.* **135**, 294 (1953).

levels are disregarded so that the degree of modulation, which represents an important source of information is usually lost. Birks and Little[23] first introduced an alternative method by which the phenomena may be transferred from the original high frequency to any desired low frequency while preserving in the latter the phase and the degree of modulation of the former. To obtain this result an alternating current voltage of the light-modulating frequency having a known fixed phase difference $\phi$ with it, is applied to one or more of the dynodes. The result is the modulation of the phototube response according to the equation

$$R(t) = a + b \cos(2\pi ft - \phi)$$

The photocurrent is then the cross-correlation product $\langle R(t)F(t)\rangle$ integrated over a period and equals

$$R(t)F(t) = aA + Bb\,\frac{\cos\delta}{2}\cos(\phi - \delta) \tag{13}$$

Birks and Little determined $\delta$ by the value of $\phi$ necessary to get maximum photocurrent. We chose to translate the phenomenon to a lower frequency $\Delta f$, by varying $\phi$ linearly with time so that $\phi = 2\pi\Delta ft$. In such case, if $\Delta f \ll f$

$$\langle R(t)F(t)\rangle_{av} = aA + Bb\,\frac{\cos\delta}{2}\cos(2\pi\Delta ft - \delta) \tag{14}$$

Thus the low frequency response preserves the phase difference $\delta$ and, the relative modulation of the high frequency photocurrent.

The advantages of the procedure utilized are several:

1. Since mixing of the high frequencies $f$ and $f + \Delta f$ takes place only inside the phototube and only the low frequency component is wanted, the photomultiplier may be used with any practical anode load or even as a current device with a great improvement in sensitivity.

2. All frequencies above $\Delta f$ may be filtered off with great increase in the signal/noise ratio and signal isolation.

3. With $\Delta f$ less than 100 Hz, numerical time-interval counting may be used to determine the phase difference between the exciting and fluorescence light. Such a procedure is particularly well adapted to increase the accuracy through averaging.

## IV. Description of the Fluorometer

The features that make for the high sensitivity and wide spectral range of the instrument are to be found in the light modulation technique, the cross-correlation electronics, and the phase-measuring device employed.

### Optics

The excitation optics is shown in Fig. 5. The light source is a direct-

[23] J. B. Birks and W. A. Little, *Proc. Phys. Soc. London* **A66**, 921 (1953).

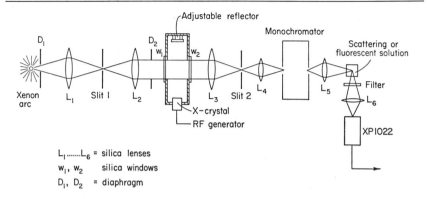

FIG. 5. Cross-correlation fluorometer optics.

current 150-watt xenon arc powered by storage batteries to reduce to a minimum the noise in the light source and to eliminate the inherent 1–5% a-c ripple present in high current power supplies.

The ultrasonics unit and lens system are similar to those described by Debye and Sears[24] and Bailey and Rollefson.[21] The ultrasonics tank is constructed of heavy stainless steel to achieve a high damping factor. An interchangeable X-cut quartz crystal is mounted at one end of the tank, and a polished stainless steel reflector at the opposite end. The plane of the reflector may be adjusted through several degrees to produce the optimum ultrasonic standing wave across the tank. Optical stability of the standing wave is provided by using a solution of 19% absolute alcohol and 81% glass-distilled water, since the temperature coefficient of the velocity of sound in this mixture at room temperature is nearly zero.[25] The major cause of phase instability appears to be unavoidable swirling suspended particles. This "phase-noise" appears to be quite random.

The ultrasonics crystal is driven in the 3rd harmonic by a 65-watt crystal-excited radio transmitter (Johnson Viking Ranger II). Two driving frequencies presently used are 7.0994 MHz and 14.1988 MHz.

Lens 1 gathers light and focuses an image of the xenon arc at slit 1. Lens 2 placed at focal length distance from this slit forms a nearly parallel beam of light through the ultrasonics tank. The ultrasonic standing wave perpendicular to the light wave front creates planes of higher refractive index, which collapse and build with the ultrasonic wave. The result is that the wave front of the emerging parallel beam is modulated in phase according to the modulation of the refractive index across the light path in the tank, just as occurs when a parallel beam strikes an ordinary grating. When this interference pattern is focused at slit 2 by lens 3 to form the image of slit 1, the image appears in the zero order as well as in several

[24] P. Debye and F. W. Sears, *Proc. Natl. Acad. Sci. U.S.* **18**, 409 (1932).
[25] W. Demtroeder, *Z. Physik.* **166**, 42 (1962).

higher orders of diffraction. Each order is modulated at twice the frequency of the standing wave, and the zero or firstorder is used by adjusting slit 2 to select the desired image. This image of modulated light is then focused upon the input slit of the monochromator so that any wavelength in the interval 240–1000 nm may be chosen for excitation of a fluorescence sample.

### Cross-Correlation Electronics (Fig. 6)

Cross correlation is accomplished by modulating the gain of the photomultiplier tube with a radiofrequency (RF) signal of 14,200,020 Hz (or 28,400,040, depending upon the excitation frequency used). Since the fluorescent light is modulated at 14,200,000 Hz (or 28,400,000 Hz), the signal detectable at the anode contains the cross-correlation frequency 20 Hz (or 40 Hz).

A Phillips XP1022 photomultiplier is used as the photodetector because of its high gain (up to $10^9$) and fast rise time of 1.7 nsec (manufacturer's specifications). This tube contains an anode screen grid which can conveniently be used to modulate the overall gain of the photomultiplier by applying to it an alternating current signal.

The key to the translation from high to low frequency output is addition of a constant known audio frequency $\Delta f$ to an RF signal $f$. The difficulty of performing it successfully is apparent when it is realized that the sum $(f + \Delta f)$ must be absolutely locked in frequency and in phase to $f$, even though they differ by only about 5 parts in $10^5$, and that other frequencies apart from $(f + \Delta f)$ must be absent.

This addition is accomplished by a Variogon Phase Shifting Transducer (Nilson Mfg. Co., Haines City, Florida), which is basically a rotating

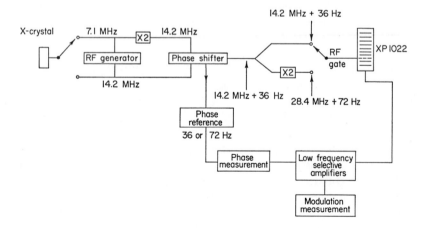

Fig. 6. Cross-correlation fluorometer electronics.

capacitor originally described by Morrison,[26] and subsequently improved.[27] The feature of the phase shifting capacitor is that the phase angle between the input and the output is linearly dependent upon the angular position of one of its plates. Thus, turning the capacitor's rotor one revolution increases the phase of the output sine wave by 360 degrees, or one full cycle.

In the fluorometer, a damped hysteresis synchronous motor drives the rotating plate of the Variogon at 20 revolutions per second, thereby adding 20 Hz to the RF signal connected to the input. The period of this audio signal is 50 msec and, with the present motor, is stable within ±0.01 msec. A phase reference signal, also 20 Hz, is obtained photoelectrically from holes in a disk mounted on the rotating shaft of the capacitor.

Figure 6 shows the electronic block diagram for measuring both phase angle and relative modulation of a fluorescence signal excited by either 14.2 MHz or 28.4 MHz. Since the phase shifting capacitor operates at only one frequency, 14.2 MHz, the reference RF signal from the ultrasonics crystal must be doubled previous to or following the phase shifter's addition of 20 Hz, depending upon the excitation frequency desired, thus producing 14,200,040 Hz or 28,400,020 Hz for the photomultiplier gating frequency.

*Phase Measurement*

The actual phase measurement is accomplished by a two-channel frequency counter (Hewlett-Packard Model 5223L) which can measure intervals to 0.01 msec. The counter is turned on by the rise of the phase reference square wave from the phase shifter, and it is turned off by the passage through zero of the audio signal from the photomultiplier tube. The counter displays or prints the time interval in milliseconds, so that the phase is calculated by

$$\alpha = \frac{\text{time interval}}{\text{period}} \times 360 \text{ degrees}$$

A phase measurement of the light scattered by a water suspension of glycogen is used to correct for phase shifts in the electronics and the optics. Therefore, the phase shift, $\delta$, due to fluorescence is

$$\delta = \alpha_{\text{fluorescence}} - \alpha_{\text{scatter}}$$

With the standard deviation of the time interval measurements within 0.01 msec, the smallest phase difference detectable (with a 40 Hz cross correlation frequency) is 0.35°. Therefore, with the 28.4 MHz excitation

[26] J. F. Morrison, *Proc. I. R. E.* **25**, 1310 (1937).
[27] E. A. Holmes, *M.I.T. Radiation Lab. Ser.* (Components Handbook) **17**, 288 (1949).

frequency, fluorescence decay times of less than 1 nsec can be measured with an expected accuracy of ±0.03 nsec.

The discrimination between lifetimes in the subnanosecond region was demonstrated by measurements in solutions of fluorescein in 0.01 $M$ NaOH, quenched by addition of KI. For an ideally collisional quenching process,[1,28] the relation $F/F_0 = \tau/\tau_0$ is expected to obtain. Here $F_0$ and $\tau_0$ are the fluorescence yield and lifetime of the solution in the absence of quencher, $F$ and $\tau$ the corresponding quantities in the presence of quencher. The classical observations of Perrin[1] on the polarization, and of Szymanovsky[29] on the actual lifetimes have substantiated the validity of this relation. The difference in lifetime between two solutions quenched to $1/11$ ($\tau = 0.49$ nsec) and $1/19$ ($\tau = 0.39$ nsec) of the initial fluorescence yield could be determined. The instrument is thus capable of determining directly a difference of 0.1 nsec between two lifetimes each less than a nanosecond long.

## V. Importance of the Determination of Lifetimes in the Study of Protein Fluorescence

From measurements of oscillator strength and fluorescence yield, it is clear that the lifetime of protein fluorescence must be in the range of 2–4 nsec. The fluorophores involved are mainly tryptophan residues, and there is a distinct possibility of detecting the heterogeneous character of their emission through measurements of $\tau$. In principle, such heterogeneity could be detected in three different ways:

1. By simultaneous measurements of phase lag and degree of modulation at a single light-modulation frequency. As set out in the introduction and graphically shown in Fig. 4, only in the case of a unique decay will the values of $\tau$ obtained by both methods be in agreement.

2. By measurements of lifetime at more than one modulating frequency. If the modulation frequency is varied systematically in the interval 1–50 MHz, the complete Fourier spectrum of the emission should be available. Present-day technical shortcomings make it very difficult to use more than a few fixed frequencies. As described, 14.2 or 28.4 MHz may be used in our fluorometer. These two ought to be sufficient to establish the character of the emission in many cases if all the lifetimes involved are in the interval of 0.1–50 nsec.

3. By measurements of the dispersion of lifetime with exciting or fluorescence light.

The practical possibility of such measurements depends upon keeping a satisfactory signal to noise ratio in spite of the inevitable loss of fluores-

[28] S. I. Wavilov, *Acta Phys. Polon.* **5**, 417 (1936).
[29] W. Szymanowsky, *Z. Physik.* **95**, 460 (1936).

cence intensity. It demands therefore the modulation of a strong light source continuous in the visible and ultraviolet regions like the xenon arc. We selected the ultrasonic technique because of the high degree of modulation attainable in all spectral regions of interest without appreciable attenuation of the exciting light. The short decay of NADH, with its low fluorescence yield of 3% was measured with monochromatic excitation of 4 nm bandwidth with very satisfactory signal to noise ratio.

LIFETIMES OF THE EXCITED STATE OF SOME COMPOUNDS
OF BIOCHEMICAL INTEREST[a]

| Substance | Solvent | Temperature (°C) | $\tau$ (nsec) |
|---|---|---|---|
| NADH | 0.1 $M$ Tris-Acetate pH 8.0 | 25 | 0.40 |
| NADH | 1.2 Propanediol $\sim 10^{-2}$ $N$ NaHCO$_3$ | 25 | 1.24 |
| NADH | Bound to LDH (1:1) | 15 | 1.50 |
| NADH | Bound to LDH (2:1) + Na oxalate | 15 | 6.53 |
| AP (DPNH) | 0.1 $M$ Tris-acetate, pH 7.0 | 15 | 0.20 |
| AP (DPNH) | Bound to LDH (2:1) | 15 | 2.09 |
| AP (DPNH) | Bound to LDH (2:1) + Na oxalate | 15 | 5.19 |
| Pyrene butyric acid | 1,2-Propanediol | 25 | 193.0[b] |
| 1-Anilino-8-naphthalene sulfonate (ANS) | 1,2-Propanediol | 25 | 8.5[b] |
| ANS | 1,2-Propanediol | −55 | 18.0[b] |
| ANS | Propanol | 25 | 10.5 |
| ANS | Bound to BSA (2:1) pH 7.0 | 25 | 16.0 |
| ANS | Water, pH 7.0 | 25 | 0.6 |
| FMN | 0.1 $M$ Phosphate pH 7.0 | 20 | 4.7 |
| FAD | 0.1 $M$ Phosphate pH 7.0 | 20 | 2.5 |
| Fluorescein (1 $\mu$g/liter) | 0.1 $M$ NaOH | 20 | 4.2 |
| Quinine sulfate | 0.1 $M$ H$_2$SO$_4$ | 20 | 19.5 |

[a] The measurements have been done with the instrumentation described. Unless otherwise specified, they were done by phase delay. The figures have been rounded to the nearest 0.1 nsec. All the values are in good agreement with published ones except those for ANS, FAD, and NADH in aqueous buffers. Their values quoted here are much shorter than those of R. F. Chen, G. G. Vurek, and N. Alexander, *Science* **156,** 949 (1967), who give ANS in water = 2.3 nsec, FAD = 4.9 nsec, NADH = 4.5 nsec; these values are up to an order of magnitude longer. Their method of measurement is, however, notoriously inaccurate for the determination of lifetimes less than about 4 nsec due to the broadness of the exciting pulse and the slow response of the 1P28 photomultiplier.

[b] Measurements by modulation at 14.2 MHz.

More recently Spencer et al.[30] have been able to resolve both the excitation and emission by introducing a second monochromator in the fluorescence beam. They were able to analyze systematically the fluorescence emitted when excitation and emission were restricted to bandwidths of 6 nm. In flavin mononucleotide the lifetime was constant within less than 0.1 nsec for all excitation and emission conditions. In anilinonaphthalene sulfonate, *small* but significant variations in lifetime across the fluorescence band were seen. The same substance showed also a small dependence of lifetime upon the exciting wavelength.[10]

[30] R. Spencer, W. Vaughan, and G. Weber, *in* "Molecular Luminescence" (E. C. Lin, ed.), p. 607. Benjamin, New York, 1969.

# Author Index

Numbers in parentheses are reference numbers and indicate that an author's work is referred to, although his name is not cited in the text.

# Subject Index